Conventional and Non-conventional Interventions in CROP IMPROVEMENT

Conventional and Non-conventional Interventions in CROP IMPROVEMENT

Edited By

C P MALIK
GULZAR S SANGHERA
SHABIR H WANI

MD PUBLICATIONS PVT LTD
NEW DELHI
www.mdppl.com

Published by :

MD Publications Pvt Ltd
"MD House", 11, Darya Ganj,
New Delhi - 110 002
Phone : +91-11-45355555
E-mail : contact@mdppl.com
Website: www.mdppl.com

ISBN : 978-81-7533-457-1

Published and Printed by Mr. Pranav Gupta on behalf of **MD Publications Pvt Ltd** at Mohit Enterprises.

PREFACE

Plant development and productivity are affected by various environmental factors limiting agricultural productivity worldwide. The loss of productivity is triggered by a series of morphological, physiological, biochemical and molecular stress induced changes. Such an unfavourable situation is contrary to the increasing global food demand. World population is increasing at an alarming rate and expected to reach more than nine billion by the end of 2050, whereas, plant productivity is being and will be seriously limited by various factors globally. More food and feed have to be grown on less land. Concomitantly agro-techniques have reached a plateau where further progress is neutral for yield. Any scientific progress helps in combating environmental constraints. Global climate pattern is getting more unpredictable with increased occurrence of different stresses. Therefore, to feed the world population, it is imperative to maintain crop productivity even under adverse environments. This is a major concern for all nations in general and developing world in particular. Large scale genome analysis and related technologies provide access to a refined understanding of crop genomes. Fortunately, molecular markers are a dependable genetic tools, accepted by the public, which helps in judging gene technology critically. Developing crop plants having ability to withstand stresses is the dire need of the day which demands modern novel strategies for thorough understanding of plants productivity mechanisms operative during stresses. Molecular breeding and genetic engineering have significantly contributed to expand basic knowledge of the cellular mechanisms involved in plant responses to stresses, exploiting new strategies to enhance productivity. Over the past few decades' research studies concerning physiological, biotechnological approaches were instrumental in enhancing our understanding of the patterns of plant responses to variable stresses and the mechanism operating. Biotechnological interventions have made well recognized impact and have revolutionized agricultural productivity. Recently, nanotechnology, which concerns with production and utilization of nanomaterials and nanoparticles , is an emerging technology of great interest and opportunity. It has wide applications in agriculture, food preservation, food packaging, development and utilization of multiplexed diagnostic kits. Further, bio-selective surfaces are the novel innovations of nanotechnology with enhanced or reduced capability to bind or hold specific molecules. There is even a tendency for classical combination breeding, together with transgenesis and marker assisted selection, physiological and nanotechnological interventions to be more effective in achieving complex breeding goals. Areas like plant genomics and molecular breeding; efficient characterization of genetic resources and search for novel alleles; Climate change and adaptation strategies by different cereals and oil seed crop species; biotic and abiotic stress tolerance are being researched by different research groups.

We have endeavored to give the reader a varied knowledge of crop improvement in all its various guises.

The present volume **'Conventional and Non-Conventional Interventions in Crop Improvement'** encompasses a collection of 10 chapters contributed by 22 experts in the field of plant breeding and genetics, physiology, biotechnology and genetic engineering, nanotechnology and crop improvement. It is a good contribution to the topic that is of prime importance to break the barriers of yield in different crops. The chapters provide the-state-of -art account of the information available on different approaches for crop improvement under diverse conditions. Chapter 1 titled **'Engineering Abiotic Stress Tolerance in Plants: Extricating Regulatory Gene Complex'** gives an insight into the different complex mechanisms that lead to the solution of the mystery of abiotic stress tolerance in plants and transcription factor genes utilized for development of transgenic stress - tolerant plants. Chapter 2 illustrates the **Potential for engineering crops with high antioxidant capacity** and describes the prospects of enhancing antioxidants in crop improvement, quantum and quality improving the nutrient capacity of economically important crops including fruits and vegetables. Chapter 3 provides a comprehensive report on current status of **Molecular markers and MAS used in sunflower (*Helianthus annuus* L.)** to accelerate their utility in various molecular biology applications like assessment of genetic relationships among genotypes, construction of linkage maps, tagging and mapping genes/QTLs of agronomic interest, development of genomic and cDNA libraries, map based cloning and marker assisted selection. Chapter 4 gives the glimpse on the **Genetics and breeding for nutritional quality aspects of grain legumes** where in systematic efforts have been made to identify factors affecting their nutritional quality and breeding approaches used for its improvement. In chapter 5 **Improvement of garlic crop using integrated approaches** like tissue culture, molecular markers and genetic transformation to enhance its nutritional quality and resistance to various biotic and abiotic stresses besides assessment of genetic diversity have been described. Chapter 6 describes **Generation of vermicompost and its utililazation in sustainable agriculture** in addition to its effects on plant growth, soil and environment. In chapter 7 attempts has been made to compile various conventional and non-conventional breeding approaches that are required for development of site specific crop varieties for **Breeding for Specific Adaptation to Increase the Yield Potential of Rice for Hill Regions .**

Further, Chapter 8 gives an account of many control mechanisms influencing **Carbon Metabolism in Transgenic Plants** formation and partitioning. Manipulation of chloroplast, the site of photosynthesis; biosynthetic pathways involved in carbon fixation, Rubisco, supercharging photosynthesis in transgenic plants. These information lead to the reassessment of ideas about the regulation of carbon metabolism and have consequences for design of bioengineering strategies to increase crop productivity. Chapter 9 adds new dimensions to crop improvement where in **Living organisms as a source of 'Nanoscale Engineering' and its Applications** are described. There are many products available in the market which shows their impressive applicability in the field of agriculture. Some of the nanodevices are CNT's , nanobiosensors , nanoparticles, etc. and have contributed to crop improvement, disease detection, disease treatment, pathogen detection, food packaging, etc. All these nanodevices have helped to increase the crop yield with high quality and also for food safety and biosecurity. Though these nanodevices have given the agriculture a new stature yet studies on their side effects on human health and environment need further analyses. Finally, Chapter 10 describes **Mutations Breeding in Genomics Era : New Opportunities and Challenges.** The advances in molecular genetics and DNA technologies that have led to revival of mutation breeding in genomic era exploring its opportunities and challenges to increase both the efficiency and efficacy of mutation techniques in crop improvement are explained.

We are sure that the present volume will prove extremely useful for those interested in exploiting different facets of crop improvement.

We gratefully appreciate all the learned contributors for their cooperation in reviewing useful information in the selected area of their interest. The opinions and text contained in their reviews are those of the authors and we have endeavored to honor their ideas.

The editors record their appreciation and sincere gratitude to Chancellor (JNU, Jaipur), Vice-Chancellor SKUAST-Kashmir (Srinagar) for encouragement, cooperation and constant support.

We record our sincere thanks to our colleagues Drs. Himakshi Bhati Kushwaha and Amandeep Hora for rendering help from time to time and in the preparation of index.

We would like to thank Jyoti Ushara and Ritesh Kaur (Ph D Scholars) for their help in proof reading and indexing the subject matter.

We are very grateful to Mr Pranav Gupta, Director of MD Publications Pvt.Ltd, New Delhi, for his cooperation and support for this book.

Editors

LIST OF CONTRIBUTORS

SHABIR H WANI

KRISHI VIGYAN KENDRA (FARM SCIENCE CENTRE) HENGBUNG, P.O. KANGPOKPI, DIST. SENAPATI, MANIPUR, 795129, INDIA

A HARIBHUSHAN

KRISHI VIGYAN KENDRA (FARM SCIENCE CENTRE) HENGBUNG, P.O. KANGPOKPI, DIST. SENAPATI, MANIPUR, 795129, INDIA

NB SINGH

ASSISTANT PROFESSOR, PLANT BREEDING AND GENETICS, CENTRAL AGRICULTURAL UNIVERSITY, IMPHAL-795004, MANIPUR, INDIA

TH RENUKA DEVI

PLANT BREEDING AND GENETICS, CENTRAL AGRICULTURAL UNIVERSITY, IMPHAL-795004, MANIPUR, INDIA

JEBERSON SM

PLANT BREEDING AND GENETICS, CENTRAL AGRICULTURAL UNIVERSITY, IMPHAL-795004, MANIPUR, INDIA

JYOTI USHAHRA

SCHOOL OF LIFE SCIENCES, JAIPUR NATIONAL UNIVERSITY, JAIPUR, RAJASTHAN, INDIA

AMANDEEP HORA

SCHOOL OF LIFE SCIENCES, JAIPUR NATIONAL UNIVERSITY, JAIPUR, RAJASTHAN, INDIA

CP MALIK

SCHOOL OF LIFE SCIENCES, JAIPUR NATIONAL UNIVERSITY, JAIPUR, RAJASTHAN, INDIA

K SUJATHA

CROP IMPROVEMENT SECTION, DIRECTORATE OF OILSEEDS RESEARCH, RAJENDRANAGAR, HYDERABAD, ANDHRA PRADESH -500 030

M SUJATHA

CROP IMPROVEMENT SECTION, DIRECTORATE OF OILSEEDS RESEARCH, RAJENDRANAGAR, HYDERABAD, ANDHRA PRADESH -500 030

ASIF M IQBAL

DIVISION OF PLANT BREEDING AND GENETICS,SHER-E-KASHMIR UNIVERSITY AND AGRICULTURAL SCIENCES AND TECHNOLOGY OF KASHMIR, SHALIMAR, 191121, J&K, INDIA.

SHAFIQ A WANI

DIVISION OF PLANT BREEDING AND GENETICS,SHER-E-KASHMIR UNIVERSITY AND AGRICULTURAL SCIENCES AND TECHNOLOGY OF KASHMIR, SHALIMAR, 191121, J&K, INDIA

AJAZ A LONE

DIVISION OF PLANT BREEDING AND GENETICS,SHER-E-KASHMIR UNIVERSITY AND AGRICULTURAL SCIENCES AND TECHNOLOGY OF KASHMIR, SHALIMAR, 191121, J&K, INDIA

ZA DAR

DIVISION OF PLANT BREEDING AND GENETICS,SHER-E-KASHMIR UNIVERSITY AND AGRICULTURAL SCIENCES AND TECHNOLOGY OF KASHMIR, SHALIMAR, 191121, J&K, INDIA.

FA NEHVI

DIVISION OF PLANT BREEDING AND GENETICS,SHER-E-KASHMIR UNIVERSITY AND AGRICULTURAL SCIENCES AND TECHNOLOGY OF KASHMIR, SHALIMAR, 191121, J&K, INDIA.

ABHILASH NAIR

SCHOOL OF LIFE SCIENCES, JAIPUR NATIONAL UNIVERSITY, JAIPUR, RAJASTHAN, INDIA

ANIL KHAR

DIRECTORATE OF ONION AND GARLIC RESEARCH, RAJGURUNAGER, PUNE

AMANDEEP HORA

SCHOOL OF LIFE SCIENCES, JAIPUR NATIONAL UNIVERSITY, JAIPUR, RAJASTHAN, INDIA

SC KASHYAP

SKUAST-KASHMIR, MOUNTAIN RESEARCH CENTRE FOR FIELD CROPS, KHUDWANI, ANANTNAG, 192102, JAMMU & KASHMIR, INDIA

GULZAR S SANGHERA

SKUAST-KASHMIR, MOUNTAIN RESEARCH CENTRE FOR FIELD CROPS, KHUDWANI, ANANTNAG, 192102, JAMMU & KASHMIR, INDIA

PUSHP SHARMA

DEPARTMENT OF PLANT BREEDING AND GENETICS, PUNJAB AGRICULTURAL UNIVERSITY, LUDHIANA-141 004, INDIA

HIMAKSHI BHATI- KUSHWAHA

SCHOOL OF LIFE SCIENCES, JAIPUR NATIONAL UNIVERSITY, JAIPUR, RAJASTHAN, INDIA

CONTENTS

1

Engineering Abiotic Stress Tolerance in Plants: Extricating Regulatory Gene Complex

Shabir H Wani, N.B. Singh, Th. Renuka Devi, A. Haribhushan and *Jeberson SM*

Environmental factors such as drought, salinity and extreme temperatures are responsible for low productivity of crop plants worldwide. Plants adapt to such situations by various mechanisms. An insight into the different complex mechanisms can lead to the solution of the mystery of abiotic stress tolerance in plants. Several researchers all over the world have tried to unravel the gene networks responsible for abiotic stress tolerance. Initially single gene delivery system was the common means of developing transgenic plants against abiotic stress tolerance. After the discovery of transcription factors, it became clear that abiotic stress tolerance is regulated by numerous genes which get up regulated during stress conditions. In this review some of the main transcription factor genes utilized for development of transgenic abiotic stress tolerant plants will be discussed.

Introduction

Drought, salinity and low temperature, are major adverse environmental factors that limit plant productivity (Boyer, 1982). Drought and salinity are already widespread in many regions and are expected to cause serious salinization of more than 50% of all arable lands by 2050 (Vinocur and Altman, 2005). The world food grain production needs to be doubled by the year 2050 to meet the ever growing demands of the population (Tilman *et al.*, 2002) which is going to reach 9 billion by that time (Virmani and Ilyas-Ahmed, 2007). Abiotic stresses present a major challenge in our quest for sustainable food production as these may reduce the potential yields as high as 70% in crop plants (Katiyar-Agarwal *et al.*, 2006). There is an increasing scarcity of fresh water and plants account for about 65% of global fresh water use (Postel *et al.*, 1996). Soil salinity limits crop production in about 20% of irrigated lands (Flowers and Yeo, 1995). Drought and salinity stresses are also limiting crop production even under irrigated conditions (Chinnusamy *et al.,* 2006). Plant's ability to tolerate water deficits (as during drought or freezing/cold) and high salt levels depends heavily on its water status which is altered in response to environmental conditions (Bray, 1993, 1997; Bohnert *et al.*, 1995; Steudle and Peterson, 1998; Blumwald, 2000; Zhu *et al.*, 2005). Therefore, improvement of abiotic stress tolerance might increase actual yields extensively in the most crops (Todake *et al.*, 2012). Understanding the mechanisms by which plants recognize and transduce the stress signals to instigate adaptive responses is crucial for engineering stress-tolerant crop plants. Various stress-induced genes have been recognized, including those encoding vital enzymes for abscisic acid (ABA) biosynthesis (Nambara and Marion-Poll, 2005), proteins such as proline, trehalose, mannitol, ectoine, glycine betaine involved in osmotic adaptation and tolerance to cellular dehydration (Wani *et al.*, 2008; Wani and Gosal, 2011; Yao *et al.*, 2011), cellular protective enzymes (Puckette *et al.*, 2007), numerous signalling proteins such as protein kinases/protein phosphatases (Zhu, 2002), and transcription factors (Mochida *et al.*, 2009, Wani *et al.*, 2008; Gosal *et al.*, 2010). Until now, we have seen that genetic engineering for stress tolerance included genes involved in encoding enzymes required for the biosyntheses of various osmo protecants (Wani *et al.*, 2008; Wani and Gosal, 2011; Yao *et al.*, 2011). Other genes that have been selected for transformation include those that encode enzymes for modifying membrane lipids, *LEA* proteins and detoxification enzymes. In all the above experiments, a single gene for a protective protein or an enzyme was constitutively expressed in the transgenic plants. Many of these genes have been shown to function in stress conditions (Wani *et al.*, 2008; Gosal *et al.*, 2010; Wani and Gosal, 2011).

The gene regulating protein factors that regulate gene expression and signal transduction and that function in stress responses may be useful for improving the tolerance of plants to abiotic stresses by gene transfer. These genes comprise of regulatory proteins i.e., transcription factors (*bZip, MYC, MYB, DREB and NACs* (*NAM, ATAF* and *CUC*) etc.), protein kinases (*MAP* kinase, CDP kinase, receptor protein kinase, ribosomal protein kinase and transcription regulation protein kinase etc.) and proteinases (phosphoesterases and phospholipase, etc) (Thomashow, 1999; Zhu, 2002; Seki *et al.*, 2003; Shinozaki *et al.*, 2003; Fujita *et al.*, 2004; Mukhopadhyay *et al.*, 2004; Katiyar-Agarwal *et al.* 2006; Wani *et al.*, 2008; Gosal *et al.*, 2010). Some of the transcription factors known to be responsible for abiotic stress tolerance and have been used for development of transgenic plants against various extreme environmental stresses are described below:

DREB (Dehydration Responsive Element Binding)

There has been a continuous and tremendous research on trancriptional control of abiotic stress in model and crop plants. So to enumerate all that work is out of the scope of this review. More emphasis will be given on work done on drought, salinity and cold tolerance in regard to transcription factors especially *DREB* family.

The transcription factors activate cascades of genes that act together in enhancing tolerance towards multiple stresses. It has been revealed that dozens of transcription factors are involved in the plant response to drought stress (Vincour and Altman, 2005; Bartels and Sunkar, 2005). The transcription activation factors interact with cis-elements present in the promoter region of various abiotic stress related genes and thus upregulating the expression of many genes resulting in imparting tolerance to abiotic stresses. Model plant *Arabidopsis* dedicates about 5.9% of its genome coding for more than 1500 transcription factors (Riechmann *et al.,* 2000). The transcription factors which are involved in stress responsive gene expression are traditionally classified in ABA dependent and ABA- independent regulatory pathways (Shinozaki and Yamaguchi Shinozaki, 2000; Thomashow, 1999; Xiong *et al.,* 2002). The DRE (dehydration responsive element) CRT (C-Repeat) was identified as a *cis*-acting element regulating gene expression in response to dehydration (salt, drought and cold stresses) in *Arabidopsis* (Yamaguchi-Shinozaki and Shinozaki, 1994). Initial attempts for genetic transformation using these *DREB* genes started with *Arabidopsis.* Over expression of *DREB1/CBF* in *Arabidopsis* resulted in the activation of expression of many stress tolerance genes and tolerance of the plant to drought, high salinity and/or freezing was improved (Jaglo-Ottosen *et al.,* 1998). Transgenic plants overexpressing *DREB1/CBF3* under the control of CaMV 35S promoter also showed increased tolerance to drought, high salinity and freezing stress (Kausga *et al.,* 1999, Gilmour *et al.,* 2000). Constitutive expression of *DREB1A* protein leads to growth retardation under normal growth conditions. Thus the use of stress inducible *rd 29A* promoter for the over expression of *DREB1A* minimizes negative effects on plant growth (Kasuga *et al.,* 1999).

In their first attempt to transform wheat with *DREB1A* transcription factor, (Pellegrineschi *et al.,* 2004) used particle gun method as the means for gene transfer. *DREB1A* was transformed under the control of stress inducible *rd29A* promoter. Under green house conditions substantial resistance to water deficit was demonstrated. Oh *et al.* (2005) ectopically expressed *Arabidopsis DREB1A /(CBF3)* in transgenic rice plants under the control of CaMV 35S promoter. *DREB1A* transgenic rice plants showed enhanced tolerance to drought and salinity but to a little extent for low temperature stress without any stunted phenotype despise its constitutive expression.

Four rice *CBF/DREB1A* orthologs, *OsDREB1A, OsDREB1B, OsDREB1C* and *OsDREB1D* have been isolated (Dubouzet *et al.,* 2003), *OsDREB1* transgenic rice plants conferred improved tolerance to drought, salt and low temperatures. On further analysis a large portion of stress inducible genes were identified. Thus giving the confirmation those DREB1/CBF cold responsive pathways is conserved in *Arabidopsis* and rice (Ito *et al.,* 2006). In another attempt constitutive or stress inducible expression of *ZmDREB2A* resulted in an improved drought stress tolerance in plants.

In addition, transgenic plants showed enhanced thermo tolerance, thus indicating that *ZmDREB2A* has a dual function mediating expression of genes responsive to both water and heat stress (Qin *et al.,* 2007). Similar results were earlier found in *Arabidopsis* plants overexpressing constitutive active *DREB2A* in which transgenic plants showed increased thermo tolerance in addition to tolerance against water stress (Sakuma *et al.,* 2006).

Stress tolerance is a complex trait so unlikely to be under single gene control. Wise strategy is to use genetic engineering to switch on a transcription factor regulating the expression of several genes related to abiotic stress. Transpiration efficiency is the most important component trait of drought. About 90% of the rooted shoots transformed with rd *29A:DREB1A* construct survived and appeared to be phenotypically normal. Fifty three *rd29A:DREB1A* and eighteen *35S:DREB1A* plants were successfully transferred to green house and their seeds collected. Transgenic peanut had higher TE than the wild type under well watered conditions.

Some transgenic events showed 70% increase over wild type JL 24, RD 2 had 40% more TE than WT JL 24 under water limited conditions (Bhatnagar-Mathur *et al.,* 2007). Transgenic over-expression of *HvCBF4* from barley in rice resulted in an increase in tolerance to drought, high-salinity and low-temperature stresses without stunting growth. Using the 60 K Rice Whole Genome microarrays, 15 rice genes were identified that were activated by *HvCBF4.* When compared with 12 target rice genes of *CBF3/DREB1A,* five genes were common to both *HvCBF4* and *CBF3/DREB1A,* and ten and seven genes were specific to *HvCBF4* and CBF3/*DREB1A,* respectively (Oh *et al.,* 2007) (Table 1).

OsDREB1F is induced by salt, drought, and cold stresses, and also ABA application (Wang *et al.,* 2008). Rice plants introduced with *OsDREB1F* gene exhibited increased stress tolerance to salt, drought, and cold stress. Overexpression of *OsDREB1F* in *Arabidopsis* enhanced expression of stress-inducible genes, such as *rd29B* and *RAB18,* that contain the DRE/CRT element in their promoter regions, which suggests that OsDREB1F might also participate in expression of osmotic stress responsive genes through the ABA-dependent signaling pathway. Overexpression of *OsDREB1G* also promotes drought tolerance (Chen *et al.,* 2008).

Transgenic *Arabidopsis* plants overexpressing the functional form of *OsDREB2B* displayed enhanced expression of DREB2A target genes and improved tolerance to drought and heat stresses (Matsukura *et al.,* 2010). These analyses indicate *OsDREB2B* is a key gene that encodes the DREB2-type TF that functions in stress-responsive gene expression in rice. Similarly, overexpression of *DREB2A CA* in transgenic *Arabidopsis* increased expression of many stress-inducible genes and resulted in enhanced tolerance to drought, salt and heat stresses (Mizoi *et al.,* 2011). The protein stability mediated by ubiquitin E3 ligases is important for activation of DREB2A (*Qin et al.,* 2011).

ZPF (Zinc finger proteins)

Zinc finger proteins (*ZFPs*) play important roles in growth and development in both animals and plants. *Arabidopsis* genes encoding distinct *ZFPs* have been identified. However, the physiological role of their homologues with putative zinc finger motif remains unclear. *ThZF1,* a novel gene was characterized from salt stressed cress (*Thellungiella halophila,* Shan Dong), encoding a functional transcription factor. *ThZF1* contains two conserved C2H2 regions and shares conserved domains, including DNA-binding motif, with *Arabidopsis thaliana ZFP* family members. The transcript of the *ThZF1* gene was induced by salinity and drought. Transient expression analysis of *ThZF1–GFP* fusion protein revealed that *ThZF1* was localized preferentially in nucleus. A gel-shift assay showed that *ThZF1* specially bind to the wild-type (WT) *EP2* element, a *cis*-element present in the promoter regions of several target genes regulated by *ZFPs.* Furthermore, a functional analysis demonstrated that *ThZF1* was able to activate HIS marker gene in yeast. Finally, ectopic expression of *ThZF1* in *Arabidopsis* mutant azf2 suggested that *ThZF1* may have similar roles as *Arabidopsis AZF2* in plant development as well as regulation of downstream gene (Xu *et al.,* 2007). In a recent observation by Chen *et al.,* 2012, transgenic rice plants overexpressing OsbZIP16 exhibited significantly improved drought resistance at both the seedling and tillering stages, which

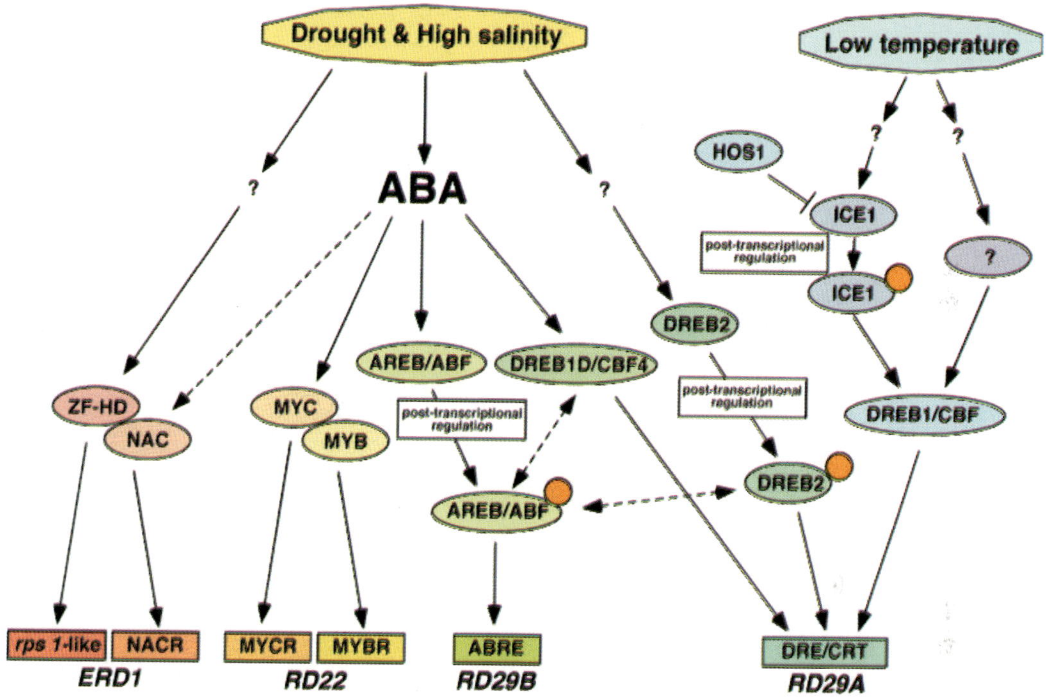

Fig 1: Regulatory network of gene expression in response to drought, high salinity and cold stresses: specificity and crosstalk of gene networks

* Cis-acting elements that are involved in stress-responsive transcription are shown in boxes. Transcription factors that control stress-inducible gene expression are shown in circles or ovals. Small circles indicate the modification of transcription factors in response to stress signals for their activation, such as phosphorylation. Dotted lines indicate possible regulation. Double arrow lines indicate possible cross talk. ABF: ABRE-binding factor, ABRE: ABA-responsive element, AREB: ABRE-binding protein, CBF: C-repeat-binding factor, CRT: C-repeat, DRE: dehydration responsive element, DREB: DRE-binding protein, ERD: early responsive to dehydration, ICE: inducer of CBF expression, MYBR: MYB recognition site, MYCR: MYC recognition site, NACR: NAC recognition site, RD: responsive to dehydration, ZF-HD: zinc finger homeodomain protein. (Source: Nakashima and Yamaguchi-shinozaki, 2005)

was positively correlated with the observed expression levels of OsbZIP16. Representative downstream drought-inducible genes were observed to have significantly higher expression levels in transgenic rice plants than in the wild type plants under drought conditions. OsbZIP16 was shown to be induced by exogenous ABA treatment, while overexpression of OsbZIP16 was observed to make transgenic plants more sensitive to ABA than wild type plants were.

NAC (NAM ATAF and CUC)

NAC transcription factors play important roles in plant growth, development and stress responses (Hao *et al.,* 2011). The NAC gene family encodes one of the largest families of plant specific transcription factors and is absent in other eukaryotes (Gao *et al.,* 2007). Rice and *Arabidopsis* genomes contain 75 and 105 putative *NAC* genes, respectively. The role of NAC gene family in abiotic stresses was discovered in *Arabidopsis*. Three NAC genes were induced under salt and/ or drought stress. Overexpression of three genes *ANACO19, ANACO55, ANACO72,* greatly enhanced drought tolerance in *Arabidopsis* (Tran *et al.,* 2004). Multiple rice transcription factors , including a NAC gene were induced in the early stages of salt stress (Chao *et al.,* 2005) *OsNAC6*, a member of *ATAF* family, was also induced by cold, salt, drought and abscisic acid (ABA) (Ohnishi *et al.,* 2005). Overexpression of stress responsive gene *SNAC1* (stress-responsive *NAC* 1) significantly enhances drought resistance in transgenic rice (22–34% higher seed setting than control) in the field under severe drought stress conditions at the reproductive stage while showing no phenotypic changes or yield penalty. The transgenic rice also shows significantly improved drought resistance and salt tolerance at the vegetative stage. Compared with wild type, the transgenic rice were more sensitive to abscisic acid and lost water more slowly by closing more stomatal pores, yet displayed no significant difference in the rate of photosynthesis. DNA chip analysis revealed that a large number of stress-related genes were up-regulated in the *SNAC1*-overexpressing rice plants. *SNAC1* holds promising utility in improving drought and salinity tolerance in rice (Hu *et al.,* 2006). Similarly several *NAC* genes are found to be induced by drought, high salinity, and cold stresses. *OsNAC6/SNAC2* encodes a NAC transcription factor in rice (Hu *et al.,* 2008). Expression of *OsNAC6*/SNAC2 is induced by cold, drought and high salinity. Transgenic rice plants overexpressing *OsNAC6* showed enhanced tolerance to dehydration and high salinity stresses, although the plants showed growth retardation and low yield under the non-stress condition. This low yield was increased in transgenic rice plants driven by a stress-inducible promoter. These results indicate OsNAC6 functions as a transcriptional activator in abiotic stress responses (Chung *et al.,* 2009). In a similar experiment *OsNAC5* expression was induced by dehydration, cold, ABA, and MeJA treatments. The growth of rice plants overexpressing *OsNAC5* was comparable to that of control plants under non-stress conditions. The *OsNAC5* overexpressors exhibited enhanced tolerance to high salinity. In another study it was reported that RNA interference (RNAi) transgenic rice plants with reduced *OsNAC5* expression exhibited a reduced amount of tolerance to abiotic stresses than control plants, whereas overexpression of *OsNAC5* improved abiotic stress tolerance. It was also revealed that accumulation of compatible solutes such as proline and soluble sugars was absolutely linked with *OsNAC5* expression level (Song *et al.,* 2011). A recent report revealed that GmNAC11 acts as a transcriptional activator, whereas GmNAC20 functions as a mild repressor; however, the C-terminal end of GmANC20 has transcriptional activation activity. Over-expression of *GmNAC20* increased salt and freezing tolerance in transgenic *Arabidopsis* plants; however, over-expression only improved salt tolerance. Over-expression of *GmNAC20* also enhanced lateral root formation (Hao *et al.*, 2011). Thus the above study revealed that GmNAC20 may regulate stress tolerance through activation of the *DREB/ CBF–COR* pathway, and may control lateral root development by altering auxin signaling-related genes. Recently, an attempt was made to introduce NAC1 gene from *E. coli* into tobacco through *Agrobacterium* mediated transformation (Ramegowda *et al.*, 2012). In another experiment, overexpression of TaNAC2 resulted in enhanced tolerances to drought, salt, and freezing stresses in *Arabidopsis*, which were simultaneously demonstrated by enhanced expression of abiotic stress-response genes and several physiological indices (Mao *et al.*, 2012). The transgenic plants resulted in increased tolerance towards abiotic stresses. The above results thus indicate that NAC genes also play an important role in the transcriptional complex of abiotic stress responses in plants.

Conclusion and Future Perspective

Significant and outstanding progress has been made in unraveling the role of transcription factor genes like DREB, NAC and ZFPs towards tolerance to abiotic stress factors such as drought, salinity and extreme temperatures. The mammoth task of development of transgenic plants harboring such transcription factor genes is still a big challenge. Almost in every study the experiment is limited to laboratory and green house conditions. But the question is whether these transgenic plants show better performance at filed level where complex abiotic stress factors are available. Also a second concern is about the selection of crop as most of the initial experiments are being done on model crops like *Arabidopsis* and tobacco. Future thrust must be laid on commercially grown crops like rice, wheat and other major cereal crops which are the most affected by abiotic stress conditions.

REFERENCES

Aharoni A, Dixit S, Jetter R, Thoenes E, Van Arkel G, Pereira A (2004) The SHINE clade of AP2 domain transcription factors activates wax biosynthesis, alters cuticle properties, and confers drought tolerance when overexpressed in *Arabidopsis*. *The Plant Cell* 16: 2463–2480

Bartels D and Sunkar R (2005) Drought and salt tolerance in plants. *Crit Rev Plant Sci* 21: 1–36

Behnam B, Kikuchi K, Celebi-Toprak F, Yamanaka S, Kasuga M, Yamaguchi Shinozaki K, Watanabe KN (2006) The *Arabidopsis* DREB1A gene driven by the stress-inducible rd29A promoter increases salt-stress tolerance in tetrasomic tetraploid potato (*Solanum tuberosum*) in proportion to its copy number. *Plant Biotechnol* 23: 169–177

Bhatnagar-Mathur P, Devi MJ, Reddy DS, Lavanya M, Vadez V, Serraj R, Yamaguchi-Shinozaki K, and Sharma KK (2007) Stress inducible expression of *At DREB1A* in transgenic peanut *(Arachis hypogaea)* increases transpiration efficiency *Plant Cell Rep* 26(12): 2071-82

Blumwald E (2000) Sodium transport and salt tolerance in plants. *Curr Opin Cell Biol* 12: 431–434

Bohnert HJ, Nelson DE, Jensen RG (1995) Adaptation to environment stress. *Plant Cell* 7: 1099–1111

Boyer JS (1982) Plant productivity and environment. *Science* 218: 443–448

Bray EA (1993) Molecular responses to water deficit. *Plant Physiol* 103: 1035–1040

Bray EA (1997) Plant responses to water deficit. *Trends Plant Sci* 2: 48–54

Chen H, Chen W, Zhou J, He H, Chen L, Chen H, Deng XW (2012) Basic leucine zipper transcription factor OsbZIP16 positively regulates drought resistance in rice. *Plant Science* 193: 8–17

Chen JQ, Meng XP, Zhang Y, Xia M, Wang XP (2008) Over-expression of OsDREB genes lead to enhanced drought tolerance in rice. *Biotechnol Lett* 30: 2191–2198

Chen JR, Lü JJ, Wang TX, Chen SY, Wang HF (2009) Activation of a DRE-binding transcription factor from *Medicago truncatula* by deleting a Ser/Thr-rich region. *In Vitro Cell Dev Biol Plant* 45: 1–11

Chen M, Xu Z, Xia L, Li L, Cheng X, Dong J, Wang Q and Ma Y (2009) Cold-induced modulation and functional analyses of theDRE-binding transcription factor gene, GmDREB3, in soybean (*Glycine max* L.). *J Exp Bot* 60: 121-135

Chinnusamy V, Xiong L and Zhu JK (2006) Use of genetic engineering and molecular biology approaches for crop improvement for stress environments. In: Ashraf M. and Harris P. J. C, editors. Abiotic Stresses, Plant Resistance through Breeding and Molecular approach. U.P. India: IBDC. p. 47-89

Chung PJ, Kim YS, Jeong JS, Park SH, Nahm BH, Kim JK (2009) The histone deacetylase OsHDAC1 epigenetically regulates the OsNAC6 gene that controls seedling root growth in rice. *Plant J* 59: 764–776

Dubouzet JG, Sakuma Y, Ito Y, Kasuga M, Dubouzet EG and Miura S (2003) OsDREB genes in rice, *Oryza sativa* L. encode transcription activators that function in drought, high-salt- and cold-responsive gene expression. *Plant J* 33: 751–763

Flowers T J and Yeo A R (1995) Breeding for salinity resistance in crop plants - where next? *Aust J Plant Physiol* 22: 875-884

Fujita M, Fujita Y, Maruyama K, Seki M, Hiratsu K, Ohme- Takagi M, Tran L, Yamaguchi-Shinozaki K, Shinozaki K (2004) A dehydration-induced NAC protein, RD26, is involved in a novel ABA dependent stress-signaling pathway. *The Plant Journal* 39: 863–876

Gao JP, Chao DY and Lin HX (2007) Understanding abiotic stress tolerance mechanisms: recent studies on stress response in rice. *J Integ Plant Biol* 49 (6): 742-750

Gilmour SJ, Sebolt AM, Salazar MP, Everard JD and Thomashow MF (2000) Overexpression of the *Arabidopsis* CBF3 transcriptional activator mimics multiple biochemical changes associated with cold acclimation. *Plant Physiol.* 124: 1854–1865

Gosal SS, Wani SH and Kang MS (2010) Biotechnology and Crop Improvement. *Journal Crop Improvement* 24: 1-65

Guo ZJ, Chen XJ, Wu XL, Ling JQ, Xu P (2004) Overexpression of the AP2/EREBP transcription factor OPBP1 enhances disease resistance and salt tolerance in tobacco. *Plant Molecular Biology* 55: 607–618

Hao YJ, Wei W, Song QX, Chen HW, Zhang YQ, Wang F, Zou HF, Lei G, Tian, AG, Zhang WK, Ma B, Zhang JS, Chen SY(2011) Soybean NAC transcription factors promote abiotic stress tolerance and lateral root formation in transgenic plants *Plant J* 68 (2): 302–313

Hsieh TH, Lee JT, Charng YY and Chan MT (2002) Tomato plants ectopically expressing *Arabidopsis CBF1* show enhanced resistance to water deficit stress. *Plant Physiol* 130: 618–626

Hu H, Dai M, Yao J, Xiao B, Li X, Zhang Q and Xiong L (2006) Overexpressing a NAM, ATAF, and CUC (NAC) transcription factor enhances drought resistance and salt tolerance in rice. *Proc Natl Acad Sci* USA 103: 12987–12992

Hu H, You J, Fang Y, Zhu X, Qi Z, Xiong L (2008) Characterization of transcription factor gene SNAC2 conferring cold and salt tolerance in rice. *Plant Mol Biol* 67: 169–181

Huang J, Sun SJ, Xu DQ, Yang X, Bao YM, Wang ZF, *et al.* (2009) Increased tolerance of rice to cold, drought and oxidative stresses mediated by the overexpression of a gene that encodes the zinc finger protein ZFP245. *Biochem Biophys Res Commun* 389: 556-561

Ito Y, Katsura K, Maruyama K, Taji T, Kobayashi M, Seki M (2006) Functional analysis of rice DREB1/CBF-type transcription factors involved in cold-responsive gene expression in transgenic rice. *Plant Cell Physiol* 47: 141–153

Jaglo-Ottosen KR, Gilmour SJ, Zarka DG, Schabenberger O and Thomashow MF (1998) *Arabidopsis* CBF1 overexpression induces COR genes and enhances freezing tolerance. *Sci* 280: 104–106

Jung J, Won SY, Suh SC, Kim HR, Wing R, Jèong Y, Hwang I, Kim M (2007) The barley ERF-type transcription factor *HvRAF* confers enhanced pathogen resistance and salt tolerance in *Arabidopsis. Planta* 225: 575–588

Karaba A, Dixit S, Greco R, *et al.* (2007) Improvement of water use efficiency in rice by expression of HARDY, an *Arabidopsis* drought and salt tolerance gene. *Proceedings of the National Academy of Sciences,* USA 104: 15270–15275

Kasuga M, Q Liu, S Miura, K Yamaguchi-Shinozaki, K Shinozaki (1999) Improving plant drought, salt, and freezing tolerance by gene transfer of a single stress inducible transcription factor. *Nat Biotechnol* 17: 287–291

Katiyar-Agarwal P, Agarwal P, Reddy M K and Sopory S K (2006) Role of DREB transcription factors in abiotic and biotic stress tolerance in plants. *Plant Cell Rep* 25: 1263–1274

Kim JC, Lee SH, Cheong YH, *et al.* (2001) A novel cold-inducible zinc finger protein from soybean, SCOF-1, enhances cold tolerance in transgenic plants. *The Plant Journal* 25: 247–259

Kobayashi F, MaetaE, Terashima A, KawauraK, Ogihara Y, Takumi S (2008) Development of abiotic stress tolerance via bZIP-type transcription factor LIP19 in common wheat. *J Exp Bot* doi: 10.1093/jxb/ern014

Lisso J, Altmann T, Mussig C (2006) The AtNFXL1 gene encodes a NF-X1 type zinc finger protein required for growth under salt stress. *FEBS Letters* 580: 4851–4856

Liu N, Zhong NQ, Wang GL, Li LJ , Liu XL, He YK, Xia GX (2007) Cloning and functional characterization of PpDBF1 gene encoding a DRE-binding transcription factor from *Physcomitrella patens. Planta* 226: 827–838

Ma Q, Dai X, Xu Y, Guo J, Liu Y, Chen N, Xiao J, Zhang D, Xu Z, Zhang X and Chong K (2009) Enhanced Tolerance to Chilling Stress in OsMYB3R-2 Transgenic Rice Is Mediated by Alteration in Cell Cycle and Ectopic Expression of Stress Genes *Plant Physiology* 150: 244–256

Mao X, Zhang H, Qian X, Li A, Zhao G and Jing R (2012) TaNAC2, a NAC-type wheat transcription factor conferring enhanced multiple abiotic stress tolerances in *Arabidopsis. Journal of Experimental Botany*, doi: 10.1093/jxb/err462

Matsukura S, Mizoi J, Yoshida T, Todaka D, Ito Y, Maruyama K, et al (2010) Comprehensive analysis of rice DREB2-type genes that encode transcription factors involved in the expression of abiotic stress-responsive genes. *Mol Genet Genomics* 283: 185–196

Mizoi J, Shinozaki K, Yamaguchi-Shinozaki K (2011) AP2/ERF family transcription factors in plant abiotic stress responses. *Biochim Biophys Acta* 1819: 86–96

Mukhopadhyay A, Vij S, Tyagi AK (2004) Overexpression of a zinc finger protein gene from rice confers tolerance to cold, dehydration, and salt stress in transgenic tobacco. *Proceedings of the National Academy of Sciences*, USA 101: 6309–6314

Nakashima K and Yamaguchi-Shinozaki K (2005) Molecular studies on stress-responsive gene expression in *Arabidopsis* and improvement of stress tolerance in crop plants by regulon biotechnology. *JARQ* 39 (4): 221-229

Nambara E, Marion-Poll A (2005) Abscisic acid biosynthesis and catabolism. *Annual Review of Plant Biology* 56: 165–185

Nelson DE, Repetti PP, Adams TR, *et al.* (2007) Plant nuclear factor Y(NF-Y) B subunits confer drought tolerance and lead to improved corn yields on water-limited acres. *Proceedings of the National Academy of Sciences*, USA 104: 16450–16455

Ning J, Li X, Hicks L M and Xiong L (2010) A Raf-Like MAPKKK Gene DSM1 Mediates Drought Resistance through Reactive Oxygen Species Scavenging in Rice. *Plant Physiology* 152: 876–890

Oh SJ, Kwon CW, Choi DW, Song SI and Kim JK (2007) Expression of barley HvCBF4 enhances tolerance to abiotic stress in transgenic rice. *Plant Biotech Journal* 5: 646–656

Oh SJ, Song SI, Kim YS, Jang HJ, Kim SY and Kim M (2005) *Arabidopsis* CBF3/DREB1A and ABF3 in transgenic rice increased tolerance to abiotic stress without stunting growth. *Plant Physiol* 138: 341–351

Oh SJ, Kim YS, Kwon CW, Park HK, Jeong JS, Kim JK (2009) Overexpression of the transcription factor AP37 in rice improves grain yield under drought conditions. *Plant Physiol* 150: 1368-1379

Ohnishi T, Sugahara S, Yamada T, Kikuchi K, Yoshiba Y and Hirano HY (2005) OsNAC6, a member of the NAC gene family, is induced by various stresses in rice. *Genes Genet Syst* 80: 135- 139

Park JM, Park CJ, Lee SB, Ham BK, Shin R, Paek KH (2001) Overexpression of the tobacco Tsi1 gene encoding an EREBP/AP2-type transcription factor enhances resistance against pathogen attack and osmotic stress in tobacco. *The Plant Cell* 13: 1035–1046

Pellegrineschi A, Reynolds M, Pacheco M, Brito RM, Almeraya R, Yamaguchi-Shinozaki K and Hoisington D (2004) Stress-induced expression in wheat of the *Arabidopsis* thaliana DREB1A gene delays water stress symptoms under greenhouse conditions. *Genome* 47: 493–500

Postel SL, Daily GC and Ehrlich PR (1996) Human appropriation of renewable fresh water. *Sci.* 271: 785-788

Puckette MC, Weng H, Mahalingam R (2007) Physiological and biochemical responses to acute ozone-induced oxidative stress in *Medicago truncatula. Plant Physiology and Biochemistry* 45: 70–79

Qin F, Shinozaki K, Yamaguchi-Shinozaki K (2011) Achievements and challenges in understanding plant abiotic stress responses and tolerance. *Plant Cell Physiol* 52: 1569–1582.

Qin F, M Kakimoto Y, Sakuma K, Maruyama Y, Osakabe LSP, Tran K, Shinozaki K, Yamaguchi-Shinozaki (2007) Regulation and functional analysis of ZmDREB2A in response to drought and heat stress in *Zea mays* L. *Plant J* 50: 54-69

Ramegowda V, Senthil-Kumar M, Nataraja KN, Reddy MK, Mysore KS, Udayakumar M (2012) Expression of a Finger Millet Transcription Factor, EcNAC1, in Tobacco Confers Abiotic Stress-Tolerance. *PLoS ONE* 7(7): e40397.

Riechmann JL, Heard J, Martin G, Reuber L, Jiang C, Keddie J, Adam L, Pineda O, Ratcliffe OJ and Samaha RR (2000) *Arabidopsis* transcription factors: genome -wide comparative analysis among eukaryotes. *Sci.* 290: 2105–2110

Sakuma Y, Maruyama K, Osakabe Y, Qin F, Seki M, Shinozaki K, Yamaguchi-Shinozaki K (2006) Functional analysis of an *Arabidopsis* transcription factor, DREB2A, involved in drought responsive gene expression. *The Plant Cell* 18: 1292–1309

Seki M, Kamei A, Yamaguchi-Shinozaki K, Shinozaki K (2003) Molecular responses to drought, salinity and frost: common and different paths for plant protection. *Current Opinion in Biotechnology* 14: 194–199

Shinozaki K and Yamaguchi-Shinozaki K (2000) Molecular responses to dehydration and low temperature: differences and cross-talk between two stress signalling pathways. *Curr Opin Plant Biol* 3: 217–223

Shinozaki K, Yamaguchi-Shinozaki K, Seki M (2003) Regulatory network of gene expression in the drought and cold stress responses. *Current Opinion in Plant Biology* 6: 410–417

Shou H, Bordallo P and Wang K (2004) Expression of the *Nicotiana* protein kinase (*NPK1*) enhanced drought tolerance in transgenic maize. *J Exp Bot* 55: 301-308

Song SY, Chen Y, Chen J, Dai XY, Zhang WH (2011) Physiological mechanisms underlying OsNAC5-dependent tolerance of rice plants to abiotic stress. *Planta* 234: 331–345.

Steudle E, Peterson CA (1998) How does water get through roots? *J Exp Bot* 49: 775–788

Suzuki N, Rizhsky L, Liang H, Shuman J, Shulaev V and Mittler R (2005) Enhanced tolerance to environmental stress in transgenic plants expressing the transcriptional coactivator multiprotein bridging factor 1. *Plant Physiol* 139: 1313–1322

Tang M, Sun J, Liu Y, Chen F, Shen S (2007) Isolation and functional characterization of the JcERF gene, a putative AP2/EREBP domain containing transcription factor, in the woody oil plant *Jatropha curcas*. *Plant Molecular Biology* 63: 419–428.

Tang W, Newton RJ, Lin JX, Charles TM (2006) Expression of a transcription factor from *Capsicum annuum* in pine calli counteracts the inhibitory effects of salt stress on adventitious shoot formation. *Molecular Genetics and Genomics* 276: 242–253

Thomashow MF (1999) Plant cold acclimation: freezing tolerance genes and regulatory mechanisms. *Annual Review of Plant Physiology and Plant Molecular Biology* 50: 571–599

Tilman D, Cassman KG, Matson PA, Naylor R and Polasky S (2002) Agricultural sustainability and intensive production practices. *Nature* 418: 671-677

Todaka D, Nakashima K, Shinozaki K, Yamaguchi-Shinozaki K (2012) Toward understanding transcriptional regulatory networks in abiotic stress responses and tolerance in rice. *Rice* 5(6): 1-9

Tran LS, Nakashima K, Sakuma Y, Simpson SD, Fujita Y and Maruyama K (2004) Isolation and functional analysis of *Arabidopsis* stress-inducible NAC transcription factors that bind to a drought-responsive cis-element in the early responsive to dehydration stress 1 promoter. *Plant Cell* **16**: 2481–2498

Vinocur B and Altman A (2005) Recent advances in engineering plant tolerance to abiotic stress: achievements and limitations. *Curr Opin Biotechnol* 16: 123–132

Virmani SS and Ilyas-Ahmed M (2007) Rice breeding for sustainable production. In: Kang M. S. and Priyadarshan P. M, editors. Breeding Major Food Staples. USA: Blackwell publ. p.141-191

Wani SH and Gosal SS (2011) Introduction of *OsglyII* gene into *Indica* rice through particle bombardment for increased salinity tolerance. *Biologia Plantarum* 55 (3): 536-540

Wani SH, Sandhu JS and Gosal SS (2008) Genetic engineering of crop plants for abiotic stress tolerance. In: Advanced Topics in Plant Biotechnology and Plant Biology. CP Malik, B Kaur and C Wadhwani, eds. MD Publications New Delhi pp. 149-183

Xiong L, Lee H, Ishitani M and Zhu JK (2002) Regulation of osmotic stress responsive gene expression by LOS6/ABA1 locus in *Arabidopsis*. *J Biol Chem* 277: 8588–8596

Xu DQ, Huang J, Guo SQ, Yang X, Bao YM, Tang HJ, et al. (2008) Overexpression of a

TFIIIA-type zinc finger protein gene ZFP252 enhances drought and salt tolerance in rice (*Oryza sativa* L.). *FEBS Lett* 582: 1037-1043

Xu SM, Wang XC and Chen J (2007) Zinc finger protein 1 (*ThZF1*) from salt cress (*Thellungiella halophila*) is a Cys-2/His-2-type transcription factor involved in drought and salt stress. *Plant Cell Rep* 26(4): 497-506

Yamaguchi-Shinozaki K and Shinozaki K (1994) A novel *cis*-acting element in an *Arabidopsis* gene is involved in responsiveness to drought, low-temperature, or high-salt stress. *Plant Cell* 6: 251-264

Yao D, Zhang X, Zhao X, *et al.* (2011) Transcriptome analysis reveals salt-stress-regulated biological processes and key pathways in roots of cotton (*Gossypium hirsutum* L.). *Genomics* 98: 47–55

Zhang X, Wollenweber B, Jiang D, Liu F, Zhao J (2008) Water deficits and heat shock effects on photosynthesis of a transgenic *Arabidopsis thaliana* mconstitutively expressing ABP9, a bZIP transcription factor. *Journal of Experimental Botany* 59: 839–848

Zhang Y, Yang C, Li Y, Zheng N, Chen H, Zhao Q, Gao T, Guo H and Xie Q (2007) SDIR1 is a ring finger E3 ligase that positively regulates stress-responsive abscisic acid signaling in *Arabidopsis*. *Plant Cell* 19: 1912-1929

Zhao H and Bughrara SS (2008) Isolation and characterization of cold-regulated transcriptional activator *LpCBF3* gene from perennial ryegrass (*Lolium perenne* L.). *Mol Genet Genomics* 279: 585–594

Zhu C, Schraut D, Hartung W, Schaffner AR (2005) Differential responses of maize MIP genes to salt stress and ABA. *J Exp Bot* 56: 2971–2981

Zhu J (2002) Salt and drought stress signal transduction in plants. *Annual Review of Plant Biology* 53: 247–273

Table 1 : List of genetically engineered plants for abiotic stress tolerance extricating regulatory gene complexes

Mechanism	*Transgene(s)*	*Plant species*	*Transformation method*	*Promoter*	*Remarks*	*Reference*
Regulatory genes	*NtTsi1*	*N. tabacum*	*Agrobacterium*	CaMV 35S	Transgenic plants showed lower loss of chlorophyll contents under high salt stress	Park *et al.*, 2001
	C2H2 zincfinger	*N. tabacum*	*Agrobacterium*	CaMV 35S	Transgenic plants exhibited faster recovery of chlorophyll content under cold stress conditions	Kim et al., 2001
	CBF1 from *A. thaliana*	*Lycopersicon esculentum*	*Agrobacterium*	CaMV 35S	Transgenic tomato plants were more resistant to water-deficit stress than the wild-type plants	Hsieh *et al.*, 2002
	SHN	*Arabidopsis*	*Agrobacterium*	CaMV 35S	Transgenic plants showed Drought tolerance and recovery with Reduced stomatal density	Aharoni *et al.*, (2004)
	NtOPBP1	*N. tabacum*	*Agrobacterium*	CaMV 35S	Under high salt stress, transgenic plants showed lower loss of chlorophyll content	Guo *et al.*, 2004
	DREB 1A from *A. thaliana*	*Triticum aestivum*	Particle gun	*rd 29 A*	Transgenic plants expressing *DREB1A* gene demonstrated substantial resistance to water stress under greenhouse conditions.	Pellegrineschi *et al.*, 2004

Mechanism	Transgene(s)	Plant species	Transformation method	Promoter	Remarks	Reference
	NPK1 from *N. tabacum*	*Zea mays*	*Agrobacterium* , Particle gun	35SC4PPDK	Transgenic maize plants showed enhanced drought tolerance. The *Agrobacterium*-derived events contained fewer than five copies of the NPK1 transgene, whereas the bombardment-derived events carried more than 20 copies of the transgene.	Shou *et al.*, 2004
	CBF3/DREB 1A from *A. thaliana*	*Oryza sativa*	-	Rice *Ubi* 1	Tolerance to drought and high salinity without growth retardation or any phenotypic alteration.	Oh *et al.*, 2005
	MBF1c from *A. thaliana*	*A. thaliana*	*Agrobacterium*	CaMV 35S	Constitutive expression of the stress-response transcriptional coactivator multiprotein bridging factor 1c (MBF1c) in *Arabidopsis* enhances the tolerance of transgenic plants to bacterial infection, heat, and osmotic stress. Moreover, the enhanced tolerance of transgenic plants to osmotic and heat stress was maintained even when these two stresses were combined.	Suzuki *et al.*, 2005
	DREB1A from *A. thaliana*	*Solanum tuberosum*	*Agrobacterium*	*rd29A*	Transgenic plants showed significant tolerance against salinity stress (1M NaCl)	Benham *et al.*, 2006
	SNAC 1 from *Oryza sativa*	*Oryza sativa*	*Agrobacterium*	CaMV 35S	Transgenic plants showed improved drought resistance under favorable conditions and strong tolerance to salt stress	Hu *et al.*, 2006

Mechanism	*Transgene(s)*	*Plant species*	*Transformation method*	*Promoter*	*Remarks*	*Reference*
	Os DREB 1A / Os DREB 1B from *Oryza sativa*	*Oryza sativa*	*Agrobacterium*	Ca MV 35S	*DREB 1* type genes are quite useful for improving stress tolerance in crop plants, including rice.	Ito *et al.*, 2006
	DREB2A from *A. thaliana*	*A. thaliana*	*Agrobacterium*	CaMV 35S	Significant drought-stress tolerance but slight freezing tolerance in transgenic *Arabidopsis* plants.	Sakuma *et al.*, 2006
	CaPF1 from Capsicum annum	*Pinus longaeva*	*Agrobacterium*	CaMV 35S	Transgenic plants exhibited Lower loss of chlorophyll contents under Oxidative stress.	Tang *et al.*, 2006
	AtNFXL1	*Arabidopsis*	*Agrobacterium*	CaMV 35S	Transgenic plants had improved maximum quantum efficiency of PSII under salt stress.	Lisso *et al.*, 2006
	JcERF	*Arabidopsis*	*Agrobacterium*	CaMV 35S	Overexprerssion of JcERF in transgenic Arabidopsis resulted in higher chlorophyll content and tolerance to high salt / freezing.	Tang *et al.*, 2007
	DREB 1A from *A. thaliana*	*Arachis hypogaea*	*Agrobacterium*	CaMV *35S, rd 29 A*	Transgenic plants showed increased transpiration efficiency, an important feature of drought tolerance.	Bhatnagar-Mathur *et al.*, 2007
	AtHRD	*Oryza sativa*	*Agrobacterium*	CaMV 35S	Transgenic rice plants showed lower stomatal conductance / enhanced photosynthesis assimilation and efficiency under Drought/ high salt stress	Karaba *et al.*, 2007

Mechanism	*Transgene(s)*	*Plant species*	*Transformation method*	*Promoter*	*Remarks*	*Reference*
	HvCBF4 from *Hordeum vulgare*	*Oryza sativa*	*Agrobacterium*	*Ubi1*	Transgenic rice resulted in an increase in tolerance to drought, high salinity and low temperature stresses without stunting growth.	Oh *et al.*, 2007
	Zm DREB 2A from *Zea mays*	*A. thaliana*	*Agrobacterium*	CaMV 35S	Drought stress tolerance in plants was improved.	Qin *et al.*, 2007
	SDIR1 from *Arabidopsis*	*Arabidopsis*	-	-	Transgenic plants showed ABA hypersensitivity and ABA-associated phenotypes, such as salt hypersensitivity in germination, enhanced ABA-induced stomatal closure, and enhanced drought tolerance.	Zhang *et al.*, 2007
	HvRAF (*Hordeum vulgare* root abundant factor)	*Arabidopsis*	*Agrobacterium*	CaMV 35S	Transgenic *Arabidopsis* plants showed enhanced resistance to *Ralstonia solanacearum*strain GMI1000, as well as seed germination and rootgrowth tolerance to high salinity indicating that HvRAF is a transcription factor that plays dual regulatory roles in response to biotic and abiotic stresses in plants.	Jung *et al.*, 2007
	PpDBF1 from Physcomitrella patens	*N. tabacum*,	*Agrobacterium*	CaMV 35S	Transgenic tobacco plants overexpressing *PpDBF1* gained higher tolerance to salt, drought and cold stresses.	Liu *et al.*, 2007

Mechanism	*Transgene(s)*	*Plant species*	*Transformation method*	*Promoter*	*Remarks*	*Reference*
	NF-Y (HAP	*Arabidopsis*	*Agrobacterium*	CaMV 35S	Under drought stress transgenic plants showed higher water potential and photosynthesis rates than controls.	Nelson *et al.*, 2007
	NF-Y (HAP	*Zea mays*	*Agrobacterium*	CaMV 35S	Transgenic plants showed higher chlorophyll index, photosynthesis rate and stomatal conductance under drought stress	Nelson *et al.*, 2007
	LpCBF3, from (Lolium perenne L.)	*Arabidopsis*	*Agrobacterium*	CaMV 35S	Transgenic *Arabidopsis* plants showed increased freezing tolerance.	Zhao and Bughrara, 2008
	Wlip 19 from wheat	*N. tabacum*	*Agrobacterium*	CaMV 35S	Wlip19-expressing transgenic tobacco showed a significant increase in abiotic stress tolerance,especially freezing tolerance.	Kobayashi *et al.*, 2008
	AtABP9	*Arabidopsis*	*Agrobacterium*	CaMV 35S	Under Drought/heat stress transgenic plants showed improved photosynthetic capacity.	Zhang *et al.*, 2008
	OsDREB1F *from Oryza sativa*	*Oryza sativa L, Arabidopsis*	*Agrobacterium*	CaMV 35S	Transgenic plants harbouring OsDREB1F gene led to enhanced tolerance to salt, drought, and low temperature in both rice and Arabidopsis.	Wang *et al.*, 2008
	ZFP 245	*Oryza sativa L.*	*Agrobacterium*	CaMV 35S	Transgenic plants overexpressing *ZFP 245* showed significant tolerance to cold and drought stresses.	Huang *et al.*, 2009

Mechanism	*Transgene(s)*	*Plant species*	*Transformation method*	*Promoter*	*Remarks*	*Reference*
	AP37	*Oryza sativa* L.	*Agrobacterium*	CaMV 35S	AP 37 expressing transgenic rice plants showed enhance tolerance towards drought stress.	Oh *et al.*, 2009
	*MtDREB2A from Medicago truncula*ta	*M. truncula*ta	*Agrobacterium*	CaMV 35S	Transgenic plants showed increased salt and drought tolerance.	Chen *et al.*, 2009
	GmDREB3 from *Glycine max*	*N. tabacum, Arabidopsis*	*Agrobacterium*	CaMV 35S, RD29A	GmDREB3 transgenic tobacco accumulated higher levels of free proline under drought stress and retained higher leaf chlorophyll levels under high salt stress than wild-type tobacco. In addition, overexpression of GmDREB3 enhanced tolerance to cold, drought, and high salt stresses in transgenic *Arabidopsis*.	Chen *et al.*, 2009
	OsMYB3R-2 from *Oryza sativa*	*Oryza sativa*	*Agrobacterium*		Transgenic plants over-expressing OsMYB3R-2 exhibited enhanced cold tolerance.	Ma *et al.*, 2009
	DSM1	*Oryza sativa*	*Agrobacterium*	CaMV 35S	Overexpression of DSM1 in rice increased the tolerance to dehydration stress at the seedling stage.	Ning *et al.*, 2010
	GmNAC11, GmNAC20 from Glycine max	*Arabidopsis*	*Agrobacterium*	CaMV 35S	Over-expression of *GmNA C20* enhanced salt and freezing tolerance in transgenic Arabidopsis plants; however, over-expression only improves salt tolerance. Over-expression of *GmNAC20* also promotes lateral root formation.	Hao *et al.*, 2011

Mechanism	*Transgene(s)*	*Plant species*	*Transformation method*	*Promoter*	*Remarks*	*Reference*
	OSNAC5	Rice	*Agrobacterium*	CaMV 35S	Overexpression of Os NAC5 in rice enhanced abiotic stress tolerance and further accumulation of proline and soluble sugars was positively correlated with *OsNAC5* expression level.	Song *et al.*, 2011
	At DREB 2A CA	*Arabidopsis*	*Agrobacterium*	CaMV 35S	Transgenic plants showed increased tolerance towards drought and salinity stress.	Mizoi *et al.*, 2011
	EcNAC1 from Escherichia coli	*N. tabaccum*	*Agrobacterium*	CaMV 35S	Transgenic tobacco plants showed tolerance towards abiotic stress.	Ramegowda *et al.*, 2012
	OsbZIP16 from Oryza sativa L.	*Oryza sativa L.*	*Agrobacterium*	Rice Actin1	Transgenic rice plants overexpressing OsbZIP16 gene exhibited significantly improved drought resistance both at seedling and tillering stages.	Chen *et al.*, 2012

2

Potential for Engineering Crops with High Antioxidant Capacity

Jyoti Ushahra, Amandeep Hora and *CP Malik*

The objectives of this review are to describe the prospective of antioxidants in crop improvement, quantum and quality. Numerous metabolic processes produce reactive oxygen species and cause cellular damage in plants. Antioxidants are valuable, plant-based systems which have the potential to defend against oxidative stress-related effects caused by undue production of reactive oxygen species (ROS) due to abiotic and biotic stresses. Natural antioxidants possibly function as reducing agents, free radical scavengers, and quenchers of the formation of singlet oxygen. Plants possess efficient enzymatic (superoxide dismutase, catalase, ascorbate peroxidase, glutathione reductase, dehydroascorbate reductase, glutathione peroxidase, glutathione reductase) and non-enzymatic (ascorbic acid, glutathione, tocopherols, carotenoids) antioxidants. The second approach deals with genetic engineering which improves the antioxidant and nutrient capacity of economically important crops including fruits and vegetables.

Introduction

Oxygen is essential component for life to perform biological functions such as catabolism of fats, proteins and carbohydrates to generate energy for growth and other activities. It involved in the generation of several reactive oxygen species (ROS) such as hydroxyl radical ($\cdot OH$), the superoxide radical ($O_2^{\cdot -}$), hydrogen peroxide (H_2O_2), singlet oxygen ($^1O_2$), nitric oxide radical ($NO\cdot$), hydroperoxyl radical ($HO_2\cdot$) and various lipid peroxides (Vellosillo *et al.*, 2010). Oxygen being extremely reactive is capable of becoming part of potentially damaging molecules commonly called "free radicals" (Pourmorad *et al.*, 2006). The main cellular components viz lipids, proteins, carbohydrates and nucleic acids are susceptible to damage by free radicals. Living organisms have evolved a tremendously efficient defensive system to protect cells and organs from free radicals that is composed of a group of compounds and enzymes called "antioxidant defensive system". In nature a wide variety of naturally occurring antioxidants, both endogenous and exogenous in origin (Al-Jaber *et al.*, 2011), that function to counteract free radicals.

Reactive oxygen species (ROS)

Free radicals are molecules or atoms of chemical class, skilled of self-regulating persistence with one or more unpaired electrons. In this process, electrons from additional sources to become stable, they instigate a chain reaction of oxidation; make by damaging DNA, varying biochemical compounds, corroding cell membranes and killing cells completely. Most commons are Reactive Oxygen Species (ROS), Reactive Nitrogen Species (RNS) or Reactive Sulfur Species (RSS).

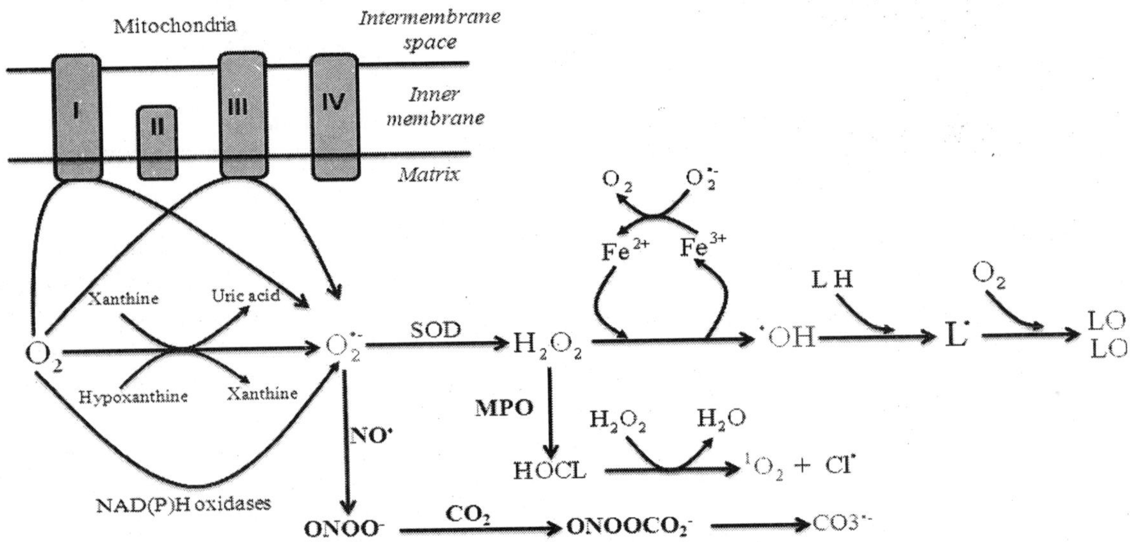

Fig 1 : Production of free radicals via different routes (Kunwar and Priyadarsini, 2011)

Sources of free radical generation

There are various resource that lead to production of free radicals in the living cells of different organisms such as: exposure of ultraviolet light, X-rays and gamma rays, product of electron transport chain, reaction catalyzed by metal, interaction of bacteria and phagocytes, addition of e^- to a natural molecule, by loss of single e^- from neutral molecule, cigarette and smoke and environmental pollution.

Detrimental effects of free radicals

ROS are hazardous for cell structures, nucleic acids, lipids and proteins at high concentration. They react with the biomolecules which cause localized injury and organ dysfunction. Major oxidation reactions are explained as follows:

DNA

Oxidative damage of DNA takes place due to interaction of DNA with ROS. Hydroxyl radical (•OH) can react with all constituents of the DNA molecule and destruction both the purine and pyrimidine bases and the deoxyribose backbone. Oxidative damage leads to the configuration of DNA lesions, which is first step associated in mutagenesis, carcinogenesis, and ageing. The interaction of pyrimidine with •OH, thymine glycol, urea residue, uracil glycol, 5-hydroxydeoxyuridine, 5-hydroxydeoxycytidine, hydantoin will genrate. Similarly, interaction of •OH with purines produces 8-hydroxy deoxyguanosine (8-OHdG), 8-hydroxyldeoxyadenosine, formamidopyrimidines etc. (Devasagayam *et al.*, 2004)

Proteins

Oxidation of proteins by free radicals (mainly •OH), involves oxidation of amino acid chains; generate stable and reactive products such as protein hydro-peroxides, which lead in fragmentation of proteins. The side chains of all the amino acid residues, particularly cysteine and methionine, are prone to oxidation by ROS. Peroxynitrite ($ONOO^-$) is involved in the oxidation of -SH group on proteins. Functionally inactive oxidized proteins are hastily removed but some can accumulate and result in the alteration of signal transduction process and transport system, ageing, atherosclerosis and many other diseases (Devasagayam *et al.*, 2004).

Lipid and lipid peroxidation

Cell membrane comprises phospholipids which have poly unsaturated fatty acids (PUFA) residues, are highly susceptible to free radical damage. When lipids are attacked by free radicals, highly damaging chain reaction of lipid peroxidation (LP) occur, which is primarily a destructive and self-prepetuating process, leading to both direct and indirect effects. Lipid peroxidation is broadly defined as "oxidative deterioration of PUFA" (Eqbal *et al.*, 2011). During lipid peroxidation a large number of toxic and mutagenic byproducts; lipid epoxides, lipid hydroperoxides, peroxyl radicals, lipid alkoxyl and unsaturated fatty acids, are formed that act as 'second messengers'. In lipid peroxidation malondialdehyde (MDA) is produced which is a good marker for free radical-mediated damage and oxidative stress. MDA can react with the free amino group of proteins, nucleic acids and phospholipid leading to structural modifications, which induce dysfunction of immune systems (Lee *et al.*, 2004).

Antioxidants

Antioxidants are first line of defense against free radical damage. A broader definition of antioxidant: "An antioxidant is a molecule that averts the transfer of electron from oxygen to organic molecules, stabilizes the free radical and terminates free radical reactions." Antioxidants have the potential to avoid the free radicals damage by preventing the formation of radicals, by stabilizing reactive oxidative species or by promoting their decomposition.

Antioxidants as defense

Free radicals derived from oxygen and nitrogen is recognized as reactive oxygen species (ROS) and reactive nitrogen species (RNS). These elements are crucial for biological metabolisms but sometimes they convert into free radicals which are highly unstable and reactive, can start chain reaction and damage biologically relevant molecules such as proteins, lipid or carbohydrates. Antioxidants avert the oxidation of other compounds. Both enzymatic and non enzymatic antioxidants such as catalase, superoxide dismutase, ascorbate peroxidase, ascorbic acid, tocopherols, carotenoids, glutathione etc. protect cells from undesired oxidations. Sarvajeet and Narendra (2010) reported that the ROS upset many cellular functions by damaging nucleic acid, oxidizing proteins and causing lipid peroxidation. Oxidative stress occurs when balance between antioxidants and ROS will disrupt (Eqbal *et al.*, 2011).

Functions of antioxidants

Antioxidants prevent the detrimental effects of free radicals and toxic products of metabolism. However, the antioxidants control free radical formation as a coordinated system. Antioxidants can act as:

- scavenge reactive oxygen species
- inhibit the process of oxidation
- terminating the chain reactions
- repairing damage
- convert the radicals to less reactive species
- stabilizing or deactivate the free radicals before they attack cells
- Hydrogen donation by the antioxidant
- Electron donation by the antioxidant
- Addition of the lipid to antioxidant
- Formation of a complex between the lipid and antioxidant
- Function as peroxide decomposers and enzyme inhibitors (Mandal *et al.*, 2009; Eqbal *et al.*, 2011; Singh and Singh, 2012)

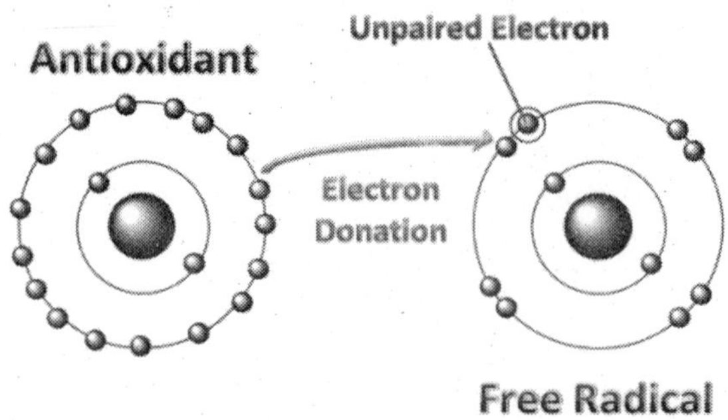

Figure 2: Reduction in formation of free radical by antioxidant

Classification of Antioxidants

Antioxidants are grouped on the bases of their synthesis:

(1) Plant-derived antioxidants (2) Secondary or synthetic antioxidants.

Plant-derived antioxidants

Plants produce an enormous variety of antioxidative molecules. These are mainly phenolics in structures. They react with radicals and convert them into more stable products (Hamid *et al.*, 2010). These are mainly: **Minerals-** co factor of antioxidant enzymes i.e. selenium, copper, iron, zinc and manganese, etc., **Vitamins-** required for metabolic functions i.e. vitamin B, vitamin C, vitamin E, and **Phytochemicals-** These are secondary metabolites i.e. anthocyanins, catechins, carotenoids, diterpene, beta carotene, curcumin, lycopene, zeaxantin (Hurrell, 2003).

Secondary or synthetic antioxidants

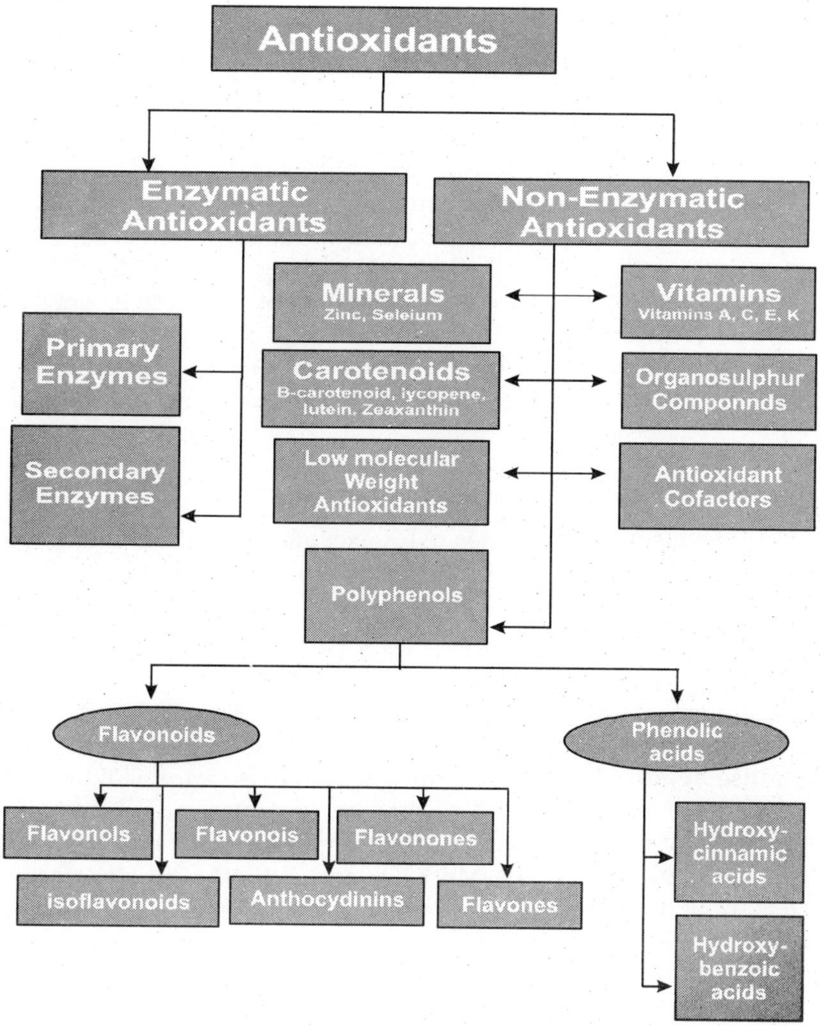

Fig 3 Flow chart showing classification of enzymztic and non-enzymatic antioxidants

Synthetic antioxidants capture free radicals and stop the chain reactions. These are phenolic compounds and used for the preservation of cosmetic and food products, in pharmaceutical etc (Kumar and Priyadarsini, 2011). These are: Butylated hydroxyl anisole (BHA), Butylated hydroxyrotoluene (BHT), Propyl gallate (PG), Nordihydro guaretic acid (NDGA) and Tertiary butyl hydroquinone (TBHQ) (Hamid *et al.*, 2010; Tavassoli and Djomeh, 2011).

Further, based on the nature of antioxidants, they are classified into two components enzymatic and non-enzymatic systems. Enzymatic antioxidants include superoxide dismutase (SOD), catalase (CAT), peroxidase (POX), ascorbate peroxidase (APX), glutathione reductase (GR) and polyphenol oxidase (PPO), Glutathione peroxidase etc. and the non- enzymatic system includes ascorbic acid (vitamin C); Ü-tocopherol, glutathione, carotenoids etc. (Mandal *et al.*, 2009). The detailed classification of antioxidants is presented in following flow chart (Fig. 3) and explained in subsequent section:

Enzymatic antioxidants

Catalase (CAT) (EC 1.11.1.6)

CAT (H_2O_2 oxidoreductase) is a haeme-containing protein based enzyme that catalyses the disproportionation of H_2O_2 into H_2O and O_2 (Karuppanapandian *et al.*, 2011). It has high capability but low affinity enzyme which destroys hydrogen peroxide. CAT promotes the redox reaction. It is a large tetrameric protein that present as multiple isoezymes encoded by nuclear genes. They are found mostly in peroxisomes and glyoxysomes, although a specific isozyme Cat3 is present in maize mitochondria. This ubiquitous enzyme has been isolated from different natural sources including plants, animal tissues and micro-organisms (Eqbal *et al.*, 2011).

Ascorbate peroxidase (APX) (EC 1.11.1.11)

CAT reduces H_2O_2 levels in peroxisomes, APX execute equivalent function in chloroplast and cytosol of plant cell. APX uses ascorbate as hydrogen donor to break down H_2O_2 to form H_2O and monodehydroascorbate (Asada, 2000).

Superoxide dismutase (SOD) (EC 1.15.1.1)

SOD is an intracellular enzyme present in every cell. It is composed of metalloprotein that catalyze the dismutation of O^-_2 superoxide radical to O_2 and H_2O_2. In cell, SOD constitutes the first line of defense against ROS. Based on the metal co-factor SOD are classified into three groups: iron SOD (Fe SOD), located in the chloroplast; manganese SOD (Mn SOD), located in the mitochondrion and peroxisome; and copper-zinc SOD (Cu-Zn SOD) located in the chloroplast (Alscher *et al.*, 2002). The MnSODs and FeSODs are structurally correlated, whereas Cu ZnSODs show no structural relationship to the other and evolved independently. It reduces superoxide into H_2O_2. When H_2O_2 concentration enhances, it inactivates Cu-Zn SOD. Catalase and glutathione peroxidase reduce H_2O_2 to conserve SOD.

Glutathione peroxidase (EC 1.11.1.9)

It is a group of enzymes containing selenium, which catalyze the degradation of hydrogen peroxide and lipid hydroperoxide, generated during lipid peroxidation, to water using reduced glutathione as substrate (Gupta and Sharma, 2006). It is cytoplasmic and mitochondrial enzyme which helps in detoxification of peroxides.

Dehydroascorbale reductase (DHAR) (EC 1.8.5.1)

DHAR play a crucial role in the oxidative stress tolerance of plants. It catalyzed the regeneration of ascorbate from dehyroascorbate (Karuppanapandian *et al.*, 2011).

Glutathione reductase (GR) (EC 1.6.4.2)

It is a low molecular weight thiol compound present in almost all cells. It acts as a disulphide reactant and regenerate ascorbate. It reacts with $^1O_2$ and OH· and defend the thiols group of enzymes. The role of GR in H_2O_2 scavenging has been well estabilished in the halliwell - asada pathway (Asada, 2000).

Non -enzymatic antioxidants

Ascorbic acid (Vitamin C)

Ascorbic acid is a water soluble, electron donor, chain breaking antioxidant; present in high concentration in many cellular environments, such as the stroma of chloroplasts. It is a monosaccharide, found in both animals and plants (Hamid *et al.*, 2010). It is capable of neutralizing ROS in the aqueous phase before lipid peroxidation is initiated. It is a reducing agent. It can directly scavenge superoxide, hydroxyl radicals and singlet oxygen and reduce H_2O_2 to water via ascorbate peroxidase reaction (Karuppanapandian *et al.*, 2011). In cells, it is maintained in its reduced form (ascorbate) by reaction with glutathione, which can be catalyzed by protein disulfide isomerase and glutaredoxins (Hamid *et al.*, 2010). Citrus fruits, tomatos, potatos, green leafy vegetables are the major source of Vitamin c.

Tocopherols (Vitamin E)

Vitamin E is a set of 8 related tocopherols (α, β, γ and δ) and tocotrienols (α, β, γ and δ) (Hamid *et a'.*, 2010). Relative antioxidant activity of the tocopherol isomers *in vivo* is α Γβ Γγ Γ δ (Blokhina *et al.*, 2003). α tocopherol has highest antioxidant activity. They are lipid soluble and chain breaking antioxidant, present in all cellular membranes protecting against lipid peroxidation and avert chain initiation and propagation of free radical reaction. It is synthesized only by plants and algae and vital component of biological membranes. In cell membrane it defends membrane fatty acids from lipid peroxidation. It reacts with RO·, ROO· and RO· derived from PUFA oxidation (Karuppanapandian *et al.*, 2011). Tocopherols act as chemical scavengers of oxygen radicals, particularly singlet oxygen, and also physical deactivators of singlet oxygen by chain transfer mechanism (Blokhina *et al.*, 2003). Nuts, seeds, grains and oils are rich sources of Tocopherols.

Glutathione (GSH)

GSH is an abundant tripeptide (γ-glutamylcysteinglycine). It is a significant water-soluble antioxidant and synthesized from the amino acids, glutamate, glycine and cysteine. ROS molecules oxidize glutathione, but the reduced form is regenerated in a redox by NADPH-dependent reductase. In cells, it is retained in the reduced form by the enzyme glutathione reductase. High reproductive potential of GSH is due to central nucleophilic cysteine residue. It scavenges cytotoxic H_2O_2, and reacts non- enzymatically with other ROS: singlet oxygen, superoxide radical and hydroxyl radical (Blokhina *et al.*, 2003). It may benefit to detoxify many inhaled oxidizing air pollutant like ozone, NO_2 and free radicals in cigarette, smoke etc. In some organisms glutathione is substituted by other thiols, such as mycothiol in the Actinomycetes, or by trypanothione in the Kinetoplastids (Hamid *et al.*, 2010).

Carotenoids

Among all carotenoids known, β carotene is the most broadly studied molecule. It is inactive form of vitamin A (retinol). These are lipolipophilic organic compounds, present in the chloroplasts and chromoplasts of plants. CARs play many functions in plant metabolism including the role in oxidative stress tolerance (Karuppanapandian *et al.*, 2011). β Carotene prevents lipid peroxidation via quenching the singlet oxygen species and scavenging the free radicals. It is an admirable scavenger of singlet oxygen.

Dietary antioxidants

According to Members of the Food and Nutrition Board of the National Research Council (USA), dietary antioxidants decrease the adverse effect of ROS and RNS (Khan and Giridhar, 2011). Foods not only supply calories but also protect body from the commencement of disease. Dietary antioxidants are exogenous. Some dietary antioxidants are:

Table 1 : Antioxidants in food and diet

S. N.	Functional types	Antioxidants	Sources
1.	Vitamins	A, B, C, E, K	Fruits, vegetables, citrus fruits, dairy products, fish, meat,
2.	Alkaloids	Capsaicinoids, berberine, caffeine	Chilli, beverages, herbs
3.	Polyamines	Putrescine, spermidine, spermine	Pea, apple, citrus fruits, spices, liver, meat
4.	Phytosterols	Campesterol, stigmasterol	Vegetables, seeds
5.	Ferulic acid ester	Oryzanol	Rice bran oil
6.	Isoprenoids derivatives	Carotenoids	carrot, tomato, palm oil
7.	Benzopyran Derivatives (polyphenols/ flavonoids)	Proanthocyanins Anthocyanins, catechins, flavonols, lignans, flavones, isoflavones, isoflavonoids, gallic acid	Grape, apple, pomegranate, banana, pineapple, brinjal, beverages, oil seeds, cucumber, olive oil, pomegranate, green tea
8.	Metals	Zinc, selenium, copper, manganese	Vegetables, meat, egg, dairy products

(Source: Khan and Giridhar, 2011)

Plants as antioxidants

Plant have many antioxidants and screening of plant antioxidants is done through various *in vitro* models like ABTS method (Joharapurkar *et al.*, 2003), DPPH methods (Mahakunakorn *et al.*, 2004), DMPD method (Rohini *et al.*, 2004), Nitric oxide method (Hazra *et al.*, 2008), ORAC method (Irshad and Chaudhuri, 2002), TBARS assay (Halliwell, 1991) etc and via various in *vivo* models (Scartezzini and speroni, 2000) using rats and mice (Gupta and Sharma, 2006).

Some plants and fruits are listed in Table 2 and 3 which possess antioxidant property with their chemical constituents and biological activity.

Table 3 : List of some fruits showing antioxidant activity

S. No.	Fruits	active constituents	References
1.	Berries	Flavanols hydroxycinammic acids, hydroxy-benzoic acids, anthocyanins	(Wang and Lin, 2000)
2.	Strawberries	Vitamin c, bioflavonoids	(Reddy *et al.*, 2003)
3.	Cherries	Hydroxycinnamicacids, anthocyanins	(Manach *et al.*, 2004)
4.	Blackgrapes	Anthocyanins, flavonols	(Belitz and Grosch, 1999)
5.	Grape	Alanine, ascorbic acid, a tocopherol, b carotene, histidine, methionine, palmitic acid, b sitosterol, selenium	(Singh and Singh, 2012)
6.	Citrus fruits	Flavanones, flavonols, phenolic acids	(Dimitrios, 2006)
7.	Olives	Polyphenols	(Singh and Singh, 2012)
8.	Plums, prunes apples, pears, kiwi	Hydroxycinnamic acids, catechins	(Manach *et al.*, 2004)

Free radicals vis-a vis environmental stresses

Plants subjected to environmental stress during their life span. It causes significant crop losses. These environmental stresses like drought, temperature extremes, radiation, light, salinity, soil pH, heavy metals, lack of essential nutrients, air pollutants, herbicides, etc. effect on the production of reactive oxygen species in plants, causing oxidative stress. Abiotic stress single or in combination induce changes in the plant metabolism, growth and general development. Certain metals, like copper, are essential for plants metabolism, but at high concentrations (depending on plant species) can be toxic. ROS can be formed during normal aerobic metabolisms like photosynthesis and respiration. Majority of ROS are produced in the mitochondria, chloroplast, peroxisomes, plasma membrane and apoplast (Ahmad *et al.*, 2008; Moller, 2001). Under certain stress conditions an imbalance in intracellular ROS content take place and this result in oxidative stress (Gill and Tuteja, 2010).

Table 2 : list of plant showing antioxidant property

S. No.	Plant Species	Common Plant	Family	Biological behaviour	Chemical/ active constituents
1.	*Acacia catechu* (L.f.) Willd.	Catechu	Mimosaceae	Diarrhoea, cough, skin eruption, asthma, check nasal bleeding dysentry, bronchitis	Catechin and quercetrine (Cartron *et al.*, 2001).
2.	*Anadenanthera macrocarpa* (Benth.) Brenan	Angico	Fabaceae	Extract of bark is taken for live problems, upper respiratory infections, angina pectoris and gonorrhea	Thiobarbituric acid (Lamarque *et al.*, 1998)
3.	*Bauhinia forticate*	Pata de vaca	Caesalpiniaceae	Hypoglycemic,blood purifier and diuretic	Kempferol-3,7-O-dirhamnoside (De Sousa *et al.*, 2004)
4.	*Dolichos biflorus* Linn.	Horse Gram	Fabaceae	Hypolipidemic effect	Flavonoids (Kottaimuthu *et al.*, 2007)
5.	*Prunues domestics*	Plums	Rosaceae	The juice from the bark applied on body swellings and contusions and oils from kernels applied on neuralgic pains	Prunetin, genistein and quercetin (Jung *et al.*, 2002).
6.	*Foeniculum vulgare* Mill.	Sauf	Apiaceae	Stimulant, aromatic, carminative, purgative, diuretic, in venereal diseases, vermicide, useful in chest, spleen and kidneys troubles, estrogenic. Fennel oil antioxidant	Volatile oil, fenchone, anethole, limonene, anisaldehyde, estragole.
7.	*Ocimum sanctum* Linn.	Tulsi	Lamiaceae	Expectorant, in catarrh, bronchitis, ringworm and other cutaneous diseases, stomachic, gastric disorders of children, in earache	Volatile oil, terpenoids, eugenol, thymol, estragole

S. No.	Plant Species	Common Plant	Family	Biological behaviour	Chemical/ active constituents
8.	*Glycyrrhiza glabra* Linn.	Mulethi	Fabaceace	Diuretic, emmenagogue, in vomiting, asthma, bronchitis, in acute conjunctivitis, for curing wounds, in peptic ulcer	Glycyrrhizin, flavonoids, liquiritin, isoliquiritin, rhamnoliquiritin, 2-methylisoflavones
9.	*Solanum nigrum* Linn.	Makoi	Solanaceae	Diuretic, laxative, dropsical affections, virulent gonorrhoea, malaria, dysentery, hepatoprotective	Polyphenolic compounds, flavonoids, steroids.
10.	*Withania somnifera* Dunal	Ashwagandha	Solanaceae	Analgesic, increase immunity, hepatoprotective	Steroidal lactone, withanolides, glycene, withanine.
11.	*Punica granatum* L.	Pomegranate	Punicaceae	Fever, cough, pimples, boils, throat disorder, dysentery, diuretic, cardiac problems, vomiting and bronchitis	Punicalin (Das *et al.*,1999; Khanbabae and Ree, 2001; Murthy *et al.*, 2002)
12.	*Mangifera indica* Linn.	Mango	Anacardiaceae	Leuchorrhea, dysentery, bronchitis, biliousness, urinary discharges, in haemorrhage from the uterus, lungs or intestine	Cyanogenetic glycosides, polyphenlos, vitamin A and C, mangiferin, β-sitosterol, quercetin, ellagic acid, gallic acid

Water stress

Water stress causes crop yield loss because it is strongly associated with transpiration and photosynthesis. It causes inhibition of protein synthesis, inactivation of several chloroplast enzymes, increase membrane permeability, impairment of electron transport, increased activity of enzymes of the H_2O_2 scavenger system and oxidative damage through generation of oxygen radicals or inhibition of antioxidant systems in plant (Zhang and Kirkham, 1996). During drought stress, high salinity and low temperature plants accumulate a range of metabolically benevolent solutes, collectively known as compatible solutes or osmolytes. Its primary function is turgor maintenance and also protective effects on macromolecules in dehydrating cells. The solutes accrued vary between species and include proline, dimethylsulfonioproprionate (DMSP), polyols (mannitol, sorbitol, and pinitol), betaines, trehalose and fructans.

In recent, years innumerable transgenic plants have been produced in which the ability has been introduced to accumulate osmolytes. Glycine betaine (GB) is accrued by a taxonomically restricted range of species. GB application in foliar sprays is reported to ameliorate the growth of water-stressed tobacco in laboratory experiments and maize, sorghum and soybean crops in the field whereas it is toxic to *Brassica napus* (Gibon *et al.*, 1997).

Salinity stress

Salinity affects significant metabolic processes located in chloroplast and mitochondria.

Higher salt concentration damage the cellular electron transport within the different subcellular compartments and lead to the generation of reactive oxygen species (ROS) such as singlet oxygen, superoxide, hydrogen peroxide and hydroxyl radicals.

Mannitol is accrued by a broad range of species under salinity stress. In transgenic tobacco its synthesizing ability was introduced by the *E. coli* mtlD gene encoding Mannitol dehydrogenase. These plants accumulated low amounts of mannitol but were more salt-tolerant. Mannitol could be act as antioxidants by scavenging hydroxyl radicals (·OH).

Chilling stress

Foremost target of chilling injury is cell membranes. As temperature is decreased, a particular temperature determined by the ratio of saturated to unsaturated fatty acids hastens the conversion of lipids of a liquid-crystalline condition into a solid condition in plant cell membranes. The alteration of fatty acid may give rise to chilling resistance at lower temperatures in the plant cells.

Dong and Chin (2000) were investigated that the antioxidant defense system and chilling stress-accelerated changes of antioxidant enzymes in the cucumber leaves. Chilling stress enhanced the activities of the ascorbate peroxidase (APX), superoxide dismutase (SOD), glutathione reductase (GR) and peroxidase specific to guaiacol, whereas it decrease catalase activity.

Metals stress

Accumulation of phytotoxic metal (Zn, Cu and Cd) in the environment results from industrial and agricultural practices, resulting in stunted growth, chlorosis and necrosis (Oncel *et al.*, 2000).

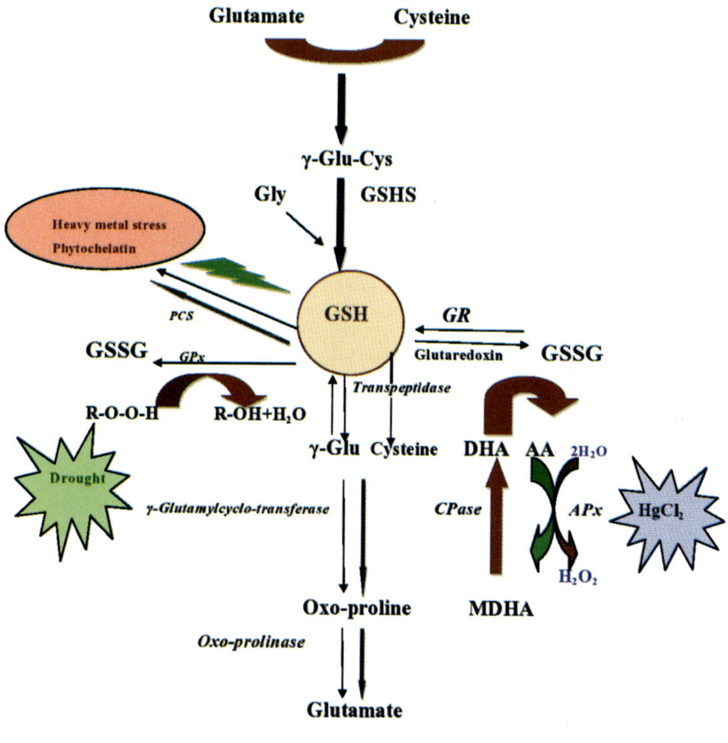

Fig. 4 : Antioxidant defense mechanism against heavy metal stress

Copper ions causes pigment bleaching and light mediated lipid peroxidation. Cadmium ions decline the chlorophyll and haeme levels of germinating mung bean seedlings by stimulation of lipoxgenase and inhibition of the antioxidative enzyme SOD and CAT. Pang *et al.* (2000) studied improving the plant ability to resist lead stress. The activities of SOD and CAT were found to be increased slightly and MDA content decreased in shoot and root of wheat seedlings under lead stress within five days after treatments.

Potential of antioxidants production in plant system

Plants comprise a wide variety of free radical scavenging molecules, enzymatic and non-enzymatic antioxidants. A vast number of plants have been studied for their potential sources of antioxidants. Plant-derived antioxidants act as reactive oxygen quenchers, peroxide decomposers and synergists.

Plants serve as one of the cheapest source of antioxidants e.g. *Olea europaea* (Silva *et al.*, 2006), *Etlingera genus* (Chan *et. al.*, 2011), *Ligustrum vulgare* (Agati *et al.*, 2009) and *Stevia rebaudiana* (Tadhani *et al.*, 2007). Bark from *Casuarina equisetifolia* (Zhang *et al.*, 2010), Acacia *confuse* (Chang *et al.*, 2001), *Populus tremuloides* Michx (Diouf *et al.*, 2009); root of *Medicago sativa* (Dalton *et al.*, 1998) and *Carissa spinarum* (Hegde and Joshi, 2010) were also reported to comprise antioxidants.

Moreover dietary sources, antioxidants can be obtained from food processing industries and agricultural-by-products. The relevant by-products are seeds, peels, bark, mill wastes and trimming wastes. The industrial-by-products, like peel and seed, are manifested to have high antioxidant level which is even higher than the flesh and other parts of the fruit with the presence of high polyphenol content, some are; *Viburnum opulus* seed (Cam *et al.*, 2007), peach peel (Chang *et al.*, 2000), mangosteen peel (Moongkarndi *et al.*, 2004), apple peel (He and Liu, 2007) and grape seed and skin (Rockenbach *et al.*, 2011).

In vitro production of antioxidants

In vitro cultured plant cells synthesize, accumulate and exudate many secondary metabolites which have significant potential as antioxidants. Antioxidant inhibits or delays the oxidation of substrates even if the compound is present in a significantly lower concentration than the oxidized substrate. The antioxidant compounds can be recycled or are irreversibly damaged in the cell, but their oxidation products are less harmful or further converted to harmless substances (Halliwell and Gutteridge, 2007). The antioxidant activity of plant secondary metabolites or antioxidants has been widely demonstrated in *in vitro* systems. *In vitro* technology offers several benefits, i.e. simpler extraction and purification from interfering matrices, independence from climatic factors and seasons, shorter and more flexible production cycles, easier fulfillment of the high-profile pharmaceutical production demands etc.

Polyphenols

Polyphenols are easily oxidizable phenolics, highly desirable as dietary and therapeutic free radical scavengers. In tissue culture excessive production of polyphenols results from unfavorable or suboptimal culture conditions. In plants polyphenols mainly derive from the shikimic acid pathway through aromatic carboxylic acids—cinnamic or benzoic. Some antioxidant polyphenols are: phenolic acids (caffeic acid derivatives), hydrolysable tannins (gallotannins and ellagitannins), lignans, stilbenes and xanthones. Phenolic acids are present in free form as glycosides in plants.

Rosmarinic acid (RA) found in many Lamiaceae and Boraginaceae species, a depside composed of two caffeic acid molecules. It can accrue in cell cultures in larger amount than in intact plants. The suspension cultures producing rosmarinic acid were generated from *Eritrichum sericeum* (Fedoreyev *et al.*, 2005), *Lithospermum erythrorrhizon* (Yamamoto *et al.*, 2000) (Boraginaceae), and *Lavandula vera* (Pavlov *et al.*, 2005a,b; Georgiev *et al.*, 2006a), *Ocimumbasilicum* (Kintzios *et al.*, 2004), *Zataria multiflora* (Mohagheghzadeh *et al.*, 2004) (Lamiaceae). Lavender cells extract accumulating rosmarinic acids which have had superior radical scavenging properties (Kovacheva *et al.*, 2006).

Lithospermic acid B is a tetrameric depside, accumulating particularly in the hairy root cultures of *H. officinalis* and elicited *S. miltiorrhiza* as well as in *Lithospermum erythrorhizon* suspension cultures (Yamamoto *et al.*, 2002). Flavonoids are tricyclic phenyl benzopyran structure but biosynthetically originate from phenylalanine and malonyl-CoA in the phenylpropanoid pathway. Subclasses of flavonoid comprise flavones, flavonols, isoflavonoids, flavanons, proanthocyanidins (catechins) and anthocyanidins. Its antioxidant properties depend on the hydroxylation pattern. Oligomeric proanthocyanidins (condensed tannins) are most powerful antioxidants of all the flavonoids due to their structure (procyanidin B1). They are responsible for the antioxidant activity of wines, teas and fruits. The generation and isolation of oligomeric proanthocyanidins have been observed in the cell cultures of *Vaccinium pahalae* (Kandil *et al.*, 2000) and *Vitis vinifera* (Decendit and Mérillon, 1996).

Another flavonoid polyphenols is anthocyanins. Anthocyanins are isolated from *in vitro* cultures of *Prunus* sp. (Blando *et al.*, 2005), *Glehnia litoralis* (Miura *et al.*, 1998), *Daucus carota* (Ravindra and Narayan, 2003), *V. pahalae* (Meyer *et al.*, 2002), *Fragaria ananassa* (Edahiro *et al.*, 2005), *Ipomaea batatas* (Terahara *et al.*, 2004), *Rudbeckia hirta* (Luczkiewicz and Cisowski, 2001), *V. vinifera* (Decendit and Mérillon, 1996) and *Ajuga reptans* (Terahara *et al.*, 2001) and others. Its Biological activity varies, because of their structural diversity.

Stilbenes are bicyclic polyphenols characterized by resveratrol. It is a phytoalexin and the red wine antioxidant. Even though it present in appreciable amount of natural sources like grapes and some Polygonaceae species, it has been also induced in elicited hairy root cultures of peanut (*Arachis hypogea*) and recovered expeditiously from the liquid medium (Medina-Bolivar *et al.*, 2002).

Isoprenoids

Isoprenoids are produced from isopentynyl pyrophosphate (IPP) and dimethylallyl pyrophosphate (DMAPP) in plant cells. They are lipophilic and play important role in protecting membrane lipids. IPP and DMAPP are subsequently fused to yield enormously variable compounds in numerous groups of terpenoids (mono, sesqui, di, and triterpenes and their derivatives), carotenoids, and steroids (Tholl, 2006). Carotenoids are most powerful antioxidants. Both β-carotene and lycopene are known as dietary chemopreventive agents, defensive against cancerogenesis. A powerful antioxidant crocin has been derived from saffron (stigmata of *Crocus sativus*) (Ochiai *et al.*, 2004). Abietane diterpenes are found in many Lamiaceae medicinal plants and in some gymnosperms. It can be obtained from *in vitro* cultures of *R. Officinalis* (Caruso *et al.*, 2000), *S. officinalis* (Grzegorczyk *et al.*, 2007) and *Salvia sclarea* (KuŸma *et al.*, 2006). Light plays a crucial role in encouraging the diterpene biosynthesis *in vitro* (Caruso *et al.*, 2000) although in natural conditions they can only be found in the underground parts of plants. Volatile terpenoids responsible for aromatic properties of plant organs serve as a multipurpose chemical defense weapon against fungal and bacterial pathogens along with long distance chemical communication e.g. attracting pollinators. Tissue culture techniques are able to produce volatile constituents of essential oils (Hrazdina, 2006).

Other antioxidants

Tocopherols broadly used in human nutrition as vitamin E and in food conservation. It is powerful free radical scavenger and defends plant cells against oxidative damage in a lipophilic environment (Kruk *et al.*, 2005). The extraction from plant oils obtained a mixture of β, γ, δ and α -tocopherols and tocotrienols; therefore plant tissue culture is an alternative source to obtain pure α -tocopherol. Several culture systems of *Helianthus annuus* (sunflower) have been established for this purpose.

Betalains is nitrogen containing compounds, widely used as non-toxic food colorants. The major benefits of betalains as dietary antioxidants are their bioavailability, which is greater than most flavonoids, and their superior constancy in comparison to anthocyanins (Stintzing and Carle, 2004). Red beetroot (*Beta vulgaris*) is a rich natural source of betacyanins. Beetroot pigments have been produced by cell suspensions and transformed roots in bioreactors (Pavlov *et al.*, 2005b). Kinobeon A is a novel, red-colored compound isolated from cell cultures of *Carthamus tinctorius*, distinctively combines the quinoid aromatic character with several conjugated double bonds. It efficiently quenches several reactive oxygen species, hinders lipid peroxidation and supports the survival of mammalian cells during oxidative stress (Kambayashi *et al.*, 2005).

Genetic engineering to enhance antioxidants activity in crop plants

The antioxidant activity of fruits vary with the species, cultivar, quantity as well as various tissues of the fruit (e.g. peel, flesh and other tissues) and consumption form (fresh, juice or processed) (Boyer and Liu, 2004; Wojdyolo *et al.*, 2008).

To increase antioxidant level of biofortify food attempts have been made by using traditional breeding strategies (Mayer *et al.*, 2008). These methods comprise crosses and back crosses for several generations which require high genetic variations of the trait and heritability, and it is a highly time consuming process (McGhie and Currie, 2008). Conventional breeding approach also has several limitations.

In cultivated tomato, the expression of chalcone isomerase (*CHI*) gene is responsible for the production of flavonoids. *CHI* gene expression is blocked in the peel and inhibited in the pulp affecting the production and accumulation of flavonoid–quercetin. Wild *Lycopersicon* accessions express the *CHI* gene. Wild *Lycopersicon* was crossed with a cultivated tomato species to develop a non-transgenic, metabolic strategy to produce flavonoid-rich hybrids (Willits *et al.*, 2005). Hybrid tomato showed 11.8-fold increase in flavonoid – quercetin in F1 hybrids. Though, further analysis of progeny was not possible because the hybrids were consistently seedless.

Genetic engineering is a powerful tool to introduce favorable modifications in metabolic pathways of plants to ameliorate the quantity and bioavailability of phytonutrients with particular attention to antioxidants (Fatima *et al.*, 2008). Various vegetable and fruit crops have been transformed to improve the concentration of a useful nutrient metabolite (Delaunois *et al.*, 2007; Giuliano *et al.*, 2008).

Tomato

Carotenoids

Tomato is a main dietary source of lycopene and comprises other antioxidants, vitamin C, flavonoids and tocopherols. Lycopene accumulation in tomato fruit is associated to a differential regulation of phytoene synthase (PSY), phytoene desaturase (PDS), lycopene beta (LCY-β) and lycopene epsilon (LCY-ε) cyclases (Ronen *et al.*, 1999). During ripening expression of the genes encoding PSY and PDS proteins, and lycopene production increase whereas LCY-β and LCY-ε, which convert lycopene to either β-or δ-carotene, declines hastily (Giuliano *et al.*, 2008). Various research groups (Wurbs *et al.*, 2007) have effectively used LCY β-cyclase gene under fruit-specific promoters to enhance the β-carotene content of tomato fruit though its silencing via antisense expression doubled the increase in lycopene (Rosati *et al.*, 2000). PDS-crtI expression also increases level of β-carotene (Ro¨mer *et al.*, 2000). Bacterial crtB gene encodes PSY gene when engineered with the fruit poly-galacturonase promoter led to 5–10 fold increase in PSY enzyme activity and a significant increase in phytoene, carotene and lycopene (Fraser *et al.*, 2002). A regulatable promoter with heterologous genes dominates endogenous control of carotenoid pathway genes, enabling higher accumulation of carotenoids during fruit ripening (Fatima *et al.*, 2008). Fruit-specific Pds promoter with lycopene β-cyclase (β-Lcy) and β-carotene hydroxylase (β-Chy) genes enhance 12-fold in β-carotene over the control (Dharamapuri *et al.*, 2002). RNAi-mediated suppression of an endogenous photo-morphogenesis regulatory gene, *DET1* increases carotenoids and flavonoids without affecting other fruit quality parameters (Davuluri *et al.*, 2005).

Polyphenols

Tomatoes contain small amounts of flavonoids, mostly located in the peel of the fruit (Muir *et al.*, 2001). Total phenolics (flavonoids and anthocyanin) in tomato have been increased successfully by using transgenic technology. Constituent expression of petunia chalcone isomerase chi-a gene in tomato fruit increase 78-fold in flavonols, mainly rutin (Muir *et al.*, 2001). Similarly, crypto chrome cry2 gene caused several-fold increase in flavonoids and non-flavonoids (Giliberto *et al.*, 2005). Cryptochrome cry2 gene is involved in the regulation of flowering time and photomorphogenesis under low fluence light in *Arabidopsis*. Isoflavone synthase GmIFS2gene of soybean catalyses the formation of genistin from substrate naringenin in isoflavone biosynthesis pathway, its constitutive expression in tomato led to an increase of genistin and quercetin glycoside in fruit peel (Shih *et al.*, 2008). The transgenic plants displayed altered antioxidant profiles. From different plant sources, genes encoding flavonoid biosynthesis pathway enzymes were used to produce transgenic tomato fruits, resulting in the accumulation of stilbenes (resveratrol and piceid), deoxychalcone (butein and isoliquirtigenin), flavones and flavonols (Schijlen *et al.*, 2006). Constitutive expression of chalcone isomerase and flavone synthase (*CHI/FNS*) gene under the control of CaMV double 35S promoter cause 3-fold higher antioxidant capacity in tomato fruit peel.

Folates

Synthetic GTP cyclohydrolase I (*gchI*) gene under the control of a fruit-specific promoter, was overexpressed in tomato, resulting in 2-fold and 3–140-fold increases, respectively, in folate and pteridine contents while p-aminobenzoate (PABA) content was reduced (Di´az *et al.*, 2004). Expression of *Arabidopsis* aminodeoxychorismate synthase (ADCS) gene resulted in 19-fold more PABA accumulation than in the controls. A genetic cross between the pteridine and PABA over-expressing transgenic lines led to 25-fold more folate in the progeny (Di´az *et al.*, 2004).

Potato

Carotenoids

The *Erwinia* PSY *crtB* gene fused to a tuber-specific promoter was introduced in potato and increase in the contents of violaxanthin, lutein and β-carotene was attained (Ducreux *et al.*, 2005). Antisense silencing of potato β-carotene hydroxylase *CHY1* and *CHY2* genes (metabolize β-carotene) led to 4–7-fold enhance in lutein but β-carotene level was enriched by 38-fold (Diretto *et al.*, 2007). *Or* gene (*orange* gene) mutation is responsible for β-carotene accumulation in cauliflower, when introduced in potato led to a 6-fold higher level of total carotenoids in transgenic against the controls (Lu *et al.*, 2006; Lopez *et al.*, 2008).

Polyphenols

Potato plants were transformed with a modified MYB transcription factor to increase the content of health promoting polyphenolic compounds such as kaempferol and chlorogenic acid. Constituent expression in potato of petunia dihydroflavonol reductase (DFR) gene (key enzyme in flavonoid biosynthesis) significantly enhanced anthocyanins and phenolics (Lukaszewicz *et al.*, 2004). The activation of the flavonoid biosynthetic pathway increases 100-fold in kaempferol content and 4-fold increase in phenolic acids (Rommens *et al.*, 2008).

Tocopherols

Tocopherols are potent antioxidants with anticarcinogenic properties and also have a potential to be used as chemotherapeutics (Constantinou *et al.*, 2008). Overexpression of *crtB* gene in potato tuber led to significant increase in the α-tocopherol content (Ducreux *et al.*, 2005).

Apple

Polyphenols

Grape stilbene synthase gene, *Vst1*, was expressed in apples for stilbene biosynthesis. This led to several-fold accumulation of piceid (*trans* resveratrol glucoside) (Ru¨hmann *et al.*, 2006). The increase in piceid consistently led to higher amounts of hydroxycinnamic acid and flavonols in the transgenic apple.

Kiwi

Piceid

Stilbene synthase gene from three different *Vitis* species when expressed constitutively in kiwi fruit led to resveratrol glucoside (piceid) (Kobayashi *et al.*, 2000).

Strawberry

Ascorbic acid

The NADPH-dependent D-galacturonate reductase (encoded by *GalUR*) converts D-galacturonic acid to ascorbic acid. Ascorbic acid content is correlated with *GalUR* expression pattern in strawberry fruits (Agius *et al.*, 2003) and overexpression of *GalUR* resulted in 2–3-fold higher levels of ascorbic acid.

Banana

Banana is a good source of vitamins B6 and C. There are wide diversity and genetic variability of banana in fruit provitamin A carotenoids, lutein and other micronutrients (Heslop-Harrison and Schwarzacher, 2007). Genetic engineering has the potential to further enhance nutritional quality and revenue, and make banana a more appreciated source of antioxidants (Davey *et al.*, 2009).

Lettuce

Tocopherols

Overexpression of *Arabidopsis* γ-tocopherol methyl transferase gene in lettuce enhances the levels of α-tocopherol compared with the wild-type control (Yusuf and Sarin, 2007).

Folates

Synthetic codon-optimized GTP cyclohydrolase I (*gchl*) based on indigenous *Gallus gallus* gene when introduced into lettuce improved folate levels. The transgenic lines had folate content that ranged from 2.1- to 8.5-fold higher than the non-transgenic lines.

Carrot

Ketocarotenoids

Carrot roots have high levels of α-carotene and β-carotene, but are scarce in ketocarotenoids. Introduction of a β-carotene ketolase gene from alga *Haematococcus pluvialis* into carrot root led to in ketocarotenoid accumulation (Jayraj *et al.*, 2008).

Canola

Carotenoids

Transformation of bacterial PSY (*crtB*) gene using seed-specific napin promoter in canola seed resulted in orange coloured seed with a 50-fold higher total carotenoid level than the wild-type (Shewmaker *et al.*, 1999). Transgenic canola seeds that co-expressed *crtI* and *crtB* gene were found to accumulate lycopene and β-carotene (Ravanello *et al.*, 2003).

Linseed /Flax

Carotenoids

Flaxseed is rich in á-linolenic acid and lignans. Lignans are sources of phytoestrogens and act as antioxidants because of the presence of α-linolenic acid, a ω-3 fatty acid, which is essential for human health (Donaldson, 2004). Overexpression of bacterial *crtB* gene in flaxseeds caused a 7.8– 18.6-fold increase in the carotenoid levels of the seeds (Fujisawa *et al.*, 2008).

Sweet Potato

Sweet potato is a rich source of vitamin A, ascorbic acid and carbohydrates. A tobacco microsomal fatty acid desaturase gene (*NtFAD3*) was overexpressed using a modified CaMV35S promoter in sweet potato to increase the levels of unsaturated fatty acids. Transgenic sweet potatoes hence obtained had increased amounts of linolenic acid (Wakita *et al.*, 2001).

Soybean

Tocopherols

Overexpression of the *Arabidopsis* genes involved in biosynthetic pathway of α-tocopherol. *VTE4* (γ-TMT–γ tocopherol methyl transferase) gene in combination with the *VTE3* (2-methyl-6-phytylbenzoquinol methyltransferase) gene, enhanced by 8-fold the α-tocopherol content in soybean seeds (Van Eenennaam *et al.*, 2003). Similarly, *Perilla frutescens* γ-TMT produced higher levels of α-tocopherol and β-tocopherol (Tavva *et al.*, 2007).

Conclusions

Cellular damage occurs due to imbalance between reactive oxygen species and antioxidant defense system. The main cellular components (DNA, protein and lipid) are susceptible to free radicals damage and cause irreversible damage that can lead to cell death. Antioxidant molecules defend the body from harmful effect of free radicals. Plants having variable phytochemical compounds possess remarkable antioxidant activity. Antioxidant activity of plants, fruits and Dietary antioxidants has been discussed. These plants exhibit significant clinical and pharmacological activity.

Table 4 : Engineered crops with enhanced target phytonutrients (antioxidants)

S. No.	GM fruit	Target nutrient	Gene introduced	Reference
1.	Tomato	β-carotene	β *Lcy/Arabidopsis*+ β *Chy*/Peppe	Dharamapuri *et al.*, 2002
			DET1/Tomato (RNAi)	Davuluri *et al.*, 2005
			β -*Lcy*/ *Arabidopsis*	Rosati *et al.*, 2000
		Lutein	*CRY2*/Tomato	Giuliano *et al.*, 2008
		Anthocyanin	*Del*/Snapdragon+*Ros1*/Snapdragon	Butelli *et al.*, 2008
		β -cryptoxanthin, zeaxanthin	B-*Lcy*/ Arabidopsis+ β *Chy*/Pepper	Dharamapuri *et al.*, 2002
		Lycopene	*ySAMDC, Spe2/Yeast*	Mehta *et al.*, 2002
		Phytoene	*CrtB/E. uredovore*	Fraser *et al.*, 2002
		isoflavone/genistin	*GmIFS2*/Soybean	Shih *et al.*, 2008
		Stilbenes	*STS/Vitis vinifera*	Schijlen *et al.*, 2006
		Flavonoid/ quercetin Kaempferol	*CHI*/Petunia	Muir *et al.*, 2001
		p-coumeric acid	*CRY2*/Tomato	Giliberto *et al.*, 2005
		Flavonoid	*DET1*/Tomato (RNAi)	Davuluri *et al.*, 2005
		Flavonol	*MYB12/Arabidopsis*	Luo *et al.*, 2008
		Campestero β-sitosterol Cycloartenol	*hmgr1/Arabidopsis*	Enfissi *et al.*, 2005
		trans resveratrol, Caffeic acid	*StSy/V. vinifera*	Giovinazzo *et al.*, 2005, Nicoletti *et al.*, 2007

S. No.	GM fruit	Target nutrient	Gene introduced	Reference
2.	Potato	Carotenoid	*Or*/cauliflower	Lu *et al.*, 2006; Lopez *et al.*, 2008
		Carotenoids	*CrtI, CrtB* and *CrtY/E. uredovora*	Diretto *et al.*, 2007
		β-carotene: carotenoids	*LCY-e*/ potato (antisense)	Diretto *et al.*, 2007
		α-tocopherol	*crtB-TP/E. uredovora*	Ducreux *et al.*, 2005; Crowell *et al.*, 2008
		Amino acids/ totalprotein	*AmA1/AmaranthusHypochondriacus*	Chakraborty *et al.*, 2000
3.	Brassica	β-carotene, Lutein	ε-*CYC/Brassica napus* (RNAi)	Yu *et al.*, 2008
		Carotenoid	*crtB*/ bacteria	Shewmaker *et al.*, 1999
		β-carotene	*crtI* +*crtb*/ bacteria	Ravanello *et al.*, 2003
			DET1 (RNAi suppression)	Wei *et al.*, 2009
4.	Canola	Carotenoid	*idi* + *crtW*+ *crtZ* (synthetic)/ marinebacteria + *crtE* + *crtB* + *crtI* +*crtY/Pantoea ananatis*	Fujisawa *et al.*, 2009
5.	Oilseed	α-tocopherol, γ tocopherol	γ-TMT/ Arabidopsis	Yusuf and Sarin, 2007
6.	Flax	Carotenoid	*crtB/P. ananatis*	Ravanello *et al.*, 2003
7.	Carrot	Ketocarotenoid	*CrtO/Haematococcus pluvialis*	Jayraj *et al.*, 2008
8.	Lettuce	Folate	*Gch1/avian*	Nunes *et al.*, 2009
		Iron	*pfe/Soybean*	Goto *et al.*, 2000
		Tocopherol	*γ-TMT/Arabidopsis*	Cho *et al.*, 2005
		Resveratrol	*STS/Parthenocissus henryana*	Liu *et al.*, 2006
9.	Soybean	α-tocopherol	*AtVTE3+AtVTE4/Arabidopsis*	Van Eenennaam *et al.*, 2003
		β-tocopherol	*γ TMT/Perilla frutescens*	Tavva *et al.*, 2007

S. No.	GM fruit	Target nutrient	Gene introduced	Reference
10.	Sweet potato	Linolenic acid	*NtFAD3*/Tobacco	Wakita *et al.*, 2001
11.	Apple	trans-piceid	*Vst1/V. vinifera*	Crowell *et al.*, 2008
12.	Kiwi	Piceid	*Stibene synthase/Vitis* spp	Ru¨hmann *et al.*, 2006
13.	Papaya	Piceid	*Vst1/V. vinifera*	Zhu *et al.*, 2004
14.	Strawberry	*p*-coumaryl alcohol *p*-coumaryl-1-acetate	*CHS/strawberry* (antisense)	Kobayashi *et al.*, 2000

(Shukla and Mattoo, 2009)

REFERENCES

Agati G, Stefano G, Biricolti S, Tattini M (2009) Mesophyll Distribution of 'Antioxidant' Flavonoid Glycosides in *Ligustrum Vulgare* Leaves under Contrasting Sunlight Irradiance. *Annals of Botany* 104(5): 853-861

Agbor GA, Kuate D, Oben JE (2007) Medicinal plants can be good source of antioxidants: Case study in Camroon. *Pakisthan Jour of Biological Sci* 10(4): 537-544

Agius F, Gonzales-Lamothe R, Caballero JL, Munoz-Blanco J, Botella MA, Valpuesta V (2003) Engineering increased vitamin C levels in plants by overexpression of a D-galacturonic acid reductase. *Nature Biotechnology* 21: 177–81

Ahmad P, Sarwat M, Sharma S (2008) Reactive oxygen species, antioxidants and signaling in plants. *Journal of Plant Biology* 51(3): 167-173

Al-Jaber NA, Awaad AS, Moses JE (2011) Review on some antioxidant plants growing in Arab world. *Journal of Saudi Chemical Society*15: 293–307

Alscher RG, Erturk N, Heath LS (2002) Role of superoxide dismutase (SODs) in controlling oxidative stress in plants. *Jour of Experimental Botany* 53(372): 1331-1341

Aly N, El-Gendy K, Mahmoud F, El-Sebae AK (2010) Protective effect of vitamin C against chlorpyrifos oxidative stress in male mice. *Pest Biochem Phys* 97: 7-12

Asada K (2000) The water-water cycle as alternative photon and electron sinks. *Phil Trans R Soc Lond B Biol Sci* 355: 1419-1431

Belitz HD, Grosch W (1999) Phenolic compounds, Food chemistry. Berlin: *Springer* 764–775

Blando F, Scardino AP, De Bellis L, Nicoletti I, Giovinazzo G (2005) Characterization of *in vitro* anthocyain-producing sour cherry (*Prunus cerasus* L.) callus cultures. *Food Res Int* 38: 937–42

Blokhina O, Virolainen E, Fagerstedt KV (2003) Antioxidants, oxidative damage and oxygen deprivation stress:a review. *Annals of Botany* 91: 179-194

Boyer J, Liu RH (2004) Apple phytochemicals and their health benefits. *Nutrition Journal* 3: 5

Butelli E, Titta L, Giorgio M, Mock HP, Matros A, Peterek S, *et al* (2008) Enrichment of tomato fruit with health-promoting anthocyanins by expression of select transcription factors. *Nature Biotechnology* 26: 1301–8

Cam M, Hisil Y, Kuscu A (2007) Organic Acid, Phenolic Content, and Antioxidant Capacity of Fruit Flesh and Seed of *Viburnum opulus. Chemistry of Natural Compounds* 43(4): 460-461

Cartron E, Carbonneau MA, Fouret G, Descomps B and Leger CL (2001) Specific antioxidant activity of caffeoyl derivatives and other natural phenolic compounds: LDL protection against oxidation and decrease in the proinflammatory lysophosphatidylcholine production. *J Nat Prod* 64: 480-486

Caruso JL, Callahan J, DeChant C, Jayasimhulu K,Winget GD (2000) Carnosic acid in green callus and regenerated shoots of *Rosmarinus officinalis. Plant Cell Rep* 19: 500–3

Chakraborty S, Chakraborty N, Datta A (2000) Increased nutritive value of transgenic potato by expressing a non allergenic seed albumin gene from *Amaranthus hypochondriacus.*

Proceedings of the National Academy of Sciences USA 97: 3724–9

Chan E, Lim Y, Wong S (2011) Antioxidant Properties of Ginger Leaves: An Overview. *Free Radicals and Antioxidants* 1(1): 6-16

Chang S, Tan C, Frankel EN, Barrett DM (2000) Low-Density Lipoprotein Antioxidant Activity of Phenolic Compounds and Polyphenol Oxidase Activity in Selected Clingstone Peach Cultivars. *Journal of Agricultural and Food Chemistry* 48(2): 147-151

Chang ST, Wu JH, Wang SY, Kang PL, Yang NS, Shyur LF (2001) Antioxidant Activity of Extracts from *Acacia Confusa* Bark and Heartwood. *Journal of Agricultural and Food Chemistry* 49(7): 3420-3424

Cho EA, Lee CA, Kim YS, Baek SH, de los Reyes B, Yun SJ (2005) Expression of g-tocopherol methyltransferase transgene improves tocopherol composition in lettuce (Latuca sativa L.). *Molecules and Cells* 19: 16–22

Constantinou C, Papas A, Constantinou AI (2008) Vitamin E and cancer: An insight into the anticancer activities of vitamin E isomers and analogs. *International Journal of Cancer* 123: 739–52

Costantini D, Verhulst S (2009) Does high antioxidant capacity indicates low oxidative stress? *Funct Ecol* 23: 506-509

Crowell EF, McGrath JM, Douches DS (2008) Accumulation of vitamin E in potato (*Solanum tuberosum*) tubers. *Transgenic Research* 17: 205–17

Dalton DA, Joyner SL, Becana M, Iturbe-Ormaetxe I, Chatfield JM (1998) Antioxidant Defenses in the Peripheral Cell Layers of Legume Root Nodules. *Plant Physiology* 116(1): 37-43

Das AK, Manda SC, Bereree S, Das J, Saha BP and Pal M (1999) Studies on antidirrhoeal activity of *Punica granathum* seed extract in rats. *J Ethnopharm* 69: 305-308

Davey MW, Van den Bergh I, Markham R, Swennen R, Keulemans J (2009) Genetic variability in Musafruit provitamin A carotenoids, lutein and mineral micronutrient contents. *Food Chemistry* 115: 806–13

Davuluri GR, van Tuinen A, Fraser PD, Manfredonia A, Newman R, Burgess D, *et al* (2005) Fruit-specific RNAi-mediated suppression of DET1 enhances carotenoid and flavonoid content in tomatoes. *Nature Biotechnology* 23: 890–5

De Sousa E, Zanatta L, Seifriz I, Creczynski-Pasa TB, Pizzolatti MG, Szpoganicz B, Silva FRMB (2004) Hypoglycemic effect and antioxidant potential of kaempferol-3, 7-O-(a)-dirhamnoside from *Bauhinia forcata* leaves. *J Nat Prod* 67: 829-832

Decendit A, Mérillon JM (1996) Condensed tannin and anthocyanin production in *Vitis vinifera* cell suspension cultures. *Plant Cell Rep* 15: 762–5

Delaunois B, Cordelier S, Conreux A, Clement C, Jeandet P (2007) Molecular engineering of resveratrol in plants. *Plant Biotechnology Journal* 7: 2–12

Devasagayam TPA, Tilak JC, Boloor KK, Sane KS, Ghaskadbi SS, Lele RD (2004) Free Radicals and Antioxidants in Human Health: Current Status and Future Prospects. *JAPI* 52: 794-804

Dharamapuri S, Rosati C, Pallara P, Aquilani R, Bouvier F, Camara B, *et al* (2002) Metabolic engineering of xanthophyll content in tomato fruits. *FEBS Letters* 519: 30–4

Di´az de la Garza R, Quinlivan EP, Klaus SM, Basset GJC, Gregory III JF, Hanson AD (2004) Folate biofortification in tomatoes by engineering the pteridine branch of folate synthesis. *Proceedings of the National Academy of Sciences* USA 101: 13720–25

Dimitrios B (2006) Sources of natural phenolic antioxidants. *Trends in Food Science & Technology* 17: 505–512

Diouf P, Stevanovic T, Cloutier A (2009) Antioxidant Properties and Polyphenol Contents of *Trembling aspen* Bark Extracts. *Wood Science and Technology* 43(5): 457-470

Diretto G, Al-Babili S, Tavazza R, Papacchioli V, Beyer P, Giuliano G (2007) Metabolic Engineering of potato carotenoid content through tuber-specific overexpression of a bacterial mini-pathway. *PLoS ONE* 2(4): 350

Diretto G, Welsch R, Tavazza R, Mourgues F, Pizzichini D, Beyer P, *et al* (2007) Silencing of beta-carotene hydroxylase increases total carotenoid and beta-carotene levels in potato tubers. *BMC Plant Biology* **7:** 11

Donaldson MS (2004) Nutrition and cancer: a review of the evidence for an anti-cancer diet. *Nutrition Journal* 3: 19

Dong HL, Chin BL (2000) Chilling stress-induced changes of antioxidant enzymes in the leaves of cucumber: in gel enzyme activity assays. *Plant Science* 159: 75–85

Ducreux LJ, Morris WL, Hedley PE, Shepherd T, Davies HV, Millam S, *et al* (2005) Metabolic engineering of high carotenoid potato tubers containing enhanced levels of β-carotene and lutein. *Journal of Experimental Botany* 56: 81–9

Edahiro JI, Nakamura M, Seki M, Furusaki S (2005) Enhanced accumulation of anthocyanin in cultured strawberry cells by repetitive feeding of L-phenylalanine into the medium. *J Biosci Bioeng* 99: 43–7

Enfissi EMA, Fraser PD, Loiș LM, Boronat A, Schuch W, Bramley PM (2005) Metabolic engineering of the mevalonate and non-mevalonate isopentenyl diphosphate-forming pathways for the production of health-promoting isoprenoids in tomato. *Plant Biotechnology Journal* 3: 17–27

Eqbal MA, Dauqan EMA, Abdullah A, Sani HA (2011) Natural antioxidants, lipid profile, lipid peroxidation, antioxidant enzymes of different vegetable oils. *Advance Jour of Food Sci and Technology* 3(4): 308-316

Fatima T, Rivera-Dominguez M, Troncoso-Rojas R, Tiznado-Hernandez MR, Handa AK, Mattoo AK (2008) Tomato. In: Kole C, Hall TC, editors. Compendium of Transgenic Crop Plants. Volume 6: Transgenic Vegetable Crops. Wiley-Blackwell Chichester UK p 1–45

Fedoreyev SA, Veselova MV, Krivoschekova OE, Mischenko NP, Denisenko VA,Dmitrenok PS, *et al* (2005) Caffeic acid metabolites from *Eritrichium sericeum* cell cultures. *Planta Med* 71: 446–51

Fraser PD, Romer S, Shipton CA, Mills PB, Kiano JW, Misawa N, *et al* (2002) Evaluation of transgenic tomato plants expressing an additional phytoene synthase in a fruit-specific manner. *Proceedings of the National Academy of Sciences* USA 99: 1092–7

Fujisawa M, Takita E, Harada H, Sakurai N, Suzuki H, Ohyama K, *et al* (2009) Pathway engineering of *Brassica napus* seeds using multiple key enzyme genes involved in ketocarotenoid formation. *Journal of Experimental Botany* 60: 1319–32

Fujisawa M, Watanabe M, Choi SK, Teramoto M, Ohyama K, Misawa N (2008) Enrichment of carotenoids in flaxseed (*Linum usitatissimum*) by metabolic engineering with introduction of bacterial phytoene synthase gene crtB. *Journal of Biosciences and Bioengineering* 105: 636–41

Gan RY, Kuang L, Xu XR, Zhang Y, Xia EQ, Song FL, Li HB (2010) Screening of Natural Antioxidants from Traditional Chinese Medicinal Plants Associated with Treatment of Rheumatic Disease. *Molecules* 15: 5988-5997

Georgiev M, Kuzeva S, Pavlov A, Kovacheva E, Ilieva M (2006a) Enhanced rosmarinic acid production by *Lavandula vera* MM cell suspension culture through elicitation with vanadyl sulfate. Z *Naturforsch* 61C: 241–4

Gibon Y, Bessieres MA, Larher F (1997) Is glycine betaine a noncompatible solute in higher plants that do not accumulate it? *Plant Cell Env* 20: 329-340

Giliberto L, Perrotta G, Pallara P, Weller JL, Fraser PD, Bramley PM,*et al* (2005) Manipulation of the blue light photoreceptor cryptochrome 2 in tomato affects vegetative development, flowering time, and fruit antioxidant content. *Plant Physiology* 137: 199–208

Gill SS, Tuteja N (2010) Reactive oxygen species and antioxidant machinery in abiotic stress tolerance in crop plants. *Plant Physiology and Biochemistry* 48(12): 909-930

Giovinazzo G, D'Amico L, Paradiso A, Bollini R, Sparvoli F, DeGara L (2005) Antioxidant metabolite profiles in tomato fruit constitutively expressing the grapevine stilbene synthase gene. *Plant Biotechnology Journal* 3: 57–69

Giuliano G, Tavazza R, Diretto G, Beyer P, Taylor MA (2008) Metabolic engineering of carotenoid biosynthesis in plants. *Trends in Biotechnology* 26: 139–45

Goto F, Yoshihara T, Saiki H (2000) Iron accumulation and enhanced growth in transgenic lettuce plants expressing the iron-binding protein ferritin. *Theoretical and Applied Genetics* 100: 658–64

Grzegorczyk I, Matkowski A, Wysokiñska H (2007) Antioxidant activity of extracts from *in vitro* cultures of *Salvia officinalis* L. *Food Chem* 104: 536–41

Gupta VK, Sharma SK (2006) Plants as natural antioxidants. *Natural Product Radiance* 5(4): 326-334

Halliwell B (1991) Drug antioxidant Effects-A basis for drug selection. *Drug* 42(4): 569-605

Halliwell B (2006) Oxidative stess and neurodegeneration: where are we now? *J Neurochem* 97: 1634-1658

Halliwell B, Gutteridge JMC (2007) Free radicals in biology and medicine. 4th edition. Oxford: Oxford University Press

Hamid AA, Aiyelaagbe OO, Usman LA, Ameen OM, Lawal A (2010) Antioxidants: Its medicinal and pharmacological applications. *African Journal of Pure and Applied Chemistry*

4(8): 142-151

Hazra B, Biswas S, Mandal N (2008) Antioxidant and free radical scavenging activity of Spondias pinnata. *BMC Complement Altern Med* 8: 63

He X, Liu RH (2007) Triterpenoids Isolated from Apple Peels Have Potent Antiproliferative Activity and May Be Partially Responsible for Apple's Anticancer Activity. *Journal of Agricultural and Food Chemistry* 55(11): 4366-4370

Hegde K, Joshi AB (2010) Hepatoprotective and Antioxidant Effect of *Carissa spinarum* Root Extract against Ccl4 and Paracetamol-Induced Hepatic Damage in Rats. *Bangladesh Journal of Pharmacology* 5(1): 73-76

Heslop-Harrison JS, Schwarzacher T (2007) Domestication, genomics and the future of the banana. *Annals of Botany* 100: 1073–84

Hrazdina G (2006) Aroma production by tissue cultures. *J Agric Food Chem* 54: 1116–23

Hurrell R (2003) Influence of vegetable protein sources on trace element and mineral bioavailability. *J Nutr* 133(9): 2973S–2977S

Irshad M, Chaudhuri PS (2002) Oxidant-antioxidant system: Role and significance in human body. *Indian Jour Exp Biol* 40: 1233-1239

Jayraj J, Devlin R, Punja Z (2008) Metabolic engineering of novel ketocarotenoid production in carrot plants. *Transgenic Research* 17: 489–501

Joharapurkar AA, Zambad SP, Wanjari MM, Umathe SN (2003) *In vivo* evaluation of antioxidant activity of alcoholic extract of rubia cordifolia Linn. And its influence on ethanol induced immunosupression. *Indian Jour Pharmacol* 35: 232-236

Jung HA, Kim AR, Chung HY and Choi JS (2002) *In vitro* antioxidant activity of some selected prunus species in Korea. *Arch Pharma Res* 25: 865-872

Kambayashi Y, Takekoshi S, Nakano M, Shibamori M, Hitomi Y, Ogino K. Kinobeon A (2005) purified from cultured safflower cells, is a novel and potent singlet oxygen quencher. *Acta Biochim Pol* 52: 903–7

Kandil F, Song L, Pezzuto J, Seigler D, Smith MAL (2000) Isolation of oligomeric proanthocyanidins from flavonoid-producing cell cultures. *In Vitro Cell Dev Biol-Plant* 6: 492–500

Karuppanapandian T, Moon JC, Kim C, Manoharan K, Kim W (2011) Reactive oxygen species in plants: their generation, signal transduction, and scavenging mechanisms. *Australian Jour of Crop Sci* 5(6): 709-725

Khan MI, Giridhar P (2011) Dietary antioxidants: the insurer of health. *Everyman's science* XLVI(4): 214-218

Khanbabae K, Ree TV (2001) Tannins: Classification and definition. *Nat Prod Rep* 18: 640-649

Kintzios S, Kollias H, Straitouris E, Makri O (2004) Scale-up micropropagation of sweet basil (*Ocimum basilicum* L.) in an airlift bioreactor and accumulation of rosmarinic acid. *Biotechnol Lett* 26: 521–3

Kobayashi S, Ding CK, Nakamura Y, Nakajima I, Matsumoto R (2000) Kiwifruits (*Actinidia deliciosa*) transformed with a *Vitis stilbens* synthase gene produce piceid (resveratrol-glucoside). *Plant Cell Reports* 19: 904–10

Kottaimuthu A, Sethupathy S, Manavalan R and Karar PK (2007) Antioxidant activity of flavonoids isolated from *Dolichos biflorus* Linn. in rabbits fed high fat diet. *Biosci Biotechnol Res. Asia,* 4: 179

Kovacheva E, Georgiev M, Pashova S, Angelova M, Ilieva M (2006) Radical quenching by rosmarinic acid from *Lavandula vera* MM cell culture. *Z Naturforsch* 61C: 517–20

Kruk J, Holländer-Czytko H, Oettmeier W, Trebst A (2005) Tocopherol as singlet oxygen scavenger in photosystem II. *J Plant Physiol* 162: 749–57

KuŸma L, Skrzypek Z, Wysokiñska H (2006) Diterpenoids and triterpenoids in hairy roots of *Salvia sclarea. Plant Cell Tissue Organ Cult* 84: 171–9

Lamarque AL, Fortunato RH and Guzman CA (1998) Seed oil alkanes from leguminosae species: Even carbon number preference. *Phytochemistry* 49: 731-736

Lee J, Koo N, Min DB (2004) Comprehensive review in food science and food safety. *Inst FoodTechnol* 3: 21-33

Liu S, Hu Y, Wang X, Zhong J, Lin Z (2006) High content of resveratrol in lettuce transformed with a stilbene synthase gene of *Parthenocissus henryana. Journal of Agricultural and Food Chemistry* 54: 8082–5

Lopez AB, Van Eck J, Conlin BJ, Paolillo DJ, O'Neill J, Li L (2008) Effect of the cauliflower OR transgene on carotenoid accumulation and chromoplast formation in transgenic potato tubers. *Journal of Experimental Botany* 59: 213–23

Lu S, Van Eck J, Zhou X, Lopez AB, O'Halloran DM, Cosman KM, *et al* (2006) The cauliflower Orgene encodes a DnaJ cysteine-rich domain-containing protein that mediates high levels of b-carotene accumulation. *Plant Cell* 18: 3594–605

Luczkiewicz M, Cisowski W (2001) Optimisation of the second phase of a two phase growth system for anthocyanin accumulation in callus cultures of *Rudbeckia hirta. Plant Cell Tissue Organ Cult* 65: 57–68

Lukaszewicz M, Matysiak-Kata I, Skala J, Feca I, Cisowski W, Szopa J (2004) Antioxidant capacity manipulation in transgenic potato tuber by changes in phenolic content. *Journal of Agricultural and Food Chemistry* 52: 1526–33

Luo J, Butelli E, Hill L, Parr A, Niggeweg R, Bailey P,*et al* (2008) AtMBY12 regulates caffeoyl quinic acid and flavonol synthesis in tomato: expression in fruits results in very high levels of both types of polyphenol. *Plant Journal* 56: 316–26

Mahakunakorn P, Tohda M, Murakami Y, Matsumoto K, Watanabe H (2004) Antioxidant and free radical-scavenging activity of Choto-san and its related constituents. *Biol Pharm Bull* 27: 38-46

Manach C, Scalbert A, Morand C, Jimenez L (2004) Polyphenols: Food sources and bioavailability. *American Journal of Clinical Nutrition* 79: 727–747

Mandal S, Yadav S, Yadav S, Nema RK (2009) Antioxidants: A Review. *Journal of Chemical and Pharmaceutical Research* 1(1): 102-104

Mayer JE, Pfeiffer WH, Beyer P (2008) Biofortified crops to alleviate micronutrient malnutrition. *Current Opinion in Plant Biology* 11: 166–70

McGhie TK, Currie AJ (2008) Effect of pre and post-harvest technologies on the health promoting properties of fruit and vegetables. In: Tomas-Barberan FA, Gil MI, editors. Improving the Health-promoting Properties of Fruit and Vegetable Products. 1st ed. CRC Press LLC, Boca Raton, FL. p. 301–25

Medina-Bolivar F, Condori J, Rimando AM, Hubstenberger J, Shelton K, O'Keefe SF, *et al.* (2002) Production and secretion of resveratrol in hairy root cultures of peanut. *Phytochemistry* 68: 1992–200

Mehta RA, Cassol T, Li N, Ali N, Handa AK, Mattoo AK (2002) Engineered polyamine accumulation in tomato enhances phytonutrient content, juice quality, and vine life. *Nature Biotechnology* 20: 613–8

Meyer JE, Pepin MF, Smith MA (2002) Anthocyanin production from *Vaccinium pahalae:* limitations of the physical microenvironment. *J Biotechnol* 93: 45–57

Mittler R (2002) Oxidative stress, antioxidants and stress tolerance. *Trends in Plant Science* 7(9): 405-410

Miura H, Kitamura Y, Ikenaga T, Mizobe K, Shimizu T, Nakamura M, *et al* (1998) Anthocyanin production of *Glehnia littoralis* callus cultures. *Phytochemistry* 48: 279–83

Moller IM (2001) Plant mitochondria and oxidative stress: Electron transport, NADPH turnover, and metabolism of reactive oxygen species. *Annual Review of Plant Physiology and Plant Molecular Biology* 52: 561-591

Moongkarndi P, Kosem N, Kaslungka S, Luanratana O, Pongpan N, Neungton N (2004) Antiproliferation, Antioxidation and Induction of Apoptosis by *Garcinia Mangostana* (Mangosteen) on Skbr3 Human Breast Cancer Cell Line. *Journal of Ethnopharmacology* 90(1): 161-166

Muir SR, Collins GJ, Robinson S, Hughes S, Bovy A, Ric De Vos CH, *et al* (2001) Overexpression of petunia chalcone isomerase in tomato results in fruit containing increased levels of flavonols. *Nature Biotechnology* 19: 470–4

Murthy CKN, Jayaprakasha GK, Singh RP (2002) Studies on antioxidant activity of pomegranate (*Punica granatum*) peel extract using *in vivo* models. *J Agric Food Chem* 50: 4791-4795

Nicoletti I, De Rossi A, Giovinazzo G, Corradini D (2007) Identification and quantification of stilbenes in fruits of transgenic tomato plants (*Lycopersicon esculentum* Mill.) by reversed phase HPLC with photodiode array and mass spectrometry detection. *Journal of Agricultural and Food Chemistry* 55: 3304–11

Nunes ACS, Kalkmann DC, Araga˜o FJL (2009) Folate biofortification of lettuce by expression of a codon optimized chicken GTP cyclohydrolase I gene. *Transgenic Research* 18: 661–67

Ochiai T, Ohno S, Soeda S, Tanaka H, Shoyama Y, Shimeno H (2004) Crocin prevents the death of rat pheochromyctoma (PC-12) cells by its antioxidant effects stronger than those of alpha-tocopherol. *Neurosci Lett* 362: 61–4

Oncel I, Keles Y, Ustun AS (2000) Interactive effects of temperature and heavy metal stress on the growth and some biochemical compounds in wheat seedlings. *Environ. Pollut* 107: 315–320

Pang X, Wang DH, Xing XY, Peng A, Zhang FS (2002) Effect of La^{+3}on the activities of antioxidant enzymes in wheat seedlings under lead stress in solution culture. *Chemosphere* 47: 1033-1039

Pavlov A, Kovatcheva P, Tuneva D, Ilieva M, Bley T (2005b) Radical scavenging activity and stability of betalains from *Beta vulgaris* hairy root culture in simulated conditions of human gastrointestinal tract. *Plant Foods Hum Nutr* 60: 43–7

Pavlov AI, Georgiev MI, Panchev IN, Ilieva MP (2005a) Optimization of rosmarinic acid pro-duction by *Lavandula vera* MM plant cell suspension in a laboratory bioreactor. *Biotechnol Prog* 21: 394–6

Pourmorad F, Hosseinimehr SJ, Shahabimajd N (2006) Antioxidant activity, phenol and flavonoid contents of some selected Iranian medicinal plants. *African Jour of Biotech* 5(11): 1142-1145

Ravanello MP, Ke D, Alvarez J, Huang B, Shewmaker CK (2003) Coordinate expression of multiple bacterial carotenoid genes in canola leading to altered carotenoid production. *Metabolic Engineering* 5: 255–63

Ravindra PV, Narayan MS (2003) Antioxidant activity of the anthocyanin from carrot (*Daucus carota*) callus culture. *Int J Food Sci Nutr* 54: 349–55

Reddy L, Odhav B, Bhoola, KD (2003) Natural products for cancer prevention: a global perspective. *Pharmacology and Therapeutics* 99: 1-13

Ro¨mer S, Fraser PD, Kiano JW, Shipton CA, Misawa N, Schuch W,*et al* (2000) Elevation of the provitamin A content of transgenic tomato plants. *Nature Biotechnology* 18: 666–9

Rockenbach II, Gonzaga LV, Rizelio VM, Goncalves AEdSS, Genovese MI, Fett R (2011) Phenolic Compounds and Antioxidant Activity of Seed and Skin Extracts of Red Grape (*Vitis Vinifera* and *Vitis Labrusca*) Pomace from Brazilian Winemaking. *Food Research International* 44(4): 897-901

Rohini G, Sabitha KE, Devi CSS (2004) *Bacopa monnieri* Linn. Extract modulates antioxidant and marker enzyme status in fibrosarcoma bearing rats. *Indian Jour Exp Biol* 42: 776-780

Rommens CM, Richael CM, Yan H, Navarre DA, Ye J, Krucker M, *et al* (2008) Engineered native pathways for high kaempferol and caffeoylquinate production in potato. *Plant Biotechnology Journal* 6: 870–86

Ronen G, Cohen M, Zamir D, Hirschberg J (1999) Regulation of carotenoid biosynthesis during tomato fruit development: expression of the gene for lycopene epsilon-cyclase is down-regulated during ripening and is elevated in the mutant Delta. *Plant Journal* 17: 341–51

Rosati C, Aquilani R, Dharamapuri S, Pallara P, Marusic C, Tavazza R,*et al* (2000)Metabolic engineering of beta-carotene and lycopene content in tomato fruit. *Plant Journal* 24: 413–9

Ru¨hmann S, Treutter D, Fritsche S, Briviba K, Szankowski I (2006) Piceid (resveratrol glucoside) synthesis in stilbene synthase transgenic apple fruit. *Journal of Agricultural and Food Chemistry* 54: 4633–40

Sarvajeet SG, Narendra T (2010) Reactive oxygen species and antioxidant machinery in abiotic stress tolerance in crop plants. *Plant Physiol Bioch* 48: 909-930

Scartezzini P, Speroni E (2000) Review on some plants of Indian traditional medicine with antioxidant activity. Jour. *Ethnopharmacol* 71: 23-43

Schijlen E, Ric de Vos CH, Jonker H, van den Broeck H, Molthoff J, van Tunen A,*et al* (2006) Pathway engineering for healthy phytochemicals leading to the production of novel flavonoids in tomato fruit. *Plant Biotechnology Journal* 4: 433–44

Shewmaker CK, Sheehy JA, Daley M, Colburn S, Ke DY (1999) Seed specific overexpression of phytoene synthase: increase in carotenoids and other metabolic effects. *Plant Journal* 20: 401–12

Shih CH, Chen Y, Wang M, Chu IK, Lo C (2008) Accumulation of isoflavone genistin in transgenic tomato plants overexpressing a soybean isoflavone synthase gene. *Journal of Agricultural and Food Chemistry* 56: 5655–61

Shukla V, Mattoo AK (2009) Potential for engineering horticultural crops with high antioxidant capacity. *Perspectives in Agriculture Veterinary Science Nutrition and Natural Resources* 4: 1-22

Silva S, Gomes L, Leitao F, Coelho AV, Boas LV (2006) Phenolic Compounds and Antioxidant Activity of *Olea Europaea* L. Fruits and Leaves. *Food Science and Technology International* 12(5): 385-395

Singh N, Singh G (2012) Antioxidants from natural source: ray of hope for oxidative damage. *Journal of pharmaceutical and scientific innovation* 1(4): 14-17

Stintzing FC, Carle R (2004) Functional properties of anthocyanins and betalains in plants, food, and in human nutrition. *Trends Food Sci Technol* 15: 19–38

Tadhani MB, Patel VH, Subhash R (2007) *In Vitro* Antioxidant Activities of *Stevia Rebaudiana* Leaves and Callus. *Journal of Food Composition and Analysis* 20(3-4): 323-329

Tavassoli S, Djomeh ZE (2011) Total Phenols, Antioxidant Potential and Antimicrobial Activity of Methanol Extract of Rosemary (*Rosmarinus officinalis* L.). *Global Veterinaria* 7(4): 337-341

Tavva VS, Kim YH, Kagan IA, Dinkins RD, Kim KH, Collins GB (2007) Increased a-tocopherol content in soybean seed overexpressing the Perilla frutscensg-tocopherol methyltransferase gene. *Plant Cell Reports* 26: 61–70

Terahara N, Callebaut A, Ohba R, Nagata T, Ohnishi-Kameyama M, Suzuki M (2001) Acylated anthocyanidin 3-sophoroside-5-glucosides from *Ajuga reptans* flowers and the corresponding cell cultures. *Phytochemistry* 58: 493–500

Terahara N, Konczak I, Ono H, Yoshimoto M, Yamakawa O (2004) Characterization of acylated anthocyanins in callus induced from storage root of purple-fleshed sweet potato, *Ipomoea batatas* L. *J Biomed Biotechnol* 5: 279–86

Tholl D (2006) Terpene synthases and the regulation, diversity and biological roles of terpene metabolism. *Curr Opin Plant Biol* 9: 297–304

Van Eenennaam AL, Lincoln K, Durrett TP, Valentin HE, Shewmaker CK, Thorne GM, *et*

al (2003) Engineering vitamin E content: from *Arabidopsis* mutant to soy oil. *Plant Cell* 15: 3007–19

Vellosillo T, Vicente J, Kulasekaran S, Hamberg M, Castresana C (2010) Emerging complexity in reactive oxygen species production and signaling during the response of plants to pathogens. *Plant Physiol* 154: 444–448

Wakita Y, Otani M, Hamada T, Mori M, Iba K, Shimada T (2001) A tobacco microsomal w-3 fatty acid desaturase gene increases the linolenic acid content in transgenic sweet potato (Ipomoea batatas). *Plant Cell Reports* 20: 244–9

Wang S Y, Lin HS (2000) Antioxidant activity in fruits and leaves of blackberry, raspberry, and strawberry varies with cultivar and development stage. *Journal of Agricultural and Food Chemistry* 48: 140–146

Wei S, Li X, Gruber MY, Li R, Zhou R, Zebarjadi A,*et al* (2009) RNAi mediated suppression of DET1 alters the levels of carotenoids and sinapate esters in seeds of Brassic napus. *Journal of Agricultural and Food Chemistry* 57: 5326–33

Willits MG, Kramer CM, Prata RTN, De Luca V, Potter BG, Steffens JC,*et al* (2005)Utilization of the genetic resources of wild species to create a non transgenic high flavonoid tomato. *Journal of Agricultural and Food Chemistry* 53: 1231–6

Wojdyolo A, Oszmianski J, Laskowski P (2008) Polyphenolic compounds and antioxidant activity of new and old apple varieties. *Journal of Agricultural and Food Chemistry* 56: 6520–30

Wurbs D, Ruf S, Bock R (2007) Contained metabolic engineering in tomatoes by expression of carotenoid biosynthesis genes from the plastid genome. *Plant Journal* 49: 276–88

Yamamoto H, Inoue K, Yazaki K (2000) Caffeic acid oligomers in *Lithospermum erythrorhizon* cell suspension cultures. *Phytochemistry* 53: 651–7

Yamamoto H, Zhao P, Yazaki K, Inoue K (2002) Regulation of lithospermic acid B and shikonin production in *Lithospermum erythrorhizon* cell suspension cultures. *Chem Pharm Bull* 50: 1086–90

Yu B, Lydiate DJ, Young LW, Scha¨fer UA, Hannoufa A (2008) Enhancing the carotenoid content of *Brassica napus* seeds by down regulating lycopene epsilon cyclase. *Transgenic Research* 17: 573–85

Yusuf MA, Sarin NB (2007) Antioxidant value addition in human diets: genetic transformation of *Brassica juncea* with g-TMT gene for increased a-tocopherol content. *Transgenic Research* 16: 109–13

Zhang J, Kirkham MB (1996) Enzymatic responses of the ascorbate glutathione cycle to drought in sorghum and sunflower plants. *Plant Sci* 113: 139-147

Zhang SJ, Lin YM, Zhou HC, Wei SD, Lin GH, Ye GF (2010) Antioxidant Tannins from Stem Bark and Fine Root of *Casuarina Equisetifolia*. *Molecules* 15(8): 5658-5670

Zhu YJ, Agbayani R, Jackson MC, Tang CS, Moore PH (2004) Expression of the grapevine stilbene synthase gene VST1 in papaya provides increased resistance against disease caused by *Phytophthora palmivora*. *Planta* 220:241–50

3

Current Status of Molecular Markers and Marker Assisted Selection in Sunflower (*Helianthus annuus* L.)

M. Sujatha and *K. Sujatha*

Molecular markers have played a significant role in accelerating sunflower breeding programmes through marker-assisted selection by virtue of their abundance, technical ease and detection at any development stage of the plant besides being not affected by environmental factors. DNA markers were developed and deployed during the last two decades in sunflower. Genomic resources are being established which should facilitate the marker-assisted breeding through direct or comparative mapping techniques. The utility of these markers in various molecular biology applications like assessment of genetic relationships among genotypes, construction of linkage maps, tagging and mapping genes/QTLs of agronomic interest, development of genomic and cDNA libraries, map based cloning and marker assisted selection are reviewed.

Introduction

Sunflower (*Helianthus annuus* L.) is one of the world's leading oilseed crops, after soybean, rapeseed, cotton, and groundnut. Sunflower oil is considered premium cooking oil for human consumption because of its high concentration of unsaturated fatty acids, oleic and linoleic, and low concentration of linolenic acid (Dorrell and Vick, 1997). Molecular breeding or marker assisted selection provides opportunities for genetic enhancement of various crops including sunflower by accelerating the plant breeding procedure. To facilitate this research, a number of molecular markers systems namely, Restriction fragment length polymorphism (RFLP) (Berry *et al* 1995; Gentzbittel *et al.*, 1995, 1998; Jan *et al.*, 1998; Bert *et al.*, 2004), Random amplified polymorphic DNA (RAPD) (Mouzeyar *et al.*, 1995; Lawson *et al.*, 1996; 1998; Horn *et al.*, 2003), Amplified fragment length polymorphism (AFLP) (Flores-Berrios, 2000a; Mokrani *et al.*, 2002; Langar *et al.*, 2003; Rachid Al-Chaarani *et al.*, 2004), Simple sequence repeats (SSRs) (Mokrani *et al.*, 2002; Bert *et al.*, 2003; Rachid Al-Chaarani *et al.*, 2004; Hu *et al.*, 2007; Kiani *et al.*, 2007a), Sequence characterized amplified region (SCAR) (Lu *et al.*, 1999; 2000; Brahm *et al.*, 2000; Bouzidi *et al.*, 2002), Cleaved amplified polymorphic sequence (CAPS) (Gedil *et al.*, 2001; Bouzidi *et al.*, 2002), insertion/deletion (INDEL) (Perez-Vich *et al.*, 2005; Garcia-Moreno *et al.*, 2006), single nucleotide polymorphism (SNP) (Lai *et al.*, 2005) and target region amplification polymorphism (TRAP) (Hu *et al.*, 2007) markers have been developed and used for various molecular applications in sunflower.

Genetic markers represent genetic differences between individual organisms or species. Generally, they do not represent the target genes themselves but act as 'signs' or 'flags'. Such markers themselves do not affect the phenotype of the trait of interest because they are located only near or 'linked' to genes controlling the trait. These are classified in to three types: (1) Morphological markers, (2) Biochemical markers, and (3) DNA markers and their application in sunflower improvement is elaborated below:

Morphological markers

Morphological markers are visually characterized phenotypic characters such as flower colour, seed shape, growth habits or pigmentation. Miller and Fick (1997) identified about 40 morphological traits controlled by one or two genes in sunflower, prior to the advent of molecular markers. The efforts to establish linkage relationships among the genes controlling various morphological traits met with limited success and as a result, no classical linkage map based on morphological markers could be developed. Leclercq (1966) discovered a tight linkage between a nuclear male sterile gene (*Ms*10) and an anthocyanin pigment gene (*T*) and proposed a scheme to use male sterility to produce sunflower hybrids without the tedious emasculation process of the seed-bearing female plants. Hockett and Knowles (1970) reported that the single gene *Y* for green versus yellow growing point color was found to be linked with the gene *Br*3 for branching with a recombination frequency of 11.6 ± 1.0%.

Biochemical markers

Characterization of sunflower genotypes was carried out using isozymes and storage proteins. Isozymes are enzymes that have different amino acid sequence but catalyze the same chemical reaction. In sunflower, isozymes have been used for investigating genetic variability in domesticated and wild populations (Torres, 1983; Quillet *et al.*, 1992; Tersac *et al.*, 1994; Dry and Burdon, 1986; Rieseberg and Seiler, 1990; Cronn *et al.*, 1997) assessing phylogenetic relationships and speciation mechanisms (Rieseberg, 1991; Rieseberg *et al.*, 1998), differentiating maintainer and restorer lines (Carrera *et al.*, 2002), determining the ploidy level (Rieseberg and Soltis, 1989) and for identifying interspecific hybrids (Carrera *et al.*, 1996). There have been few reports on mapping of isozyme loci on to various linkage groups of sunflower (Kahler and Lay, 1985; Lay *et al.*, 1988; Quillet *et al.*, 1995; Fambrini *et al.*, 1997; Mestries *et al.*, 1998; Carrera *et al.*, 2002).

The storage proteins namely albumins were used to differentiate sunflower genotypes (Anisimova *et al.*, 1995; Serre *et al.*, 2001). The biochemical markers have been used to a limited extent in sunflower due to their low frequency, poor genome coverage, stage specific expression and low levels of polymorphism. Pizarro *et al.* (2000) and Popov *et al.* (2002) compared isozyme and RAPD markers to assess genetic diversity in 21 and 30 sunflower inbred lines, respectively, and demonstrated the greater discriminatory power of the DNA-based marker systems.

Owing to the disadvantages of morphological and biochemical markers, DNA markers have assumed greater utility and importance in sunflower genomics.

DNA-based markers

DNA markers are the most widely used marker predominantly due to their abundance. They arise from different classes of DNA mutations such as substitution mutations (point mutations), rearrangements (insertions or deletions) or errors in replication of tandemly repeated DNA (Paterson, 1996a). These markers are selectively neutral because they are usually located in non-coding regions of DNA. DNA markers are practically unlimited in number and are not affected by environmental factors and/or the developmental stage of the plant (Winter and Kahl, 1995). DNA markers are broadly divided into three classes based on the method of their detection: (1) Hybridization-based; (2) Polymerase chain reaction (PCR)-based and (3) DNA sequence-based. The DNA marker techniques developed and used for various applications in sunflower are reviewed.

Hybridization-based markers

RFLP (Restriction fragment length polymorphism)

It is the first DNA marker technique developed for genome mapping in humans (Botstein *et al.*, 1980). This technique detects the gain or loss of a restriction enzyme recognition site and the insertion or deletion of a DNA fragment between two restriction sites. Using RFLP markers on 17 inbred lines, Gentzbittel *et al.*, (1994) reported a lower genetic variability in cultivated sunflower, suggesting that efforts to introgress new genes from wild sunflower species should be increased. RFLPs consistently separated sunflower inbred lines into sterility maintainer (B-line) and fertility restorer (R-line) groups (Berry *et al.*, 1994; Gentzbittel *et al.*, 1994; Zhang *et al.*, 1995), reflecting the breeding strategies that maximize heterosis. Zhang *et al.* (1995) used RFLPs to screen four inbred lines for intra-line polymorphisms, and found RFLPs within them.

Rieseberg and Seiler (1990) used RFLP analysis of chloroplast DNA along with isozymes, for an evolutionary study of cultivated sunflower in relation to wild *Helianthus annuus*. They found extensive isozyme and chloroplast DNA polymorphism in the wild accessions and a virtual monomorphism in cultivated lines, concluding that domesticated sunflower must have evolved from a very limited gene pool.

The open reading frame, *orfH522*, in the 3′-flanking region of the mitochondrial (mt) *atpA* gene was associated with the CMS phenotype (Kohler *et al.*, 1991; Laver *et al.*, 1991) and the gene product of *orfH522*, a 16-kDa protein, differed between the male fertile and male sterile lines (Horn *et al.*, 1991; Monéger *et al.*, 1994). Mitochondrial RFLP studies of 28 male sterility-inducing cytoplasms and one male fertile cytoplasm of *Helianthus* using nine mtDNA genes and three probes for the *orfH522* clearly distinguished the CMS sources (Horn, 2002).

Three classical linkage maps have been constructed in sunflower using RFLPs. The first published linkage map of cultivated sunflower (Berry *et al.*, 1995) spanned 1,380 cM on 17 linkage groups and contained 234 RFLP loci detected by 213 cDNA probes in an F_2 population of 289 individuals derived from a cross between two inbred lines, HA 89 and ZENB8. This map was later "joined" with maps constructed from eight other F_2 populations to form a high density, composite map with 635 marker loci and was 1,472 cM in length (Berry *et al.*, 1996). The second linkage map (Gentzbittel *et al.*, 1995) of 237 loci spanning 1,150 cM and consisting of 16 linkage groups of more than 20 cM and seven groups of less than 20 cM was constructed from three F_2 and two BC_1 populations. The third map published by the USDA sunflower research group (Jan *et al.*, 1998) had 271 marker loci detected by 232 cDNA probes and covered 1,164 cM in 20 linkage groups.

RFLPs were used to map the downy mildew resistance genes *Pl1, Pl2, Pl6* (Mouzeyar *et al.*, 1995; Roeckel-Drevet *et al.*, 1996; Vear *et al.*, 1997) and *Sclerotinia sclerotiorum* resistance QTLs (Mestries *et al.*, 1998). Some of the mapped single or low-copy probes were recently used to identify chromosome-specific bacterial artificial chromosome (BAC) clones (Feng *et al.*, 2006) and for assessing SNP frequencies, nucleotide diversity, and linkage disequilibrium (LD) in modern cultivated sunflower lines (Kolkman *et al.*, 2007).

AFLP (Amplified fragment length polymorphism)

This is a multi locus dominant marker technique based on PCR (Vos *et al.*, 1995). It converts the detection of RFLP from the hybridization-based system to an amplification-based system. In other words, AFLP methodology employs PCR to detect RFLP indirectly and collectively. This involves digestion of DNA with restriction enzymes, ligation of the restricted fragments with synthetic oligonucleotides (adapters), pre-amplification with primers complementary to that of the adapters and selective amplification of sets of restriction fragments, achieved by adding one to three selective nucleotides to the 3' ends of the primers. AFLP has proven to be a powerful genotyping technique for almost all organisms with different levels of genome complexity. In Sunflower, AFLPs were used for fingerprinting germplasm (Hongtrakul *et al.*, 1997; Dong *et al.*, 2007) mapping downy mildew resistance gene *Pl2* (Brahm *et al.*, 2000), fertility restorer locus *Rf1* (Kusterer *et al.*, 2005) and for quantitative trait loci (QTL) analysis of various traits such as *in vitro* organogenesis (Flores Berrios *et al.*, 2000a; 2000b), partial resistance to downy mildew (Rachiu Al- Chaarani *et al.*, 2002), leaf chlorophyll concentration and stomatal conductance (Hervé *et al.*, 2001).

Four linkage maps have been constructed using AFLPs and in combination with other markers. Peerbolte and Peleman (1996) used 11 primer combinations on the DNA of 92 individuals from two CARTISOL F_2 populations, CX x RHA 266 and PAC2 x RHA 266 and mapped 241 (188 co-dominant) and 282 AFLP (243 co-dominant) markers in the two populations, respectively. Gedil *et al.* (2001) reported an integrated RFLP-AFLP linkage map constructed from the 180 F_2 progeny derived from two public sunflower lines, HA 370 and HA 372. This map consisted of 296 AFLP and 104 RFLP markers in 17 linkage groups and was 1,326 cM long. Kusterer *et al.* (2004) constructed a linkage map from 183 F_2 individuals from the cross RHA 325 (cms) × HA 342. The map covered 1,751.5 cM with 202 AFLP and 19 SSR markers in 18 linkage groups. Langar *et al.* (2003) used both AFLP and DALP for constructing a linkage map for an RIL population derived from a cross of HA 89 x LR4. The map contained seven allele-specific PCR markers, 64 DALP, and 301 AFLP markers, and covered 2,168.6 cM in 18 linkage groups.

PCR based markers

RAPD

Random amplified polymorphic DNA is a polymorphism assay based on the amplification of random DNA segments with single primers of arbitrary nucleotide sequence (Welsh and McClelland, 1990; Williams *et al.*, 1990). The amplification depends on the correct orientation and proper distances of the randomly distributed sequences complementary to the decanucleotide primers being used. Using arbitrarily primed-PCR (AP-PCR) the genetic variability to mid-stem infection of *Sclerotinia sclerotiorum* (white mold) among interspecific sunflower hybrids was demonstrated (Köhler and Friedt, 1999). Interspecific hybrid progenies exhibited substantial genetic distance from their parental sunflower inbred lines, and hybrids having reduced *Sclerotinia* infection were identified.

RAPDs were used to study the ancestral relationships between sunflower inbred lines, wild populations, and land races (Arias and Rieseberg, 1995; Iqbal *et al.*, 2008), for tagging resistance genes to rust (Lawson *et al.*, 1998), broom rape (Lu *et al.*, 2000) and downy mildew resistance gene *Pl2* (Brahm *et al.*, 2000). In recent years, RAPD has given way to other marker techniques because of its dominant nature, and lower reproducibility (Jones *et al.*, 1997).

SSR

Simple sequence repeats, also known as microsatellites or short tandem repeats (STR) is a class of DNA sequence stretches characterized by highly repeated di-, tri-, tetra-, or penta-nucleotide motifs and spread throughout the genomes. SSR polymorphisms are assayed by specifically amplifying a small DNA fragment that contains the repeats via PCR and then by sizing the amplified fragments on agarose/ polyacrylamide gels or capillary electrophoresis. The number of repeats varies among genotypes under investigation and results in a length variation of the amplified fragments. Variation in SSR length occurs primarily due to slipped-strand mispairing during DNA replication (Toth *et al.*, 2000; Katti *et al.*, 2001; Li *et al.*, 2002), and mutations of this sort occur at a much higher frequency than do point mutations and insertions/deletions (Rossetto, 2001). Thus, SSRs reveal much higher levels of polymorphism than do most other marker systems. SSRs have been a common source of markers for genetic mapping, molecular breeding, and population genetic analyses in a wide variety of species including sunflower (Chase *et al.*, 1996; Xiong *et al.*, 1999; Matsuoka *et al.*, 2002; Tang *et al.*, 2003; Neeraja *et al.*, 2007). Being highly polymorphic, highly abundant, co-dominant inheritance, analytically simple and readily transferable, SSRs are the markers of choice in Sunflower genomic studies. Yu *et al.* (2002) developed 131 unique SSRs from 970 clones isolated from the (CA)*n*–, (CT)*n*–, (CAA)*n*–, (CATA)*n*–, or (GATA)*n*– enriched genomic DNA libraries. The dinucleotide, trinucleotide, and tetranucleotide repeats, were very informative depicting polymorphic information content (PIC) scores of 0.53, 0.53 and 0.83, respectively.

SSRs were used to characterize the genetic diversity in 19 elite inbred lines and 28 domestic and wild germplasm accessions, including Native American landraces (Tang and Knapp,

2003) using 122 SSRs distributed throughout the sunflower genome. An extraordinary allelic diversity in the Native American landraces and wild populations, and less diversity in the cultivated germplasm was observed. The potential of SSR markers in identification and assessment of genetic relationships between inbred lines (Paniego *et al.*, 2002; Darvishzadeh *et al.*, 2010) and backcross derived inbred lines obtained from interspecific crosses (Sujatha *et al.*, 2008) was demonstrated. Zhang *et al.* (2005) identified 78 SSR markers from the 1,111 primer pairs provided by CARTISOL as an effective set of SSR markers for diversity assessment, purity testing, hybrid formula verification, and varietal identification in the process of seed certification. These molecular markers could also play an important role in protecting plant breeders' rights in sunflower and in assessing whether or not an inbred line was essentially derived [UPOV (International Union for the Protection of New Varieties of Plants) Convention revised in 1991].

Wills *et al.* (2005) investigated the mode of organellar inheritance using chloroplast SSRs and proposed that it is strictly maternal. Ellis *et al.* (2006) used nuclear and chloroplast SSRs to investigate the levels of genetic diversity and population structure of an extremely rare sunflower species, *Helianthus verticillatus* in comparison to a common congener, *H. angustifolius.* Despite its rarity, *H. verticillatus* possessed significantly higher levels of genetic diversity than *H. angustifolius* at nuclear loci and equivalent levels of chloroplast diversity. SSR markers were successively used to map the fertility restorer genes *Rf3* (Liu *et al.*, 2011), *Rf4* (Feng *et al.*, 2008), male sterility gene (Perezwich *et al.*, 2005) and *Tph1* gene controlling β-tocopherol in seeds (Vera-Ruiz *et al.*, 2006).

The development of an SSR linkage map (Tang *et al.*, 2002) marked the most significant milestone in sunflower genome mapping. The first sunflower SSR map was constructed from a mapping population of 94 RILs derived from a cross between two public inbred lines: a confectionery, RHA 280 (Fick *et al.*, 1974), and a single-headed oilseed restorer line, RHA 801 (Roath *et al.*, 1981). The map consisted of 459 loci coalesced into 17 linkage groups with an accumulated length of 1,368 cM. This public SSR map was soon expanded to 1,432 cM adding 118 new SSR and INDEL markers and was aligned with other proprietary maps constructed from the PHA × PHB RIL population and from the HA 370 × HA 372 F_2 map with 80 RFLP marker loci (Yu *et al.*, 2003). The mapped SSR markers have been used by numerous researchers in their mapping projects and made it possible for all published maps to have the same linkage group numbers. Recently, the original RHA 280/RHA 801 RIL map has been expanded to 1,161 SSR loci.

SCARs

Sequence characterised amplified region markers are an extension of RAPD markers, wherein the RAPD fragment linked to the trait of interest is sequenced and longer primers (20–25 bp) are synthesized so that PCR amplification is performed under high stringent conditions rather than the less stringent RAPD conditions to improve specificity and reproducibility (Paran and Michelmore, 1993). In sunflower, SCAR markers were developed for the rust resistance genes R_{adv} and R_1 (Lawson *et al.*, 2010), the downy mildew resistance gene *Pl2* (Brahm *et al.*, 2000), and *Or5* gene which confers resistance to race E of broomrape *orobanche cumana* (Lu *et al.*, 2000).

CAPS

Cleaved amplified polymorphic sequences are also called PCR based RFLPs. A CAPS marker obtained by linked RAPD fragment sequencing namely H13 and depicting polymorphism when restricted with the enzyme *Hinf*I was used to map the fertility restorer gene *Rf1* (Kusterer *et al.,* 2005) in sunflower. Pankovic *et al.* (2007; 2009) developed CAPS markers for the downy mildew resistance gene *Pl6.*

DALP

Direct amplification of length polymorphism was proposed by Desmarais *et al.* (1998). In this technique, several selective primers are made by extending the 3′ end of the M13–40 universal sequencing primer with two or four nucleotides. These selective primers can be used with another universal sequencing primer M13 to generate multi-banded DNA fingerprints resolved with a sequencing gel. A different fingerprint can be obtained when a different selective primer is used. This technique is very easy to perform and offers the possibility to sequence each of the amplified fragments with the universal sequencing primers. Since the procedure requires the use of radioisotope to label one of the primers, this technique was not widely used. Langer *et al.* (2003) used it in combination with AFLP for construction of a linkage map.

TRAP

Target region amplification polymorphism is a technique (Hu and Vick, 2003) designed to harness the existing sequence information and explores the bioinformatics tools to develop markers associated with traits of interest. It involves fixed primers designed against known sequences of annotated putative genes and used in combination with other primers of arbitrary sequence to trap polymorphism in the targeted genomic region. Since the fixed primers target the candidate gene sequences, it offers a potentially higher probability of amplifying markers linked to the phenotype under investigation than RAPD and AFLP, which generate random polymorphic markers across the genome. This technique has been successful in assessing the genetic relationships of 16 perennial *Helianthus* species (Hu *et al.,* 2003) and among 177 USDA-ARS released public sunflower inbred lines (Yue *et al.,* 2009).

TRAPs were used to develop markers for a gene conferring resistance to several races of downy mildew (Hu *et al.,* 2004b), a recessive gene for apical branching (Rojas-Barros *et al.,* 2008), a recessive gene governing complete male sterility (Chen *et al.,* 2006), a recessive gene governing light green leaf color, a gene controlling lemon ray flower color (Yue *et al,* 2008b) and QTLs for resistance to *Sclerotinia* head rot (Yue *et al.,* 2008a). Yue *et al.* (2007) has demonstrated the utility of a TRAP marker developed from a sunflower EST that showed homology to a *Petunia* fertility restorer gene to recover *rf1/rf1* genotypes with a greater efficiency in progeny tests.

Hu *et al.* (2004a) developed a linkage map constructed with TRAP markers. Using 38 primer combinations, a map consisting of 160 markers on 17 linkage groups spanning a length of 1,140 cM was constructed with 129 RILs derived from the cross 83HR4/ RHA 345. Hu (2006) added 183 TRAP markers to the public SSR map (Tang *et al.,* 2002) and defined 21 of the 34 linkage group ends by using 94 individuals of the RIL mapping population developed by Steven Knapp's laboratory and nine fixed primers containing the conserved *Arabidopsis* type telomeric repeats. An additional 220 TRAP markers were integrated to the same map and expanded the length from 1,523 to 1,920 cM (Hu *et al.,* 2007). A brief summary of map information of different traits and linked markers in sunflower is presented in Table1.

Table 1 : Map information of different traits in sunflower

Trait	Gene	Markers used	Closest genetic distance(cM)	Reference
Downy mildew	*Pl1*	RFLP, RAPD	–	Mouzeyar *et al.,* 1995
	Pl2	RAPD, AFLP, SCAR	2cM	Brahm *et al.,* 2000
	Pl2, Pl5, Pl7, Pl8	RFLP, AFLP	–	Bert *et al.,* 2001
	Pl6	RAPD, SCAR	3.0	Bouzidi *et al.,* 2002
	Pl1	RFLP, CAPS	–	Gedil *et al.,* 2001
	Pl1	IFLP, NBS-LRR, SSR	2.4	Slabaugh *et al.,* 2003
	Pl5/ Pl8	TIR-NBS-LRR	–	Radwan *et al.,* 2003
	Pl5/Pl8	CC-NBR-LRR-sequence	–	Radwan *et al.,* 2004
	Pl5/Pl8	SSR	–	Duble *et al* , 2004
	Pl5/Pl8	STS	–	Radwan *et al.,* 2004
	Pl5/Pl8	EST	–	Bouzidi *et al.,* 2007
	Pl6	SSR, TIR-NBS-LRR, CAPS	–	Bouzidi *et al.,* 2002, Pankovic *et al.,* 2007; 2009
	Pl7	SSR, RFLP	–	Mandy *et al.,* 2008
	Pl13	SSR	0.9	Sujatha *et al.,* 2009
	Plarg	SSR	–	Duble *et al.,* 2004
	QTL	SSR, AFLP	–	Rachid Al-Chaarani *et al.,* 2002
	QTL	RFLP, SSR	–	Vear *et al.,* 2008
Orobanche	*Or_5*	RFLP, SCAR	–	Lu *et al.,* 1999
	Or_5	RAPD, SCAR	5.6	Lu *et al.,* 2000
	Or_5	SSR, AFLP	5.6	Tang *et al.,* 2003
	QTL	RFLP, SSR	–	Perez-Vich *et al.,* 2004
Sclerotinia midstalk-rot	–	RFLP	–	Bert *et al.,* 2002
	QTL	RFLP	–	Bert *et al.,* 2004
	QTL	SSR	–	Micic *et al.,* 2004

Trait	Gene	Markers used	Closest genetic distance(cM)	Reference
	QTL	SSR	–	Micic *et al.*, 2005a
	–	SSR	–	Micic *et al.*, 2005b
Black stem	*QTL*	AFLP, SSR	–	Rachid Al-Charani *et al.*, 2002
	QTL	RFLP, SSR	–	Bert *et al.*, 2004
Chlorotic mottle virus	*Rcmo-1*	SSR	4.0	Lenardon *et al.*, 2005
Phoma basal stem and root necrosis	*QTL*	SSR, AFLP	–	Alfadil *et al.*, 2007
Rust (*Puccinia helianthi*)	*R1 and R (Adv)*	RAPD	–	Lawson *et al.*, 1996
	R1	RAPD, SCAR	0.0,	Lawson *et al.*, 1998
	R2	SSR	1.6	Lawson *et al.*, 2011
Oil content	*QTL*	RFLP, AFLP	–	Leon *et al.*, 1995a
	QTL	AFLP, SSR	20.9	Mokrani *et al.*, 2002
Beta-tocopherol content	*Tph1*	SSR	–	Vera-ruiz *et al.*, 2006
Gamma-tocopherol content	*Tph2*	SSR, INDEL	3.6	Garcia-Moreno *et al.*, 2006
High stearic acid content	*Es1, Es2 and Es3*	SSR	3.9	Perez-Vich *et al.*, 2006
Seed hypodermis color	*Hyp*	RFLP	–	Leon *et al.*, 1996
Seed-quality	*QTL*	SSR	–	Ebrahimi *et al.*, 2008
Self-pollination and seed dormancy	*QTL*	SSR	–	Gandhi *et al.*, 2005
Restoring pollen	*Rf1*	RAPD, SCAR, AFLP	0.3 0.1	Horn *et al.*, 2003; Kusterer *et al.*, 2005
	Rf4	RAPD, AFLP	0.9	Feng and Jan 2008

Trait	Gene	Markers used	Closest genetic distance(cM)	Reference
	Rf3	SSR and RFLP-derived STS-markers	0.7	Liu *et al.*, 2012
Nuclear male-sterility	*Ms10, ms11* *ms9*	RFLP, SSR, INDEL SSR, TRAP	1.0 1.2	Perez-vich *et al.*, 2005 Chen *et al.*, 2006
Plant height or flowering dates	*QTL* *QTL* *QTL*	RFLP SSR AFLP, SSR	5.9 – –	Leon *et al.*, 2001 Bert *et al.*, 2003 Rachid *et al.*, 2004
Agronomic characters	*QTL* *QTL*	AFLP, SSR AFLP, SSR	– –	Rachid *et al.*, 2004 Haddadi *et al.*, 2011
Seed morphological traits	*QTL*	TRAP, SSR	–	Yue *et al.*, 2009
Germination and seedling development	*QTL*	AFLP, SSR	–	Rachid Al-Chaarani *et al.*, 2005
Lemon ray flower color	*Yf,* *Yf1*	SSR, EST	2.3, 1.5	Yue *et al.*, 2008b
Branching and pericarp pigment	*Hyp*	SSR	–	Tang *et al.*, 2006a
Early domestication	*QTL*	SSR	–	Wills and Burke, 2007
Chlorophyll fluorescence	*QTL*	SSR	–	Kiani *et al.*, 2008
Chlorophyll deficiency	*Yl*	SSR, TRAP	4.2	Yue *et al.*, 2009
Photosynthesis and water status	*QTL*	AFLP	6.0	Herve *et al.*, 2001
Growing degree days and photoperiod response	*QTL* *QTL*	RFLP SSR	– –	Leon *et al.*, 2001 Fonts *et al.*, 2008

Trait	Gene	Markers used	Closest genetic distance(cM)	Reference
In vitro organogenesis Somatic embryogenesis	*QTL*	AFLP	–	Flores Berrios *et al.*, 2000a; Flores Berrios *et al.*, 2000b
Plant water status and osmotic adjustment	*QTL*	AFLP, SSR	3.7	Kiani *et al.*, 2007
Drought or salinity stress	*QTL*	SSR	–	Xianan and Baird 2003
Chilling and salt stresses	*QTL*	EST	–	Paula *et al.*, 2008

Markers from DNA sequence databases

The advent of high-throughput sequencing technology produced a large volume of DNA sequence data, which provides a rich source for DNA marker development. The Compositae Genome Program (CGP; *http://cgpdb.ucdavis.edu/)*, supported jointly by grants from the National Science Foundation Plant Genome Program and the United States Department of Agriculture Plant Genome Program, has generated over two million ESTs for many biologically and economically important plant species in the Compositae (Asteraceae) family. The majority of the 284,745 sunflower ESTs in GenBank were deposited by the CGP.

EST SSRs

The EST approach has been useful in taxa whose whole genome sequences remain unavailable as in the family Compositae. Although the utility of SSR markers has been well-established, their *de novo* development is costly and a time consuming endeavor on a locus-by-locus basis (Squirrell *et al.*, 2003). Consequently, expressed-sequence tag (EST) databases have become an increasingly valuable resource for SSR marker development (Pinto *et al.*, 2004; Feingold *et al.*, 2005; Pashley *et al.*, 2006; Yi *et al.*, 2006; Becher, 2007; Hanai *et al.*, 2007; Laurent *et al.*, 2007; Heesacker *et al.*, 2008). EST databases are now available for more than 140 biologically and economically important plant species (Heesacker *et al.*, 2008), and SSRs are typically abundant within these sequence collections (Ellis and Burke, 2007). A key advantage of EST-SSRs is that they are more transferable across species as compared to the genomic SSRs developed from sequence information from non-coding sequences (Cordeiro *et al.*, 2001; Yu *et al.*, 2004; Pashley *et al.*, 2006).

Heesacker *et al.* (2008) mined the sunflower EST database for SSRs and insertions-deletions (INDELs) and developed a catalog of 16,643 EST-SSRs and a collection 484 EST-SSR and 43 INDEL markers for comparative mapping and other genomics applications in sunflower. Ellis *et al.* (2006) used EST-SSRs derived from the cultivated sunflower to investigate levels of genetic diversity in an extremely rare sunflower *(H. verticillatus*) and a more common congener (*H. angustifolius*). Based on a simple comparison of the mean level of genetic diversity present within each taxon, the two species are statistically indistinguishable.

Cross transferability

Several studies have shown the feasibility of transferring SSR markers from one plant species to another belonging to the same or different genera in cereals (Kuleung *et al.*, 2004), legumes (Peakall *et al.*, 1998), and Cruciferae (Plieske and Struss, 2001). Garcýa-Moreno *et al.* (2010) evaluated the transferability of sunflower markers in safflower and reported that the non-genic SSR markers (17.6%) had lower transferability rates when compared to the gene based markers namely intron fragment length polymorphism markers (56.2%) and resistant gene candidate based markers (73.7%). Pashley *et al.* (2006) reported that EST derived SSRs were three times more transferable across species of the genus *Helianthus* than anonymous SSRs. Heesacker *et al.* (2008) studied the transferability of 466 EST-SSR or INDEL markers and reported 413 (88.6%) markers to be amplifying from one or more wild species (*H. argophyllus, H. tuberosus, H. anomalus, H. paradoxus* and *H. deserticola*), 69 (14.8%) amplifying in safflower (*Carthamus tinctorius*) and 67 (14.4%) in lettuce (*Lactuca sativa*).

The transferability between the two economically important genera of the family compositae namely *H. annuus* and the two species of lettuce *Lactuca sativa* and *L. serriola* was tested with respect to chloroplast SSRs (Wills *et al.*, 2005). Thirty out of 36 primers amplified in both the species indicating the highly conserved nature of chloroplast genome and considerable homology between these two species. EST SSRs developed from safflower showed a transferability of 18% with lettuce, 19% with Gerbera and 25% with sunflower (Chapman *et al.*, 2009). Thus the EST-SSRs represent a valuable resource for the genetic analysis of sunflower and related species, facilitating comparative map-based analyses across a broader array of taxa in the family Asteraceae.

SNP

Single nucleotide polymorphism (SNP) refers to a DNA sequence variation at a single nucleotide position in the genome. Theoretically SNP variations involve four different nucleotides at a particular site, but only two of these four possibilities are mostly observed. Thus, in practice, SNPs are biallelic markers and so the information content on a single SNP is limited compared to other markers. This disadvantage is overcome by the relatively larger abundance and stability of SNP loci compared to the most popular SSR marker system. In plant genomes, the usual frequency of SNPs is one in every 100–300 bp (Gupta *et al.*, 2001). In sunflower, a relatively high frequency of SNPs has been reported. Liu and Burke (2006) investigated nine genomic loci of a total length of 8,207 bp in 32 individuals sampled from wild populations, primitive and improved germplasm accessions. They found 444 polymorphic sites among the samples with an average of 1 SNP for every 16.8 bp of sequence. The wild sunflowers harbored 392 polymorphic sites (1 SNP/19.1 bp), whereas the cultivated sunflower harbored 194 polymorphic sites (1 SNP/38.8 bp). Kolkman *et al.* (2007) surveyed the nucleotide diversity of 82 previously mapped RFLP marker loci in two wild and 10 elite inbred lines. They identified 1,078 SNPs in 49.4 kbp of DNA/ genotype with an average of 1 SNP/45.7 bp. Non-coding sequences harbored two-fold more SNPs (1/32.1 bp) than coding sequences (1/

62.8 bp). Fusari *et al.* (2008) reported that one SNP was found in every 69 bp in the 14,348-bp aligned sequences of 28 candidate genes related to biotic and abiotic stresses in 19 sunflower inbred lines. To assess single nucleotide polymorphism (SNP) frequencies, nucleotide diversity, and linkage disequilibrium (LD) in sunflower cultivars, alleles were resequenced from 81 genic loci distributed throughout the sunflower genome. DNA polymorphisms were abundant; 1,078 SNPs (1/45.7 bp) and 178 insertions-deletions (INDELs) (1/277.0 bp) were identified in 49.4 kbp of DNA/genotype. SNPs were two-fold more frequent in noncoding (1/32.1 bp) than coding (1/62.8 bp) sequences. Nucleotide diversity was only slightly lower in inbred lines (u ¼ 0.0094) than wild populations (u ¼ 0.0128). When extrapolated across the genome, sunflower was predicted to harbor at least 76.4 million common SNPs among modern cultivar alleles (Kolkman *et al.,* 2007). Lai *et al.* (2005) integrated the classical SSR linkage map (Tang *et al.,* 2002) with 243 SNP markers mapped to the 17 linkage groups.

The abundance, ubiquity and interspersed nature of SNPs together with the potential of automatic high-throughput analysis make them ideal candidates as molecular markers for construction of high density genetic maps, QTL fine mapping, marker-assisted plant breeding and association mapping studies. SNPs located in known genes provide a fast alternative to analyze the fate of agronomically important alleles in breeding populations, thus providing functional markers.

Functional markers, also called "perfect markers" detect the causal DNA sequence differences between alleles of a gene underlying a given phenotype. These markers provide perfect marker-trait association and 100% accurate prediction of the phenotype in all genetic backgrounds and can be used directly for marker-aided selection and map-based cloning. Through marker-trait association, these markers have been developed for a variety of traits in a large number of plant systems. For instance, Paris *et al.* (2002) developed a high throughput assay for selecting barley seedlings carrying superior alleles of β-amylase (a key enzyme involved in the degradation of starch). In rice, SNPs in the waxy gene also proved useful for MAS leading to the production of cultivars with superior eating, cooking and processing qualities (Larkin and Park, 2003).

Functional markers in sunflower are rare, but they have been developed for traits determining oil quality (Tang *et al.,* 2006b) and herbicide resistance (Kolkman *et al.,* 2004). For oil quality, Tang *et al.* (2006b) determined that a non-lethal knockout mutation in a 2-methyl -6-phytyl-1,4-benzoquinone/2-methyl-6-solanyl-1,4-benzoquinone methyl transferase locus (MT-1) caused by the insertion of a 5.2 kb *Ty3/gypsy*-like retrotransposon was underlying β-tocopherol accumulation in sunflower seeds, and robust STS markers diagnostic for wild type and mutant MT-1 alleles were developed. For herbicide resistance, Kolkman *et al.* (2004) identified mutations in codons 197 and 205 in the acetohydroxyacid synthase gene *AHAS-1* that confers resistance to sulfonylurea (SU) and imidiazolonone (IMI) herbicides, and developed a SNP genotyping assay diagnostic for the codon 205 mutation.

Marker-assisted selection (MAS)

MAS involve the selection of superior individuals in segregating progenies based on the linked marker genotype rather than the phenotype as done in conventional plant breeding. Although significant strides have been made in crop improvement through phenotypic selections for agronomically important traits, considerable difficulties are often encountered during this process, primarily due to genotype–environment interactions. Besides, testing procedures may be many times difficult, unreliable or expensive due to the nature of the target traits (e.g. abiotic stresses) or the target environment. With the development and availability of an array of molecular markers and dense molecular genetic maps in crop plants, MAS has become possible for traits both governed by major genes as well as quantitative trait loci (QTLs).

Marker-assisted selection was initially proposed by Young and Tanksley (1989) and experimented by various researchers (Hospital, 1992; Visscher *et al.*, 1996; Frisch *et al.*, 1999). It gained considerable importance as it improved the efficiency of plant breeding through precise transfer of genomic regions of interest (foreground selection) and accelerated the recovery of the recurrent parent genome (background selection). It has been more widely employed for simply inherited traits than for polygenic traits, although there are a few success stories in improving quantitative traits through MAS. The success of MAS depends upon several critical factors, including the number of target genes to be transferred, the distance between the flanking markers and the target gene, the number of genotypes selected in each breeding generation, the nature of germplasm and the technical options available at the marker level. At present, MAS in sunflower is limited as the genetic distances between the identified linked markers and target genes is high.

MAB in sunflower

Marker-assisted backcross breeding in sunflower is mainly being carried out in private companies to accelerate the introgression of target genes into elite germplasm. Traits such as downy mildew resistance, high oleic acid content in the seed oil, and herbicide resistance are currently the main targets, although complex traits such as resistance to *Sclerotinia, Phoma* and *Phomopsis* are also undertaken. Marker-assisted backcross breeding is also a very effective way of transferring genes or QTLs determining these traits from wild donor genotypes into elite breeding lines by reducing both the time needed to produce a commercial cultivar and the risk of undesirable linkage drag. Advanced backcross QTL analysis" (AB-QTL) has been proposed for transferring QTLs for agronomically important traits from unadapted donor lines (wild species, landraces) into established elite inbred lines (Tanksley and Nelson, 1996). This approach has been used in sunflower to improve populations derived from interspecific crosses for resistance to *Sclerotinia sclerotiorum* or Phomopsis (Serieys *et al.*, 2000).

Gene pyramiding is a genotype building strategy for simultaneous manipulation of several genes or favorable alleles at QTL when they are originally present in multiple parents (Hospital, 2003). For downy mildew control, the preliminary results reported by Tourvieille de Labrouhe *et al.* (2004) comparing different methods for obtaining durable resistance indicated that gene pyramids were less effective in reducing the appearance of new races compared to other control methods such as the use of combinations of resistance *Pl* genes by alternation or in "multi-hybrids". Vear (2004) suggested that in order to increase the durability of these major genes, they need to be backed up by quantitative, non-race specific resistance QTLs. Markers will be essential for this strategy because it would be impossible to select for these QTLs phenotypically when they are combined with major resistance genes. Additionally, molecular markers will be useful for pyramiding tightly-linked resistance genes within the same RGC cluster (Slabaugh *et al.*, 2003). For partial resistances such as *Sclerotinia* and Phomopsis, a very important step towards the improvement of the levels of resistance is the use of MAS to combine different resistance QTL. Evolution of new races of *P. halstedii* emerged in the cultivation areas of sunflower (Tourvieille de Labrouhe *et al.*, 2000; Molinero-Ruiz *et al.*, 2002). Moreover, some races developed tolerance to the fungicide metalaxyl (Albourie *et al.*, 1998; Spring *et al.*, 2006), which is the only effective plant protecting agent against this disease. The discovery of linkage between resistance genes and molecular markers in order to be able to use these purposefully and durably in plant breeding assumes importance for pyramiding several resistance genes (Brahm *et al.*, 2000).

Genomic resources

The genomic resources available are the public databases (NCBI) and the Compositae Genome Project Database (CGPDB) which contain sequence information for approximately 93,500 cultivated sunflower EST sequences corresponding to approximately 31,600 unique genes. The different types of markers designed and the immortal RIL populations developed by various researchers also constitute a potential resource for genomic studies in sunflower.

Mapping Populations

All the published sunflower linkage maps were constructed from two types of mapping populations: F_2 and RIL (recombinant inbred line) populations each having advantages that are useful in different situations for different purposes. The two parental plants were usually selected by maximum polymorphism based on molecular marker screening or by the difference in the phenotypic trait of interest. After creating the F_1 hybrid, an F_2 or a RIL population can be generated and used for molecular mapping. An F_2 population is developed by selfing the F_1 hybrids. The major disadvantage of using F_2 populations is that the populations are not immortal. Therefore, the sample tissue to isolate DNA for genotyping is finite. Another problem is that each individual plant in the F_2 population possesses one unique genotype and cannot be replicated in measuring quantitative traits. Since the F_2 populations are easy to develop, they were used in the construction of the early sunflower linkage maps (Berry *et al.*, 1995; Gentzbittel *et al.*, 1995; 1999; Jan *et al.*, 1998).

RIL populations are developed by self-pollinating individual F_2 plants followed by additional cycles of single-seed descent (SSD) to advance the generation to F_7. Since each of the resulting RIL can be traced back to the individual F_2 plants, these lines are also called F_2-derived lines. Therefore, an RIL population needs at least four years to develop. After several generations of self-pollination, a RIL population becomes an eternal mapping population since each individual line possesses a practically homozygous genotype and can be propagated generation after generation as a clone. Doubled haploid (DH) populations are produced by regenerating plants by the induction of chromosome doubling from pollen grains of the F_1. Homozygous lines can be obtained in just two seasons by this method however; the production of DH populations is possible in those species that are amenable for tissue culture.

RILs and DH populations increase the mapping accuracy of the high-throughput, dominant markers such as AFLP and TRAP, which cannot easily differentiate heterozygous from homozygous genotypes in the F_2 populations. These populations are also powerful tools for analyzing quantitative traits because lines of identical genotypes can be used in replicated trials to generate quantitative trait data, which is used to determine the chromosomal location of the molecular markers closely associated with the traits of interest. There are a number of sunflower linkage maps constructed with RIL populations (Flores Berrios *et al.*, 2000a; Tang *et al.*, 2002; Yu *et al.*, 2003; Langer *et al.*, 2003; Hu *et al.*, 2004a; Kiani *et al.*, 2007). However, the use of DH populations is limited in sunflower. Backcross populations or NILs (Near isogenic lines) developed by crossing the F_1 with the recipient parent for 5-6 generations and selfed were routinely used for mapping genes and individual QTLs from wild species and unadapted germplasm. In the recent past, a novel method of linkage analysis called Association Mapping is employed as it does not necessitate the development of mapping populations for mapping target genes.

Association mapping

Association mapping, also known as linkage disequilibrium (LD) mapping, is an emerging field of genetic mapping that has the potential to reach resolution to the level of individual genes (alleles) underlying quantitative traits. It has emerged as a tool to resolve complex trait variation down to the sequence level by exploiting historical and evolutionary recombination events at the population level (Nordborg and Tavare, 2002; Risch and Merikangas, 1996) rather than marker trait linkages in segregating populations. Thus, it circumvents the time taking and cumbersome step of development and analysis of segregating progenies for mapping target genes. As a new alternative to traditional linkage analysis, association mapping offers the advantages of increased mapping resolution, reduced research time, and greater allele number (Yu and Buckler, 2006). Association mapping identifies quantitative trait loci (QTLs) by examining the marker-trait associations that can be attributed to the strength of linkage disequilibrium between markers and functional polymorphisms across a set of diverse germplasm. The phenotypic variation of many complex traits of agricultural or evolutionary importance is influenced by multiple quantitative trait loci (QTLs), their interaction, the environment, and the interaction between QTL and environment. Linkage analysis and association mapping are the two most commonly used tools for dissecting complex traits. Linkage analysis in plants typically localizes QTLs to 10 to 20 cM intervals because of the limited number of recombination events that occur during the construction of mapping populations and the cost for propagating and evaluating a large number of lines (Doerge, 2002; Holland, 2007). While hundreds of linkage analysis studies have been conducted in various plant species over the past two decades (Holland, 2007; Kearsey and Farquhar, 1998), only a limited number of identified QTLs were cloned or tagged at the gene level (Price, 2006). Association mapping has been done for different traits in the plant species namely Maize (Andersen *et al.*, 2005; Camus-Kulandaivelu *et al.*, 2006; Harjes *et al.*, 2008), *Arabidopsis* (Ehrenreich *et al.*, 2007), sorghum (Casa *et al.*, 2008), wheat (Breseghello and Sorrells, 2006b), barley (Kraakman *et al.*, 2006), potato (Malosetti *et al.*, 2007) and rice (Agrama *et al.*, 2007). The SNP frequencies and LD decay were studied in modern sunflower cultivars from 82 previously mapped RFLP marker loci distributed throughout the sunflower genome (Kolkman *et al.*, 2007). Liu and Burke (2006) studied LD at nine genic loci in wild populations and exotic germplasm accessions namely Native American land races and prehybrid era open-pollinated confectionery and oilseed cultivars which help in high-density genetic mapping and high-resolution association mapping.

Physical mapping

To characterize and isolate genes corresponding to desirable traits, physical map or ordered array of genomic DNA fragments in the desired region containing the gene is required. For this purpose, yeast artificial chromosome (YAC), bacterial artificial chromosome (BAC) and cosmids are appropriate vectors for cloning large segments of DNA molecules. In sunflower, six libraries were constructed in different vector systems and available for physical mapping (Table 2).

Table 2 : List of genomic libraries developed in sunflower

S. No.	Genotype	No. of clones	Enzyme	Vector	Avg. insert size (Kb)	Genome coverage (X)	Reference
1 BAC	HA821	158,208	*Hind*III	pBelo- BAC11	80	4-5	Gentzbittel *et al.*, 2002
2 BAC	RHA325	100,000	*Hind*III	pBelo- BAC11	60	1.9	Ozdemir *et al.*, 2004
3 BAC	HA383	202,752	*Hind*III	pIndigoBAC536	125	8.3	Knapp(CUG1)
4 BAC	YDQ	147,456	*Hind*III	pBAC(*FRT-oriV*)	118	5	Bouzidi *et al*., 2006
5 BAC	HA89	107,136	*Bam*HI	pECBAC1	140	5	Feng *et al.*, 2006
6 BIBAC	HA89	84,864	*Hind*III	pCLD04541	137	3.9	Feng *et al.*, 2006

Cloning

There are two approaches for cloning target genes. One is Map based cloning and the other is a candidate gene approach.

Map-based cloning

It is being employed to isolate gene(s) corresponding to desirable traits. Through genetic mapping, a fine scale linkage map can be obtained using DNA markers and then a genomic DNA library of large-sized fragments can be screened with molecular markers linked to the desired trait. Combination of genetic analysis and molecular tools makes it possible to isolate genes corresponding to the target traits. This type of gene cloning is called map-based cloning. After tagging the trait, linked DNA markers are used for screening an appropriate YAC or BAC library to isolate putative clones. This strategy is reviewed by Tanksley *et al.* (1995) and has been applied successfully for the isolation of *Pto* gene in tomato and *Xa-1* gene in rice (Martin *et al.*, 1993; Yoshimura *et al.*, 1996).

Efforts for Map-based cloning of some of the genes conferring nuclear male sterility, fertility restoration, downy mildew resistance in sunflower have been unsuccessful. This is primarily due to the large genome size, the high content of repetitive elements, the lack of publicly available large mapping populations and large insert genomic libraries. With regard to the relationship of genetic maps with physical maps, 1 cM map distance represents about 2 Mb in sunflower genome assuming an estimated haploid genome size (1C) of about 3,000 Mb to 3,500 Mb (Arumuganathan and Earle, 1991; Baack *et al.*, 2005) and a map size for the whole genome of about 1,500 cM (Micic *et al.*, 2004; Hu *et al.*, 2007).

Candidate gene approach

Candidate gene approach requires that the gene of interest has been isolated in another species and sequences are available to be used as probes or to design primers for PCR amplification in sunflower. The major problem in this approach is the degree of homology between genes in different species, which either allows to start with specific primers from conserved regions or degenerate primers if the sequence homology proves only moderate between species. Primers are then used to amplify the candidate gene either from genomic DNA or CDNA. The large amount of EST data now available for sunflower will be very helpful for employing candidate gene approach in sunflower. Candidate gene approach, although limited to genes of known function, has been very successful in sunflower in the recent years.

Gentzbittel *et al.* (1998) used degenerate primers designed from conserved NBS domains of *N* from tobacco (Whitham *et al.*, 1994), *RPS* from *A. thaliana* (Mindrinos *et al.*, 1994) and *L6* from flax (Lawrence *et al.*, 1995) to clone resistance gene analogs (RGAs). One of the clones was sequenced and mapped close to the *Pl6* gene. Gedil *et al.* (2001a) placed *Pl1* and six RGCs of the NBS-LRR type on the HA370 x HA372 public sunflower map. Mapping of resistance gene candidates produced by degenerate primers pointed to the clustering of disease resistance genes in the *Pl1-Pl2-Pl6* region (Vear *et al.*, 1997). Genetic mapping of these RGAs on two segregating F_2 populations showed that the non-TIR-NBS-LRR RGAs are clustered and linked to the *Pl5/Pl8* locus for resistance to downy mildew (Radwan *et al.*, 2003). Radwan *et al.* (2004) cloned full-length cDNA and genomic sequences from two of these RGAs, which were named as *Ha-NTIR11g* (accession no. AY490793) and *Ha-NTIR3a* (accession no. AY490791). Primers flanking an intron between the TIR and NBS encoding region of the *HaRGC1* sequence amplify a large RGC family (Slabaugh *et al.*, 2003). These RGC paralogs showed differences in intron length between 90 to >800 nucleotides. Twenty-four HaRGC1 loci were mapped to a 2–4 cM region of LG 8. Nine haplotypes (based on these 24 RGAs) were identified among elite inbred lines and were correlated to known downy mildew resistance specificities (Slabaugh *et al.*, 2003).

Radwan *et al.* (2008) mined the sunflower EST database and used comparative genomics to develop a deeper understanding of the diversity and distribution of NBS-LRR homologs in the sunflower genome. Collectively, 630 NBS-LRR homologs were identified, 88 by mining a database of 284,241 sunflower ESTs and 542 by sequencing 1,248 genomic DNA clones isolated from common and wild sunflower species. These genomic sequences were obtained by using degenerate primers against four conserved NBS amino acid sequences. DNA markers were developed from 196 unique NBS-LRR sequences and facilitated the mapping of 167

NBS-LRR loci. The latter were distributed throughout the sunflower genome in 44 clusters and singletons (Radwan *et al.*, 2008). The largest and most complex cluster of NBS-LRR loci discovered so far in sunflower is located on linkage group 8 of the public sunflower map. It spans a segment of 36 cM in three subclusters containing 54 NBS-LRR loci. This region also harbors the *Pl1* and other downy mildew R-genes. The second largest NBS-LRR cluster was identified on the lower end of LG 13, with 27 loci in two subclusters and a few singletons spanning a segment of 31 cM. This is the segment that harbors downy mildew (*Pl5* and *Pl8*) and black rust (*RADV*) R-genes (Lawson *et al.*, 1998; Radwan *et al.*, 2003; Slabaugh *et al.*, 2003; Yu *et al.*, 2003). This cluster harbors TIR- and non-TIR-NBS-LRR encoding genes.

Conclusion and future prospects

Recent developments in DNA marker technology together with the concept of marker-assisted selection provide new solutions for selecting and maintaining desirable genotypes. Once molecular markers closely linked to desirable traits are identified, marker assisted selection can be performed in early segregating populations and at early stages of plant development. Marker-assisted selection can be used to pyramid the major genes including resistance genes, with the ultimate goal of producing varieties with more desirable characters. Thus, with MAS it is now possible for the breeder to conduct many rounds of selection in a year. Molecular marker technology is now integrated into existing plant breeding programmes all over the world in order to allow researchers to access, transfer and combine genes at a rate and with a precision not previously possible. However, one of the major drawbacks in most of the species including sunflower is when the linked marker used for selection is at a distance away from the gene of interest, leading to cross-overs between the marker and the gene. This produces a high percentage of false-positives/negatives in the screening process. In the final analysis, the success depends on identifying marker(s) as close to the gene as possible for its utility across all populations. The availability of functional markers will influence future plant breeding both through making available important genes to be used for production of transgenic crops and also through MAS and MAB for more effective and speedy progress in conventional plant breeding.

Standardised techniques for anther/pollen culture have to be developed for production of DH populations which yields homozygous lines in two seasons when compared to RILs to accelerate the mapping studies. Development of gene specific markers especially for closely linked and clustered downy mildew resistance genes is a necessity for pyramiding them to achieve durable resistance in sunflower. An effort to study syntenic relationships of sunflower with sequenced model genomes like *Arabidopsis* has to be made so that the orthologous sequence information can be used to design more markers for saturation of linkage maps in sunflower. The high degree of cross transferability of safflower markers offers a potential for use of those markers in sunflower.

Advances in high-throughput genotyping and sequencing technologies have markedly reduced the cost per data point of molecular markers, particularly single nucleotide polymorphisms (SNPs). Marker-assisted selection has been successful for introgressing and pyramiding major-effect genes. Use of Association mapping strategy with the abundant SSR/SNP markers to dissect minor QTLs because of enhanced chance of recombination in natural populations is advocated. However many challenges remain to be resolved before MAS can routinely provide added value for breeding very complex traits. The rate, scale, and scope of uptake of MAS in public crop breeding programs has continually lagged behind expectations. There are many technical and logistical factors that have hindered the speed and scope of MAS uptake. These include the unit cost and scalability of DNA extraction systems, the capital costs associated with high throughput genotyping equipment, the prolonged and labor-intensive methods for identifying polymorphism. Strategic use of different marker systems can go a long way in practicing MAS in sunflower.

REFERENCES

Alfadil AT, Kiani PS, Dechamp-Guillaume G, Gentzbittel L and Sarrafi A (2007) QTL mapping of partial resistance to *Phoma* basal stem and root necrosis in sunflower (*Helianthus annuus* L.). *Plant Sci* 172: 815–823

Agrama HA, Eizenga GC and Yan W (2007) Association mapping of yield and its components in rice cultivars. *Mol Breed* 19: 341–356

Albourie JM, Tourvieille J and Tourvieille de Labrouhe D (1998) Resistance to metalaxyl in and *Phoma macdonaldii. Theor Appl Genet* 109: 865–874

Andersen JR, Schrag T, Melchinger AE, Zein I and Lubberstedt T (2005) Validation of Dwarf8 polymorphisms associated with flowering time in elite European inbred lines of maize (*Zea mays* L.). *Theor Appl Genet* 111: 206–217

Anisimova IN, Fido RJ, Tatham AS and Shewry PR (1995) Genotypic variation and polymorphism of 2S albumins of sunflower. *Euphytica* 83: 15–23

Arias DM and Rieseberg LH (1995) Genetic relationships among domesticated and wild sunflower (*Helianthus annuus*, Asteraceae). *Econ Bot* 49: 239–248

Arumuganathan K and Earle ED (1991) Nuclear DNA content of some important plant species. *Plant Mol Biol Rep* 9: 208–218

Baack E, Whitney KD and Rieseberg LH (2005) Hybridization and genomic size evolution: timing and magnitude of nuclear DNA content increases in *Helianthus* homoploid species. *New Phytol* 167: 623–630

Becher R (2007) EST-derived microsatellites as a rich source of molecular markers for oats. *Plant Breed* 126: 274–278

Berry ST, Allen RJ, Barnes SR and Caligari PDS (1994) Molecular marker analysis of *Helianthus annuus* L. 1. Restriction fragment length polymorphism between inbred lines of cultivated sunflower. *Theor Appl Genet* 89: 435–441

Berry ST, Leon AJ, Hanfrey CC, Challis P, Burkholz A, Barnes SR, Rufener GK, Lee M and Caligari PDS (1995) Molecular-marker analysis of *Helianthus annuus* L. 2. Construction of an RFLP map for cultivated sunflower. *Theor Appl Genet* 91: 195–199

Berry ST, Leon AJ, Chalis P, Livin C, Jones R, Hanfrey CC, Griffiths S and Roberts A (1996) Construction of a high density composite RFLP linkage map for cultivated sunflower *H.annuus* L. Proceeding of the 14th International Sunflower Conference, Beijing 2: 1150- 1160

Bert PF, Dechamp-Guillaume G, Serre F, Jouan I, Tourvieille de Labrouhe D, Nicolas P and Vear F (2004) Comparative genetic analysis of quantitative traits in sunflower (*Helianthus annuus* L.). 3. Characterisation of QTL involved in resistance to *Sclerotinia sclerotiorum* and *Phoma macdonaldii. Theor Appl Genet* 109: 865–874

Bert PF, Jouan I, de Labrouhe DT, Serre F, Philippon J, Nicolas P and Vear F (2003) Comparative genetic analysis of quantitative traits in sunflower (*Helianthus annuus* L.) 2. Characterisation of QTL involved in developmental and agronomic traits. *Theor Appl Genet* 107: 181–189

Bert PF, Jouan I, Tourvieille de Labrouhe D, Serre F, Nicolas P and Vear F (2002) Comparative genetic analysis of quantitative traits in sunflower (*Helianthus annuus* L.) l. QTL involved in resistance to *Sclerotinia sclerotiorum* and *Diaporthe helianthi. Theor Appl Genet* 105: 985–993

Bert PF, Tourvieille de Labrouhe D, Philippon J, Mouzeyar S, Jouan I, Nicolas P and Vear F (2001) Identification of a second linkage group carrying genes controlling resistance to downy mildew (*Plasmopara halstedii*) in sunflower (*Helianthus annuus* L.). *Theor Appl Genet* 103: 992–997

Botstein D, White RL, Skolnick M and Davis RW (1980) Construction of a genetic linkage map in man using restriction fragment length polymorphisms. *Am J Hum Genet* 32: 314–331

Bouzidi MF, Franchel J, Tao Q, Stormo K, Mraz A, Nicolas P and Mouzeyar S (2006) A sunflower BAC library suitable for PCR screening and physical mapping of targeted genomic regions. *Theor Appl Genet* 113: 81–89

Bouzidi MF, Badaoui S, Cambon F, Vear F, Tourvielle D, Labrouhe D, Nicolas P and Mouzeyar S (2002) Molecular analysis of a major locus for resistance to downy mildew in sunflower with specific PCR-based markers. *Theor Appl Genet* 104: 592–600

Bouzidi MF, Parlange F, Nicolas P and Mouzeyar S (2007) Expressed Sequence Tags from the oomycete *Plasmopara halstedii*, an obligate parasite of the sunflower. *BMC Microbiol* 7: 110

Brahm L, Rocher T and Friedt W (2000) PCR-Based markers facilitating marker assisted selection in sunflower for resistance to downy mildew. *Crop Sci* 40: 676–682

Breseghello F and Sorrells ME (2006b) Association mapping of kernel size and milling quality in wheat (*Triticum aestivum* L.) cultivars. *Genetics* 172: 1165–1177

Burke JM, Tang S, Knapp S and Rieseberg LH (2002b) Genetic analysis of sunflower domestication. *Genet* 162: 1257–1267

Camus-Kulandaivelu L, Veyrieras JB, Madur D, Combes V, Fourmann M, Barraud S, Dubreuil P, Gouesnard B, Manicacci D and Charcosset A (2006) Maize adaptation to temperate climate: Relationship between population structure and polymorphism in the dwarf 8 gene. *Genet* 172: 2449–2463

Carrera A, Pizarro G, Poverene M, Feingold SE, León A and Berry ST (2002) The genetic mapping of five isozyme loci on the public sunflower RFLP linkage map. *Genet Mol Biol* 25: 65–72

Carrera A, Poverene M and Rodriguez RH (1996) Isozyme variability *in Helianthus argophyllus.* Its application in crosses with cultivated sunflower. *Helia* 19: 19–28

Casa AM, Pressoira G, Brown PJ, Mitchell SE, Rooney WL, Tuinstrac MR, Franks CD and Kresovicha S (2008) Community resources and strategies for association mapping in sorghum. *Crop Sci* 48: 30–40

Chapman MA, Hvala J, Strever J, Matvienko M, Kozik A, Michelmore RW, Tang S, Knapp SJ and Burke JM (2009) Development, polymorphism, and cross-taxon utility of EST-SSR markers from safflower (*Carthamus tinctorius* L.). *Theor Appl Genet* 120: 85–91

Chase MR, Moller C, Kesseli R and Bawa KS (1996) Distant gene flow in tropical trees. *Nature* 383: 398–399

Chen J, Hu J, Vick BA and Jan CC (2006) Molecular mapping of a nuclear male-sterility gene in sunflower (*Helianthus annuus* L.) using TRAP and SSR markers. *Theor Appl Genet* 113: 122–127

Cordeiro GM, Casu R, McIntyre CL, Manners JM and Henry RJ (2001) Microsatellite markers from sugarcane (*Saccharum* spp.) ESTs cross transferable to erianthus and sorghum. Plant Sci 160: 1115–1123 major mechanism for DNA-sequence evolution. *Mol Biol Evol* 4: 203–221

Cronn R, Brothers ME, Klier K, Wendel JF and Bretting PK (1997) Allozyme variation in domesticated annual sunflower and its wild relatives. *Theor Appl Genet* 95: 532–545

Darvishzadeh R, Azizi M, Hatami-maleki H, Bernousi I, Mandoulakani BA, Jafari M and Sarrafi A (2010) Molecular characterization and similarity relationships among sunflower (*Helianthus annuus* L.) inbred lines using some mapped simple sequence repeats. *Afr J Biotechnol* 9: 7280-7288

Desmarais E, Lanneluc I and Lagnel J (1998) Direct amplification of length polymorphisms (DALPs) or how to get and characterize new genetic markers in many species. *Nucl Acids Res* 26: 1458-1465

Doerge RW (2002) Mapping and analysis of quantitative trait loci in experimental populations. *Nat Rev Genet* 3: 43–52

Dong GJ, Liu GS and Li KF (2007) Studying genetic diversity in the core germplasm of confectionary sunflower (*Helianthus annuus* L.) in China based on AFLP and morphological analysis. *Russ J Genet* 43: 627–635

Dorrell DG and Vick BA (1997) Properties and processing of oilseed sunflower. In: AA Schneiter (ed) Sunflower Technology and Production. CSSA, Madison, WI, USA, pp 709–745

Dry PJ and Burdon JJ (1986) Genetic structure of natural populations of wild sunflowers (*Helianthus annuus* L.) in Australia. *Aust J Biol Sci* 39: 255–270

Dußle CM, Hahn V, Knapp SJ and Bauer E (2004) *PlArg* from *Helianthus argophyllus* is unlinked to other known downy mildew resistance genes in sunflower. *Theor Appl Genet* 109: 1083–1086

Ebrahimi A, Maury P, Berger M, Kiani PS, Nabipour A, Shariati F, Grieu P and Sarrafi A (2008) QTL mapping of seed-quality traits in sunflower recombinant inbred lines under different water regimes. *Genome* 51: 599–615

Ehrenreich IM, Stafford PA and Purugganan MD (2007) The genetic architecture of shoot branching in *Arabidopsis thaliana*: A comparative assessment of candidate gene associations vs. quantitative trait locus mapping. *Genet* 176: 1223–1236

Ellis JR and Burke JM (2007) EST-SSRs as a resource for population genetic analyses. *Heredity* 99: 125–132

Ellis JR, Pashley CH, Burke JM and Mc cauley DE (2006) High genetic diversity in a rare and endangered sunflower as compared to a common congener. *Mol Ecol* 15: 2345-2355

Fambrini M, Sebastiani L, Rossi VD, Cavallini A and Pugliesi C (1997) Genetic analysis of an electrophoretic variant for the chloroplast-associated form of Cu/Zn superoxide dismutase in sunflower (*Helianthus annuus* L.). *J Exptl Bot* 48: 1143–1146

Feingold S, Lloyd J, Norero N, Bonierbale M and Lorenzen J (2005) Mapping and characterization of new EST-derived microsatellites for potato (*Solanum tuberosum* L.). *Theor Appl Genet* 111: 456–466

Feng J and Jan CC (2008) Introgression and molecular tagging of *Rf4*, a new male fertility restoration gene from wild sunflower *Helianthus maximiliani* L. *Theor Appl Genet* 117: 241–249

Feng J, Brady AV, Lee MK, Zhang HB and Jan CC (2006) Construction of BAC and BIBAC libraries from sunflower and identification of linkage group-specific clones by overgo hybridization. *Theor Appl Genet* 113: 23–32

Fick GN, Zimmer DE and Kinman ML (1974) Registration of six sunflower parental lines. *Crop Sci* 14: 912

Flores-Berrios E, Gentzbittel L, Kayyal H, Alibert G and Sarrafi A (2000a) AFLP mapping of QTLs for *in vitro* organogenesis traits using recombinant inbred lines in sunflower (*Hetianthus annuus* L.). *Theor Appl Genet* 101: 1299–1306.

Flores-Berrios E, Sarrafi A, Fabre F, Alibert G and Gentzbittel L (2000b) Genotypic variation and chromosomal location of QTLs for somatic embryogenesis revealed by epidermal layers culture of recombinant inbred lines in the sunflower (*Helianthus annuus* L.). *Theor Appl Genet* 101: 1307–1312

Frisch M, Bohn M and Melchinger AE (1999) Comparison of selection strategies for marker assisted back crossing of a gene. *Crop Sci* 39: 1295–1301

Gandhi SD, Heesacker AF, Freemann CA, Argyris J, Bradford K and Knapp SJ (2005) The self-incompatibility locus (S) and quantitative trait loci for self-pollination and seed dormancy in sunflower. *Theor Appl Genet* 111: 619–629

Garcia-Moreno MJ, Vera-Ruiz EM, Fernandez-Martinez JM, Velasco L and Perez-Vich B (2006) Genetic and molecular analysis of high gamma-tocopherol content in sunflower. *Crop Sci.* 46: 2015–2021

Garcia-Moreno MJ, Velasco L and Perez-Vich B (2010) Transferability of non-genic microsatellite and gene-based sunflower markers to safflower. *Euphytica* 175: 145-150

Gedil MA, Slabaugh MB, Berry S, Johnson R, Michelmore R, Miller J, Gulya T and Knapp SJ (2001a) Candidate disease resistance genes in sunflower cloned using conserved nucleotide-binding site motifs: Genetic mapping and linkage to the downy mildew resistance gene *Pl1*. *Genome* 44: 205–212

Gedil MA, Wye C, Berry S, Egers B, Peelman J, Jones R, Leon A, Slabaugh MB and Knapp SJ (2001) An integrated RFLP linkage map for cultivated sunflower. *Genome* 44: 213-221

Gentzbittel L, Abbott A, Galaud JP, Georgi L, Fabre F, Liboz T and Alibert G (2002) A bacterial artificial chromosome (BAC) library for sun.ower, and identi.cation of clones containing genes for putative transmembrane receptors. *Mol Genet Genomics* 266: 979–987

Gentzbittel L, Mestries E, Mouzeyar S, Mazeyrat F, Badaoui S, Vear F, Labrouhe DTD and Nicolas P (1999) A composite map of expressed sequences and phenotypic traits of the sunflower (*Helianthus annuus* L.) genome. *Theor Appl Genet* 99: 218–234

Gentzbittel L, Mouzeyar S, Badaoui S, Mestries E, Vear F, Tourvieille de Labrouhe D and Nicolas P (1998) Cloning of molecular markers for disease resistance in sunflower, *Helianthus annuus* L. *Theor Appl Genet* 96: 519–525

Gentzbittel L, Vear F, Zhang Y-X, Bervillé A and Nicolas P (1995) Development of a consensus linkage RFLP map of cultivated sunflower (*Helianthus annuus* L.). *Theor Appl Genet* 90: 1079–1086

Gentzbittel L, Zhang Y-X, Vear F and Griveau B (1994) RFLP studies of genetic relationships among inbred lines of the cultivated sunflower, *Helianthus annuus* L.: evidence for distinct restorer and maintainer germplasm pools. *Theor Appl Genet* 89: 419–425

Gupta PK, Roy JK and Prasad M (2001) Single nucleotide polymorphisms: A new paradigm for molecular marker technology and DNA polymorphism detection with emphasis on their use in plants. *Curr Sci* 80: 4524-4535

Haddadi P, Yazdi-samadi B, Naghavi M, Kalantari A, Maury P and Sarrafi A (2011) QTL analysis of agronomic traits in recombinant inbred lines of sunflower under partial irrigation. *Plant Biotechnol Rep* 5: 135-146

Hanai LR, de Campos T, Camargo LEA, Benchimol LL, de Souza AP, Melotto M, Carbonell SAM, Chioratto AF, Consoli L, Formighieri EF, Siqueira M, Tsai SM and Vieira MLC (2007) Development, characterization, and comparative analysis of polymorphism at common bean SSR loci isolated from genic and genomic sources. *Genome* 50: 266–277

Harjes CE, Rocheford TR, Bai L, Brutnell TP, Kandianis CB, Sowinski SG, Stapleton AE, Vallabhaneni R, Williams M, Wurtzel ET, Yan J and Buckler ES (2008) Natural genetic variation in lycopene epsilon cyclase tapped for maize biofortification. *Science* 319: 330–333

Heesacker A, Kishore VK, Gao W, Tang S, Kolkman JM, Gingle A, Matvienko M, Kozik A, Michelmore RM, Lai Z, Rieseberg LH and Knapp SJ (2008) SSRs and INDELs mined from the sunflower EST database: Abundance, polymorphisms, and cross-taxa utility. *Theor Appl Genet* 117: 1021–1029

Hervé D, Fabre F, Flores Berrios E, Ieroux N, Al Chaarani GR, Planchon C, Sarrafi A and Gentzbittel L (2001) QTL analysis of photosynthesis and water status traits in sunflower (*Helianthus annuus* L.) under greenhouse conditions. *J Exptl Bot* 52: 1857– 1864

Hockett EA and Knowles PF (1970) Inheritance of branching in sunflowers, *Helianthus annuus* L. *Crop Sci* 10: 432–436

Holland JB (2007) Genetic architecture of complex traits in plants. *Curr Opin Plant Biol* 10: 156–161

Hongtrakul V, Huestis GM and Knapp SJ (1997) Amplified fragment length polymorphisms as a tool for DNA fingerprinting sunflower germplasm: genetic diversity among oilseed inbred lines. *Theor Appl Genet* 95: 400–407

Horn R (2002) Molecular diversity of male sterility inducing and male-fertile cytoplasms in the genus *Helianthus*. *Theor Appl Genet* 104: 562–570

Horn R, Köhler RH and Zetsche K (1991) A mitochondrial 16 kDa protein is associated with cytoplasmic male sterility in sunflower. *Plant Mol Biol* 17: 29-36

Horn R, Kusterer B, Lazarescu E, Prüfe M and Friedt W (2003) Molecular mapping of the *Rf1* gene restoring pollen fertility in PET1-based F1 hybrids in sunflower (*Helianthus annuus* L.). *Theor Appl Genet* 106: 599–606

Hospital F (2003) Marker-assisted breeding. In: HJ Newbury (ed) Plant Molecular Breeding. Blackwell Scientific Publ, London, UK, pp 30–56

Hospital F (1992) Using markers in gene introgression breeding programs. *Genet* 132: 1199–1210

Hu J (2006) Defining the sunflower (*Helianthus annuus* L.) linkage group ends with the *Arabidopsis*-type telomere sequence repeat-derived markers. *Chromosome Res* 14: 535–548

Hu J, Chen J, Berville A and Vick BA (2004a) High potential of TRAP markers in sunflower genome mapping. Proc 16th Intl Sunflower Conf, Aug 29–Sept 2, Fargo, ND, USA, pp 665–671

Hu J, Chen J, Gulya TJ and Miller JF (2004b) TRAP markers for a sunflower downy mildew resistance gene from a new *Helianthus annuus* source, PI468435. Proc 16th Intl Sunflower Conf, Aug 29–Sept 2, Fargo, ND, USA, pp 623–629

Hu J, Miller JF and Vick BA (2006) Registration of a tricotyledon sunflower genetic stock. *Crop Sci* 46: 2734–2735

Hu J, Seiler GJ, Jan CC and Vick BA (2003) Assessing genetic variability among sixteen perennial *Helianthus* species using PCR-based TRAP markers. Proc 25th Sunflower Res Forum, Fargo, ND, USA

Hu J and Vick BA (2003) Target region amplified polymorphism, a novel marker technique for plant genotyping. *Plant Mol Biol Rep* 21: 289–294

Hu J, Yue B and Vick BA (2007) Integration of TRAP markers onto a sunflower SSR marker linkage map constructed from 92 recombinant inbred lines. *Helia* 30: 25-36

Iqbal MA, Sadaqat HA and Khan IA (2008) Estimation of genetic diversity among sunflower genotypes through random amplified polymorphic DNA analysis. *Genet Mol Res* 7: 1408-1413

Jan CC, Vick BA, Miller JF, Kahler AL and Butler ET (1998) Construction of an RFLP linkage map for cultivated sunflower. *Theor Appl Genet* 96: 15–22

Jones CJ, Edwards KJ, Castaglione S, Winfiefd MO, Sala F, vande Wiel C, Bredemeijer G, Vosman B, Matthes M, Daly A, Brettschneider R, Bettini P, Buiatti M, Maestri E, Malcevschi N, Aert R, Volckaert G, Rueda J, Linacero R, Vazquez A and Karp A (1997) Reproducibility testing of RAPD, AFLP and SSR markers in plants by a network of European laboratories. *Mol Breed* 3: 381–390

Kahler AL and Lay CL (1985) Genetics of electrophoretic variants in the annual sunflower. *J Hered* 76: 335–340

Katti MV, Ranjekar PK and Gupta VS (2001) Differential distribution of simple sequence repeats in eukaryotic genome sequences. *Mol Biol Evol* 18: 1161–1167

Kearsey MJ and Farquhar AG (1998) QTL analysis in plants; where are we now? *Heredity* 80: 137–142

Kiani S, Grieu P, Maury P, Hewezi T, Gentzbittel L and Sarrafi A (2007) Genetic variability for physiological traits under drought conditions and differential expression of water stress-associated genes in sunflower (*Helianthus annuus* L.). *Theor Appl Genet* 114: 193–207

Kiani S, Maury P, Sarrafi A and Grieu P (2008) QTL analysis of chlorophyll fluorescence parameters in sunflower (*Helianthus annuus* L.) under well-watered and water-stressed conditions. *Plant Sci* 175: 565–573

Köhler RH and Friedt W (1999) Genetic variability as identified by AP-PCR and reaction to mid-stem infection of *Sclerotinia sclerotiorum* among interspecific sunflower (*Helianthus annuus* L.) hybrid progenies. *Crop Sci* 39: 1456–1463

Köhler RH, Horn R, Lössl A and Zetsche K (1991) Cytoplasmic male sterility in sunflower is correlated with the co-transcription of a new open reading frame with the *atpA* gene. *Mol Gen Genet* 227: 369–376

Kolkman JM, Berry ST, Leon AJ, Slabaugh MB, Tang S, Gao W, Shintani DK, Burke JM and Knapp SJ (2007) Single nucleotide polymorphisms and linkage disequilibrium in sunflower. *Genet* 177: 457–468

Kolkman JM, Slabaugh MB, Bruniard JM, Berry ST, Bushman SB, Olungu C, Maes N, Abratti G, Zambelli A, Miller JF, Leon A and Knapp SJ (2004) Acetohydroxyacid synthase mutations conferring resistance to imidazolinone or sulfonylurea herbicides in wild sunflower biotypes. *Theor Appl Genet* 109: 1147–1159

Kraakman ATW, Martinez F, Mussiraliev B, Eeuwijk FAV and Niks RE (2006) Linkage disequilibrium mapping of morphological, resistance, and other agronomically relevant traits in modern spring barley cultivars. *Mol Breed* 17: 41–58

Kuleung C, Baenziger PS and Dweikat I (2004) Transferability of SSR markers among wheat, rye, and triticale. *Theor Appl Genet* 108: 1147–1150

Kusterer B, Horn R and Friedt W (2005) Molecular mapping of the fertility restoration locus

Rf1 in sunflower and development of diagnostic markers for the restorer gene. *Euphytica* 143: 35–43

Lai Z, Livingstone K, Zou Y, Church SA, Knapp SJ, Andrews J and Rieseberg LH (2005) Identification and mapping of SNPs from ESTs in sunflower. *Theor Appl Genet* 111: 1532–1544

Langar K, Lorieux M, Desmarais E, Griveau Y, Gentzbittel L and Berville A (2003) Combined mapping of DALP and AFLP markers in cultivated sunflower using F9 recombinant inbred lines. *Theor Appl Genet* 106: 1068–1074

Larkin PD and Park WD (2003) Association of waxy gene single nucleotide polymorphisms with starch characteristics in rice (*Oryza sativa* L.). *Mol Breed* 12: 335–339

Laurent V, Devaux P, Thiel T, Viard F, Mielordt S, Touzet P and Quillet MC (2007) Comparative effectiveness of sugar beet microsatellite markers isolated from genomic libraries and GenBank ESTs to map the sugar beet genome. *Theor Appl Genet* 115: 793–805

Laver HK, Reynolds SJ, Moneger F and Leaver CJ (1991) Mitochondrial genome organization and expression associated with cytoplasmic male sterility in sunflower (*Helianthus annuus* L.). *Plant J* 1: 185–193

Lawrence GJ, Finnegan EJ, Ayliffe MA and Ellis JG (1995) The *L6* gene for flax rust resistance is related to the *Arabidopsis* bacterial resistance gene *RPS2* and the tobacco viral resistance gene *N*. *Plant Cell* 7: 1195–1206

Lawson WR, Goulter KC, Henry RJ, Kong GA and Kochman JK (1998) Marker-assisted selection for two rust resistance genes in sunflower. *Mol Breed* 4: 227–234

Lawson WR, Jan CC, Shatte T, Smith L, Kong GA and Kochman JK (2011) DNA markers linked to the R2 rust resistance gene in sunflower (*Helianthus annuus* L.) facilitate anticipatory breeding for this disease variant. *Mol Breed* 28: 569-576

Lawson WR, Goulter KC, Henry RJ, Kong GA and Kochman JK (1996) RAPD markers for a sunflower rust resistance gene. *Aust J Agric Res* 47: 395–401

Lay C, Gerdes J, Kahler A and Whalen R (1988) Inheritance and linkage of nine electrophoretic variants in *Helianthus annuus* L. In: Proc 12th Intl Sunflower Conf, Novi Sad, Yugoslavia, pp 411–445

Leclercq P (1969) Une stérilité mâle chez le tournesol. *Ann Amélior Plant* 19: 99–106

Lenardon SL, Bazzalo ME, Abratti G, Cimmino CJ, Galella MT, Grondona M and Giolitti F (2005) Screening sunflower for resistance to sunflower chlorotic mottle virus and mapping the *Rcmo-1* resistance gene. *Crop Sci* 45: 735–739

Leon AJ, Lee M and Andrade FH (2001) Quantitative trait loci for growing degree days to flowering and photoperiod response in sunflower (*Helianthus annuus* L.) *Theor Appl Genet* 102: 497–503

Leon AJ, Lee M, Rufener GK, Berry ST and Mowers RP (1995a) Use of RFLP markers for genetic linkage analysis of oil percentage in sunflower. *Crop Sci* 35: 558–564

Leon AJ, Lee M, Rufener GK, Berry ST and Mowers RP (1996) Genetic mapping of a locus (*Hyp*) affecting seed hypodermis color in sunflower. *Crop Sci* 36: 1666–1668

Li YC, Korol AB, Fahima T, Beiles A and Nevo E (2002) Microsatellites: genomic distribution, putative functions and mutational mechanisms: a review. *Mol Ecol* 11: 2453–2465

Liu Z, Sujatha M, Feng J, Vick BA and Jan CC (2012) Molecular mapping of the *Rf3* fertility restoration gene to facilitate its utilization in breeding confection sunflower. *Mol Breed* 29: 275-284

Liu A and Burke JM (2006) Patterns of nucleotide diversity in wild and cultivated sunflower. *Genet* 173: 321–330

Lu YH, Gagne G, Grezes-Besset B and Blanchard P (1999) Integration of a molecular linkage group containing the broomrape resistance gene *Or5* into an RFLP map in sunflower. *Genome* 42: 453–456

Lu YH, Melero-Varo JM, Garcia-Tejada JA and Blanchard P (2000) Development of SCAR markers linked to the gene *Or5* conferring resistance to broomrape (*O. cumana* Wallr.) *Theor Appl Genet* 100: 625-632

Luoras M and Vrânceanu AV (1998) Search for RAPD markers associated with recessive branching gene *b* in three sunflower near-isogenic lines. *Romanian Agril Res* 9-10: 1-4

Malosetti M, van der Linden CG, Vosman B and van Eeuwijk FA (2007) A mixed-model approach to association mapping using pedigree information with an illustration of resistance to *phytophthora infestans* in potato. *Genet* 175: 879–889

Martin GB, Brommonschenkel SH, Chunwongse J, Frary A, Ganal MW, Spivey R, Wu T, Earle ED and Tanksley SD (1993) Map based cloning of a protein kinase gene conferring disease resistance in tomato. *Science* 262: 1432–1436

Matsuoka Y, Vigouroux Y, Goodman MM, Sanchez GJ, Buckler E and Doebley J (2002) A single domestication for maize is shown by multilocus microsatellite genotyping. *Proc Natl Acad Sci USA* 99: 6080–6084

Mestries E, Gentzbittel L, Tourvieille de Labrouhe D, Nicolas P and Vear F (1998) Analysis of quantitative trait loci associated with resistance to *Sclerotinia sclerotiorum* in sunflower (*Helianthus annuus* L.) using molecular markers. Mol Breed 4: 215–226

Micic Z, Hahn V, Bauer E, Melchinger AE, Knapp SJ, Tang S and Schön CC (2005b) Identification and validation of QTL for *Sclerotinia* midstalk rot resistance in sunflower by selective genotyping. *Theor Appl Genet* 111: 233–242

Micic Z, Hahn V, Bauer E, Schon CC, Knapp SJ, Tan S and Melchinger AE (2004) QTL mapping of Sclerotinia midstalk-rot resistance in sunflower. *Theor Appl Genet* 109: 1474–1484

Micic Z, Hahn V, Bauer E, Schön CC and Melchinger AE (2005a) QTL mapping of resistance to *Sclerotinia* midstalk rot in RIL of sunflower population NDBLOSsel × CM625. *Theor Appl Genet* 110: 1490–1498

Miller JF and Fick GN (1997) The genetics of sunflower. In: AA Schneiter (ed) Sunflower Technology and Production. CSSA, Madison, WI, USA, pp 441–495

Mindrinos M, Katagiri F, Yu GL and Ausubel FM (1994) The *A. thaliana* disease-resistance gene *RPS2* encodes a protein containing a nucleotide binding site and leucine-rich repeats. *Cell* 78: 1089–1099

Mokrani L, Gentzbittel L, Azanza E, Fitamant L, Al-Chaarani G and Sarrafi A (2002) Mapping and analysis of quantitative trait loci for grain oil and agronomic traits using AFLP and SSR in sunflower (*Helianthus annuus* L.). *Theor Appl Genet* 106: 149–156

Molinero-Ruiz ML, Dominguez J, Melero-Vara JM (2002) Races of isolates of *Plasmopara halstedii* from Spain and studies on their virulence. *Plant Dis* 86: 736–740

Monéger F, Smart CJ and Leaver CJ (1994) Nuclear restoration of cytoplasmic male sterility in sunflower is associated with the tissue-specific regulation of a novel mitochondrial gene. *EMBO J* 13: 8–17

Mouzeyar S, Roeckel-Drevet P, Gentzbittel L, Philippon J, Tourvieille de Labrouhe D, Vear F and Nicolas P (1995) RFLP and RAPD mapping of the sunflower *Pl1* locus for resistance to *Plasmopara halstedii* race 1. *Theor Appl Genet* 91: 733–737

Neeraja CN, Maghirang-Rodriguez R, Pamplona A, Heuer S, Collard BCY, Septiningsih EM, Vergara G, Sanchez D, Xu K, Ismail AM and Mackill DJ (2007) A marker-assisted backcross approach for developing submergence-tolerant rice cultivars. *Theor Appl Genet* 115: 767–776

Nordborg M and Tavare S (2002) Linkage disequilibrium: What history has to tell us. *Trends Genet* 18: 83–90

Ozdemir N, Horn R and Friedt W (2004) Construction and characterization of a BAC library for sunflower (*Helianthus annuus* L.). *Euphytica* 138: 177–183

Paniego N, Echaide M, Munoz M, Fernandes L, Torales S and Facio P (2002) Microsatellite variation and characterisation in sunflower (*Helianthus annuus* L.). *Genome* 45: 34-43

Pankovic D, Radovanovic N, Jocic S, Satovic Z and Skoric D (2007) Development of co-dominant amplified polymorphic sequence markers for resistance of sunflower to downy mildew race 730. *Plant Breed* 126: 440–444

Pankovic DS, Šatovic Z, Radovanovic N, Jocic S and Miklic V (2009) Positioning of CAPS markers for resistance to downy mildew on linkage maps as determined in three sunflower mapping populations. Proceedings of the 43rd Croatian and 3rd International Symposium on Agriculture. Opatija, Croatia: 357-361

Paran I and Michelmore RW (1993) Development of reliable PCR based markers linked to downy mildew resistance genes in lettuce. *Theor Appl Genet* 85: 985–993

Paris M, Jones MGK and Eglinton JK (2002) Genotyping single nucleotide polymorphism for selection of barley â-amylase alleles. *Plant Mol Biol Rep* 20: 149–159

Pashley CH, Ellis JR, McCauley DE and Burke JM (2006) EST databases as a source for molecular markers: lessons from *Helianthus*. *J Hered* 97: 381–388

Paterson AH (1996a) Making genetic maps. In: A.H. Paterson (Ed.), Genome Mapping in Plants, pp. 23–39. R.G. Landes Company, San Diego, California; Academic Press, Austin, Texas

Peakall R, Gilmore S, Keys W, Morgante M and Rafalski A (1998) Cross-species amplification of soybean (*Glycine max* L.) simple sequence repeats (SSRs) within the genus and other legume genera: implications for the transferability of SSRs in plants. *Mol Biol Evol* 15: 1275–1287

Peerbolte RP and Peleman J (1996) The CARTISOL sunflower RFLP map (146 loci) extended with 291 AFLP markers. In: Proceedings of the 18th sunflower research forum, 11–12 January 1996, Fargo, ND. National Sunflower Association, Bismarck, ND pp 174–178

Perez-Vich B, Achoutch B, Knapp SJ, Leon AJ, Velasco L, Fernandez-Martinez JM and Berry ST (2004) Quantitative trait loci for broomrape (*Orobanche cumana* Wallr.) resistance in sunflower. *Theor Appl Genet* 109: 92-102

Perez-Vich B, Berry ST, Velasco L, Fernandez-Martinez JM, Gandhi S, Freeman C, Heesacker A, Knapp SJ and Leon AJ (2005) Molecular mapping of nuclear male sterility genes in sunflower. *Crop Sci* 45: 1851–1857

Pérez-Vich B, Velasco L, Grondona M, Leon AJ and Fernández-Martínez JM (2006) Molecular analysis of the high stearic acid content in sunflower mutant CAS-14. *Theor Appl Genet* 112: 867–875

Pinto LR, Oliveira KM, Ulian EC, Garcia AAF and de Souza AP (2004) Survey in the sugarcane expressed sequence tag database (SUCEST) for simple sequence repeats. *Genome* 47: 795–804

Pizarro G, Carrera AD and Poverene M (2000) Comparative analysis of genetic relationships in sunflower inbred lines based on isozymic, RAPD, and pedigree data. In: Proc 15th Intl Sunflower Conf, 12–15 June, Toulouse, France, vol 2, pp E111–E116

Plieske J and Struss D (2001) Microsatellite markers for genome analysis in *Brassica*. I. Development in *Brassica napus* and abundance in Brassicaceae species. *Theor Appl Genet* 102: 689–694

Popov VN, Yu Urbanovich O, Kirichenko VV (2002) Studying genetic diversity in inbred sunflower lines by RAPD and isozyme analyses. *Russ J Genet* 38: 785–790

Price AH (2006) Believe it or not, QTLs are accurate! *Trends Plant Sci* 11: 213–216

Quillet MC, Madjidian N, Griveau Y, Serieys H, Tersac M, Lorieux M and Bervillé A (1995) Mapping genetic factors controlling pollen viability in an interspecific cross in *Helianthus* sect. *Helianthus*. *Theor Appl Genet* 91: 1195–1202

Quillet MC, Vear F and Branlard G (1992) The use of isoenzyme polymorphism for identification of sunflower (*Helianthus annuus*) inbred lines. *J Genet Breed* 46: 295–304

Rachid Al-Chaarani G, Gentzbittel L, Huang X and Sarrafi A (2004) Genotypic variation and identification of QTLs for agronomic traits using AFLP and SSR in recombinant inbred lines of sunflower (*Helianthus annuus* L). *Theor Appl Genet* 109: 1353–1360

Rachid Al-Chaarani G, Gentzbittel L, Wedzony M and Sarrafi A (2005) Identification of QTLs for germination and seedling development in sunflower (*Helianthus annuus* L.) *Plant Sci* 169: 221–227

Rachid Al-Chaarani G, Roustae L, Gentzbittel L, Mokrani L, Barrault G, Dechamp-Guillaume G and Sarrafi A (2002) A QTL analysis of sunflower partial resistance to downy mildew (*Plasmopara halstedii*) and black stem (*Phoma macdonaldii*) by the use of recombinant inbred lines (RILs). *Theor Appl Genet* 104: 490–496

Radwan O, Bouzidi MF, Nicolas P and Mouzeyar S (2004) Development of PCR markers for the *Pl5*/*Pl8* locus for resistance to *Plasmopara halstedii* in sunflower, *Helianthus annuus*

L. from complete CC-NBS-LRR sequences. *Theor Appl Genet* 109: 176–185

Radwan O, Bouzidi MF, Vear F, Philippon J, Tourvieille de Labrouhe D, Nicolas P and Mouzeyar S (2003) Identification of non-TIR-NBS-LRR markers linked to the *Pl5/ Pl8* locus for resistance to downy mildew in sunflower. *Theor Appl Genet* 106: 1438–1446

Radwan O, Gandhi S, Heesacker A, Whitaker B, Taylor C, Plocik A, Kesseli R, Kozik A, Michelmore RW and Knapp SJ (2008) Genetic diversity and genomic distribution of homologs encoding NBS-LRR disease resistance proteins in sunflower. *Mol Genet Genom* 280: 111–125

Radwan O, Bouzidi MF, Nicolas P and Mouzeyar S (2004) Development of PCR markers for the *Pl5/Pl8* locus for resistance to *Plasmopara halstedii* in sunflower, *Helianthus annuus* L. from complete CC-NBS-LRR sequences. *Theor Appl Genet* 109: 176–185

Rieseberg L (1991) Phylogenetic and systematic inferences from chloroplast DNA and isozyme variation in *Helianthus* sect. *Helianthus* (*Asteraceae*). *Syst Bot* 16: 50–76

Rieseberg L, Baird SJE and Desrochers AM (1998) Patterns of mating in wild sunflower hybrid zones. *Evolution* 52: 713–726

Rieseberg L and Seiler G (1990) Molecular evidence and the origin and development of the domesticated sunflower (*Helianthus annuus, Asteraceae*). *Econ Bot* 44: 79–91

Rieseberg LH and Soltis DE (1989) Assessing the utility of isozyme number for determining ploidal level: evidence from *Helianthus* and *Heliomeris* (Asteraceae). *Aliso* 12 : 277–286

Risch N and Merikangas K (1996) The future of genetic studies of complex human diseases. *Science* 273: 1516–1517

Roath WW, Miller JF and Gulya TJ (1981) Registration of RHA 801 sunflower germplasm. *Crop Sci* 21: 479

Roeckel-Drevet P, Gagne G, Mouzeyar S, Gentzbittel L, Philippon J, Nicolas P, Tourvieille de Labrouhe D and Vear F (1996) Colocation of downy mildew (*Plasmopara halstedii*) resistance genes in sunflower (*Helianthus annuus* L.). *Euphytica* 91: 225–228

Rojas-Barros P, Hu J and Jan CC (2008) Molecular mapping of an apical branching gene of cultivated sunflower (*Helianthus annuus* L.). *Theor Appl Genet* 117: 19–28

Rossetto M (2001) Sourcing of SSR markers from related plant species. In: Henry RJ, editor. Plant genotyping: the DNA fingerprinting of plants. Wallingford, CT: CABI Publishing. p 211–224.

Serieys H, Tagmount A, Kaan F, Griveau Y, Tersac M, André T, Grave H, Pinochet X and Bervillé A (2000) Evaluation of an interspecific sunflower population issued from the perennial species *H. occidentalis* ssp. *plantagineus* for resistance to *Diaporthe helianthi* and *Sclerotinia sclerotiorum* in relation with phenotypic and molecular traits. In: Proc 15th Intl Sunflower Conf, June 12–15, 2000, Toulouse, France, pp 25–30

Serre M, Feingold S, Salaberry T, Leon A and Berry S (2001) The genetic map position of the locus encoding the 2S albumin seed storage proteins in cultivated sunflower (*Helianthus annuus* L.). *Euphytica* 121: 273–278

Slabaugh MB, Yu JK, Tang X, Heesacker A, Hu X, Lu G, Bidney D, Han F and Knapp SJ (2003) Haplotyping and mapping large cluster of downy mildew resistance gene candidates in sunflower using multilocus intron fragment length polymorphisms. *Plant Biotechnol J* 1: 167–185

Spring O, Zipper R and Heller-Dohmen M (2006) First report of metalaxyl resistant isolates of *Plasmopara halstedii* on cultivated sunflower in Germany. *J Plant Dis Protect* 113: 224

Squirrell J, Hollingsworth PM, Woodhead M, Russell J, Lowe AJ, Gibby M and Powell W (2003) How much effort is required to isolate nuclear microsatellites from plants? *Mol Ecol* 12: 1339–1348

Sujatha M, Prabakaran AJ, Dwivedi SL and Chandra S (2008) Cytomorphological and molecular diversity inbackcross-derived inbred lines of sunflower (*Helianthus annuus* L.) *Genome* 51: 284-293

Sujatha M, Zhao L, Feng J, Gulya TJ and Jan CC (2009) Inheritance and molecular mapping of a downy mildew resistance gene, *Pl13* in cultivated sunflower (*Helianthus annuus* L.) *Theor Appl Genet* 119: 795–803

Tang S, Heesacher A, Kishore VK, Fernández A, Sadik ES, Cole G and Knapp SJ (2003) Genetic mapping of the *Or5* gene for resistance to *Orobanche* race E in sunflower. *Crop Sci* 43: 1021–1028

Tang S, Kishore VK and Knapp SJ (2003) PCR-multiplexes for a genome-wide framework of simple sequence repeat marker loci in cultivated sunflower. *Theor Appl Genet* 107: 6–19

Tang S and Knapp SJ (2003) Microsatellites uncover extraordinary diversity in Native American land races and wild populations of cultivated sunflower. *Theor Appl Genet* 106: 990–1003

Tang S, Leon A, Bridges WC and Knapp SJ (2006a) Quantitative trait loci for genetically correlated seed traits are tightly linked to branching and pericarp pigment loci in sunflower. *Crop Sci* 46: 721–734

Tang S, Hass C and Knapp SJ (2006b) Ty3/gypsy-like retrotransposon knockout of a 2-methyl-6-phytyl-1,4-benzoquinone methyltransferase is non-lethal, unmasks a cryptic paralogous mutation, and produces novel tocopherol (vitamin E) profiles in sunflower. *Theor Appl Genet* 113: 783–799

Tang S, Yu JK, Slabaugh MB, Sintani DK and Knapp SJ (2002) Simple sequence repeat map of the sunflower genome. *Theor Appl Genet* 105: 1124-1136

Tanksley SD, Ganal MW and Martin GB (1995) Chromosomal landing: a paradigm for map-based gene cloning in plants with large genomes. *Trends Genet* 11: 63–68

Tanksley SD and Nelson JC (1996) Advanced backcross QTL analysis: A method for the simultaneous discovery and transfer of valuable QTLs from unadapted germplasm into elite breeding lines. *Theor Appl Genet* 92: 191–203

Tersac M, Blanchard P, Brunel D and Vincourt P (1994) Relationships between heterosis and enzymatic polymorphism in populations of cultivated sunflowers (*Helianthus annuus* L.). *Theor Appl Genet* 88: 49–55

Torres A (1983) Sunflower (*Helianthus annuus* L.). In: SD Tanksley, TJ Orton (eds) Isozymes in Plant Genetics and Breeding, part B. Elsevier Science Publ, Amsterdam

Toth G, Gaspari Z and Jurka J (2000) Microsatellites in different eukaryotic genomes: survey and analysis. *Genome Res* 10: 967–981

Tourvieille de Labrouhe D, Lafon S, Walser P and Raulic I (2000) A new race of *Plasmopara halstedii*, sunflower downy mildew. *Oléagineux, Corps Gras, Lipides* 7: 404–405

Tourvieille de Labrouhe D, Walser P, Mestries E, Gillot L, Penaud A, Tardin MC and Pauchet I (2004) Sunflower downy mildew resistance gene pyramiding alternation and mixture: First results comparing the effects of different varietal structures on changes in the pathogen. In: Proc 16th Intl Sunflower Conf, 29 Aug–2 Sept 2004, Fargo, ND, USA, pp 111–116

Vear F (2004) Breeding for durable resistance to the main diseases of sunflower. Proc 16th Intl Sunflower Conf, Fargo, ND, USA, Aug 29 to Sept 3, 2004, vol 1, pp 15–28

Vear F, Serre F, Dufournel IJ, Bert PF, Roche S, Walser P, De Labrouhe DT and Vincourt P (2008) Inheritance of quantitative resistance to downy mildew (*Plasmopara halstedii*) in sunflower (*Helianthus annuus* L.). *Euphytica* 164: 561–570

Vear F, Gentzbittel L, Philippon J, Mouzeyar S, Mestrie E, Roeckel–Drevet P, Tourvieille de Labrouhe D and Nicolas P (1997) The genetics of resistance to five races of downy mildew (*Plasmopara halstedii*) in sunflower (*Helianthus annuus* L.). *Theor Appl Genet* 95: 584–589

Vera-Ruiz EM, Velasco L, Leon AJ, Fernandez-Martinez JM and Perez-Vich B (2006) Genetic mapping of the *Tph1* gene controlling beta-tocopherol accumulation in sunflower seeds. *Mol Breed* 17: 291–296

Visscher PM, Haley CS and Thompson R (1996) Marker assisted introgression in back cross breeding programs. *Genet* 144: 1923–1932

Vos P, Hogers R, Bleeker M, Reijans M, van de Lee T, Hornes M, Frijters A, Pot J, Peleman J, Kuiper M and Zabeau M (1995) AFLP: a new technique for DNA fingerprinting. *Nucl Acids Res* 23: 4407-4414

Welsh J and McClelland M (1990) Fingerprinting genomes using PCR with consensus tRNA gene primers. *Nucl Acids Res* 19: 861-866

Whitham S, Dinesh Kumar SP, Choi D, Hehl R, Corr C and Baker B (1994) The product of the tobacco mosaic virus resistance gene *N*: Similarity to toll and the interleukin-1 receptor. *Cell* 78: 1101–1115

Williams JG, Kubelik AR, Livak KJ, Rafalski JA and Tingey SV (1990) DNA polymorphisms amplified by arbitrary primers are useful as genetic markers. *Nucl Acids Res* 18: 6531-6535

Wills DM, Hester ML, Liu A and Burke JM (2005) Chloroplast SSR polymorphisms in the compositae and the mode of organellar inheritance in *Helianthus annuus* L. *Theor Appl Genet* 110: 941–947

Winter P and Kahl G (1995) Molecular marker technologies for plant improvement. *World J Microbiol Biotechnol* 11: 438–448

Xiong LX, Liu KD, Dai XK, Xu CG and Zhang QF (1999) Identification of genetic factors controlling domestication-related traits of rice using an F2 population of a cross between *Oryza sativa* and *O. rufipogon*. *Theor Appl Genet* 98: 243–251

Yi GB, Lee JM, Lee S, Choi D and Kim BD (2006) Exploitation of pepper EST-SSRs and an SSR-based linkage map. *Theor Appl Genet* 114: 113–130

Yoshimura S, Umehara Y, Kurata N, Nagamura Y, Sasaki T, Minobe Y and Iwata N (1996) Identification of aYAC clone carrying the *Xa-1* allele, a bacterial blight resistance gene in rice. *Theor Appl Genet* 93: 117–122

Young ND and Tanksley SD (1989) RFLP analysis of the size of chromosomal segments retained around Tm-2 locus of tomato during back cross breeding. *Theor Appl Genet* 77: 353–359

Yu JK, Tang S, Slabaugh MB, Heesacker A, Cole G, Herring M, Soper J, Han F, Chu WC, Webb DM, Thompson L, Ewards KJ, Berry S, Leon A, Olungu C, Maes N and Knapp S (2003) Towards a saturated molecular genetic linkage map for cultivated sunflower. *Crop Sci* 43: 367-387

Yu JK, Dake TM, Singh S, Benscher D, Li WL, Gill B and Sorrells ME (2004) Development and mapping of EST-derived simple sequence repeat markers for hexaploid wheat. *Genome* 47: 805–818

Yu JK, Mangor J, Thompson L, Edwards KJ, Slabaugh MB, Knapp SJ (2002) Allelic diversity of simple sequence repeat markers among elite inbred lines in cultivated sunflower. *Genome* 45: 652–660

Yu J and Buckler ES (2006) Genetic association mapping and genome organization of maize. *Curr Opin Biotechnol* 17: 155–160

Yu JK, Tang S, Slabaugh MB, Heesacker A, Cole G, Herring M, Soper J, Han F, Chu WC, Webb DM, Thompson L, Edwards KJ, Berry S, Leon AJ, Grondona M, Olungu C, Maes N and Knapp SJ (2003) Towards a saturated molecular genetic linkage map for cultivated sunflower. *Crop Sci* 43: 367–387

Yue B, Miller JF and Hu J (2007) Experimenting with marker-assisted selection in confection sunflower germplasm enhancement. In: Proc 29th Sunflower Res Workshop, 10–11 Jan, 2007, Fargo, ND, USA

Yue B, Radi SA, Vick BA, Cai X, Tang S, Knapp SJ, Gulya TJ, Miller JF and Hu J (2008a) Identifying quantitative trait loci for resistance to Sclerotinia head rot in two USDA sunflower germplasms. *Phytopathol* 98: 926–931

Yue B, Vick BA, Yuan W and Hu J (2008b) Mapping of one of the 2 genes controlling lemon ray flower color in sunflower (*Helianthus annuus* L.). *J Hered* 99: 564–567

Yue B, Cai X, Yuan W, Vick B and Hu J (2009) Mapping quantitative trait loci (QTL) controlling seed morphology and disk diameter in sunflower (*Helianthus annuus* L.) *Helia* 32: 17-36

Zhang LS, Le Clerc V, Li S and Zhang D (2005) Establishment of an effective set of simple sequence repeat markers for sunflower variety identification and diversity assessment. *Can J Bot* 83: 66–72

Zhang YX, Gentzbittel L, Vear F and Nicolas P (1995) Assessment of inter- and intra-inbred line variability in sunflower (*Helianthus annuus*) by RFLPs. *Genome* 38: 1040–1048

4

Breeding for Quality Traits in Grain Legumes

Asif M Iqbal, Shafiq A Wani, Ajaz A Lone, Z A Dar, Shabir H Wani and *F A Nehvi*

Introduction

Quality means fitness for the purpose. The nutritional importance of a seed constituent must be related to the purpose for which the seed meal is being used, i.e. human food, animal feed etc. and also to the other feed components or to the diet. Under nutrition and malnutrition have disastrous consequences on human health, as revealed by chemical investigations. To elevate protein energy malnutrition a minimum of 50 g of pulses per capita per day should be available in addition to other sources of proteins such as cereals, milk meat and eggs. In grain legumes there is a deficiency in the sulphur containing amino acid content of legume seeds. Therefore, breeding for nutritional quality entails an improvement primarily in protein quantity and quality which are of paramount significance. Grain legumes have historically complemented cereals in both human diets and production systems, especially in efficient use of land and water resources and sustainability of agriculture systems globally. They also are good sources of animal feed in crop-livestock mixed farming systems, and contribute significantly to income growth, poverty alleviation and employment generation. Grain legumes are commonly known as "Poor man's meat" because of their protein content, and thus have widespread access to the poor. Chickpea, common bean, cowpea, faba bean, groundnut, lentil, pigeonpea and soybean are the eight major legumes that contribute to food, nutrition and feed security in Asia, Latin America, Sub-Saharan and North Africa. Tepary bean, lima bean, grass pea and bambara groundnut are four "neglected crops" with excellent drought and heat tolerance. Despite their importance in human nutrition and role in sustainability of agriculture systems, grain legume yields are low and unstable across seasons and environments. With declining per capita availability of grain legumes, there is an immediate need to address their production constraints to raise productivity, quality and stability of production to ensure their increased availability to the poor at affordable prices (Battisti and Naylor, 2009).

Pulses are grown primarily for their protein rich seeds. In India pulses were grown in 26.30 m ha with production of about 18.10 m ha during 2010-11 (Anonymous, 2011-). These leguminous crops are grown on marginal lands with inadequate rainfall (Kumar *et al.*, 2009). The food value of legumes seeds is high. The calorie value per unit weight is comparable to cereals. Additionally the protein content is double besides being fair source of minerals and vitamins. Legume protein is cheaper than animal protein and is important in developing countries for nutritional security. However the legume seeds are deficient in sulphur containing amino acids. The vegetarian diets combining cereal and pulses provide balanced amino acids. The crude fiber, protein and lipid components of pulses have exhibited a hyperchloresteronic effect. In the recent years there is more emphasis on vegetarianism even in western countries and lot of efforts is being made to improve nutritional value of pulses. The pulses are being developed as neutraceuticals and the big change in the health care system will be a change of focus *from* "TREATMENT *to* PREVENTION". Information on the genetics and breeding for nutritional quality aspects of grain legumes is scanty and very limited systematic efforts have been made to identify factors affecting their nutritional quality.

Protein

The protein in pulses ranged from 14.9-34.6% (Table 1). The most proteins are located in cotyledons and embryonic axis and seed coat contains very little protein. Protein of pulses is of two types: metabolic protein (enzymatic and structural) and storage protein. Metabolic protein is responsible for normal cellular activities. The storage protein occurs in cell in discrete protein bodies. Wide range of variation has been recorded in protein content of pulses. In chickpea protein content ranged from 14.9-29.6%. The variability in protein content is observed in seed collected from different branches.

Table 1 : Protein content of pulses

Crop	Range	Crop	Range
Chickpea	14.9-29.6	Pea	21.2-32.9
Pigeonpea	18.8-28.5	Lathyrus	22.7-29.6
Mungbean	20.8-33.1	Moth bean	21-31.3
Urdbean	21.2-31.3	Cowpea	20.9-34.6
Lentil	20.4-30.5		

The pulses are deficient in sulphur containing amino acids and tryptophan, but are rich in lysine (deficient in cereals).The cotyledons account for 93% of methionine and tryptophan of whole seed (Table 2). Embryo contributes only 2.5% protein. The composition of amino acids in seed is influenced by phosphorous, molybdenum and phosphorous level in soil (Chen *et al.*, 2010). The application of sulphur containing fertilizer enhances cystine content of chickpea protein.

Table 2 : Essential amino acid composition of different pulses

Crop	Lysine	Threonine	Valine	Leucine	Histidine	Iso Leucine	Methio-nine	Phenyla-lanine	Trypto-phan	Argnine
Chickpea	6.3	3.4	5.5	8.2	2.3	6.0	1.2	4.9	0.8	6.9
Pigeonpea	6.8	3.8	4.8	6.8	3.4	5.7	1.1	9.0	0.8	5.4
Mungbean	7.3	3.4	6.9	7.7	2.7	6.3	1.5	5.3	0.4	6.9
Urdbean	6.5	3.9	5.6	7.2	2.7	5.8	1.1	5.5	0.5	5.7
Lentil	5.1	3.0	5.1	5.5	2.1	5.8	0.6	4.0	0.6	7.0
Lathyrus	7.4	2.3	4.7	6.6	2.5	6.7	0.6	4.2	0.4	7.8
Pea	8.9	4.2	6.5	9.5	2.7	7.4	1.3	4.6	0.7	13.4

The nutritive value of pulse protein has been studied extensively. The biological values of pulses presented in Table 3 indicate wide variation and range from 32-78%.Low biological value of pulses is due to low methionine content. The pulse protein has poor digestibility. The low biological value and digestibility of pulse protein is attributed to presence of protease inhibitors and anti nutritional factors. The availability and utilization of protein are influenced by the amino acid composition, digestibility of proteins and presence of anti nutritional factors.

Table 3 : Biological value of grain legumes

Crop	Biological value (%)	Crop	Biological value (%)
Chickpea	52-78	Urdbean	60-64
Pigeonpea	46-74	Lentil	32-58
Mungbean	39-66	Pea	48-49

Carbohydrates

The total carbohydrate of pulses ranges from 53.3-68%. Carbohydrates include mono and oligosaccharides, starch and other polysaccharides. Starch is most abundant carbohydrate and varies from 31.5-53.6%. Total sugars (mono and oligosaccharides) represent small portion of total carbohydrate in pulses (Table 4). The amount of amylase in the starch influences starch solubility, lipid binding and other functional properties. Amylopectin is responsible for solubility of starch granules. Amylose and amylopectin are responsible for the structural form of starch granule. Ingestion of large of large quantity of pulses is known to cause flatulence in human beings. Accumulation of flatus in intestinal tract results in discomfort, abdominal rumbling, cramps, pain and diarrhea. Members of raffinose family of sugars are not digested by man because human mucosa lacks the hydrolytic enzyme α -1, 6-galactosidase. These oligosaccharides produce large amount of carbon dioxide, hydrogen and methane causing flatus effect. Pulse varieties with low raffinose family oligosaccharides can be developed through genetics manipulations. Soaking of seeds in water also lowers oligosaccharide content considerably.

Table 4 : Carbohydrate % of pulses

Pulses	Starch	Amylose	Total
Chickpea	32.2-50.0	31.8-45.8	60.1-61.2
Pigeonpea	40.4-48.2	38.6	57.3-58.7
Mungbean	37.0-53.6	13.8-35.0	53.1-61.2
Urdbean	32.2-47.9	43.9	56.5-63.7
Lentil	34.7-52.8	20.7-45.5	59.7
Pea	36.9-48.6	23.5-33.1	56.6
Cowpea	31.5-48.0	-	56.0-68.0

Among sugars oligosaccharides of raffinose family of sugars (raffinose, stachyose, verbascose and ajugose) predominate in most pulses accounting for 31.1-76% of soluble sugars. Verbascose is major oligosaccharide in mungbean, urdbean and pigeonpea. Stachyose is major oligosaccharide in chickpea, lentil and cowpea. Ajugose is found in peas. Raffinose is maderate in all pulses (Table 5).

Table 5 : Oligosaccharides % of pulses

Pulses	Sucrose	Raffinose	Stachyose	Verbascose	Total
Chickpea	0.7-2.9	0.7-2.4	2.1-2.6	0.4- 4.5	3.5-9.0
Pigeonpea	2.7	1.0-1.1	2.7-3.0	4.0-4.1	3.5-10.2
Mungbean	0.3-0.2	0.3-2.6	1.2-2.8	1.7-2.8	3.9-7.2
Urdbean	0.7-1.5	0.0-1.3	0.9-3.0	3.4-3.5	3.0-7.1
Lentil	1.8-2.5	0.4-1.0	1.9-2.7	1.0-3.1	4.2-6.1
Pea	2.3-2.4	0.3-0.9	2.2-2.9	1.7-3.2	5.3-8.7
Cowpea	1.8-3.1	0.4-1.2	2.0-3.6	0.6-3.1	6.0-13.0

Crude fiber consists of cellulose, hemicelluloses, lignin, pectic and cutin substance. Pulses contain good amount of crude fiber (1.2-13.5%). Large variation was recorded for crude fibre in chickpea, pigeonpea, mungbean and urdbean (Saxena *et al.*, 2010). Cellulose is major component in chickpea and peas whereas hemicelluloses is important for pigeonpea, urdbean and lentil (Table 6).

Table 6 : Crude fiber % of pulses

Pulses	Lignin	Cellulose	Hemicellulose	Total
Chickpea	2.9-7.1	1.1-13.7	0.6-8.4	1.2-13.5
Pigeonpea	2.9	7.3	10.1	1.2-8.1
Mungbean	2.2-7.2	2.5-4.6	0.3-9.1	1.2-12.8
Urdbean	3.8	5.0	10.7	1.2-7.1
Lentil	2.6	4.1	6.0	3.8-4.6
Pea	0.5-0.9	0.9-4.9	1.0-5.1	4.6-7.0
Cowpea	0.6-1.8	-		1.7-4.0

Lipids

Lipids include free fatty acids, mono, di and triacyl glycerol, phospholipids, sterols, sterol esters, glycolipids and lipoproteins. Total lipid content of pulses range from 1.0 - 4.99%. Total lipid content varies with variety, origin, location and climate. These are three classes of lipids viz. neutral lipids, phospholipids and glycolipids. Neutral lipids consist of triacyl glycerol accompanied by smaller proportion of di and monoacylglycerol, free fatty acids, sterols and sterol esters. Fatty acids are the main components of neutral lipids, phospholipids and glycolipids. The main fatty acids are palmitic, stearic, oleic, linoleic and linolenic acid present in varying proportion. High amount of either linoleic or linolenic acid are associated with pulses containing insignificant amount of lipids. Palmitic acid ranges from 9.22-25%. The time of harvest and maturity plays role in fatty acid composition. Unsaturated fatty acids esterify cholesterol.

Problems and prospects associated with breeding for nutritional quality

Negative correlation has been recorded in many legume species between grain yield and protein content and also between protein content and sulphur containing amino acids. the main problems include: negative correlation between yield and protein content, negative correlation between protein and sulphur containing amino acids and lack of proper field screening technique. However, some exceptions showing positive correlations (or absence of correlation) have been reported which indicate the possibility of simultaneous improvement through genetic recombination/mutagenesis followed by selection.

Negative correlation between yield and protein can be altered by breeding

A concurrent choice of both yield and protein will largely mitigate the situation. Variation for both the traits coexists between crop species, and between varieties within a crop (Boulter, 1982). Availability of mutant genes responsible for a higher amount of protein content in grains (such as opaque-2 in maize) could further enlarge the spectrum of genetic variation. With such a range of variability, making inter-family selection for high yield, followed by intra-family selection for high protein content from selected productive families, would thus offset the ill - effects of the negative correlation between yield and protein.

Negative correlation between protein quality and quantity can be manipulated

In legumes the negative correlations between protein content and methionine and cystein content can be modified as have been done in case of cereals. This apart, as in cereals, several mutants higher in protein content and better in amino acid profile have been also identified in legumes. Appropriate genetic manipulation of these mutants may bring about an appreciable improvement in both protein quantity and quality.

Better screening techniques are now available

Many procedures for a chemical analysis of the protein and amino acid profile have been developed. Some of them are remarkably rapid and reliable but all these require double screening (first in the field and second in the laboratory) hence the final selection is a time - consuming process. Moreover, the field selection is likely to be attended by rejection of genotypes which otherwise highly nutritive. These techniques involve both rapid chemical analysis of protein and amino acid profile of grains in the laboratory, and improved field techniques including the use of genetic markers associated with high protein content etc. However, the best choice would be availability of molecular markers tightly linked to these traits. Suitable methods for screening for protein content are dye binding coupled with automated N determination or by using IR reflectance. Fast screening methods also exist for the sulphur amino acids, e.g. IR reflectance. In screening for sulphur amino acids both methionine and cysteine must be taken into account since the latter can spare the former. Total sulphur analysis, with prior removal of aqueous alcohol soluble sulphur components is often satisfactory (Evans and Boulter, 1974). This must be followed by a secondary protein screen, to ensure, for example, that the high sulphur content is not due to increased trypsin inhibitor content.

Genetic variability for quality traits in grain legumes

Sufficient variation exists for the main components in legume seeds since oil concentrations vary from 1 to more than 40 percent, carbohydrate from 2-3 to 60% and protein from about 15 to 50 percent of the dry weight. Manipulation of the proportion of these components, however, may not be accomplished easily since the different combinations of the major components are each associated with a complex biochemical machinery about which very little is known, e.g. what is the biological regulation whereby soya beans are oil seeds while broad beans mainly store starch (Boulter, 1982).

Broad sense heritability

Broad sense heritability estimates for protein content in various grain legumes have been presented in following Table 7. Reasonable scope for genetic improvement through selection was indicated by these heritability estimates.

Table 7 : Heritability estimates of protein content in grain legumes

S.No.	Crop	Broad sense Heritability	Reference
1	White lupin	0.83	Green *et al.*, (1977)
2	Soybean	0.4-0.8	Johnson *et al.*, 1955)
3	Chickpea	0.75	Sandhu *et al.*, (1968)
4	Dry beans	0.3-0.64	Leleji *et al.*, (1972)
6	Fababean	0.54	Bond, (1977)
7	Beans	49.3% (Broad sense) 20% (narrow sense)	Dickson and Hackler, (1973)
8	Kabuli Chickpea	High	Singh *et al.*, (1986)
9	Fababean	High	Robertson *et al.*, (1986)

Genetics of protein content

Genetics of protein content was determined in several legume species and has been presented in following Table 8.

Table 8 Genetic control of protein content

S.N.	Crop	Genetics	References
1	Dry beans	Low partially dominant to high	Leleji *et al.*, (1972)
2	Fababean	Low partially dominant to high	Bond, (1977)
3	Peas	Low content partially dominant to high	Cousin, (1979)
4	Soybean	High partially dominant to low	Weber, (1950)
5	Soybean	Additive > dominance gene action	Takagi and Keshikawa, (1988)
6	Mungbean	High governed by dominant genes; mostly non-additive variance (Biparental mating system and modified diallel selective mating system)	Malhotra *et al.*, (1979)
7	Pea	Non additive gene action, both additive and non-additive gene effects	Singh and Singh, (1996) Chauhan and Singh, 1995
8	Soybean	Maternal effect	Hue *et al.*, (1984)
9	Soybean	mainly additive effects	Song *et al.*,(1991)

S.N.	Crop	Genetics	References
10	Chickpea	Predominance of additive effects for protein content and tryptophan content but for sulphur content non-additive effects	Sandhu *et al.*, (1984)
11	Pea Chickpea	Heterosis for protein content pea, chickpea and soybean GCA effects > SCA	Singh and Santoshi, (1989) Salimath *et al.*, (1988)

Correlation between protein content and grain yield

The correlation between protein content and grain yield has been worked out in some studies of grain legume. The results have been presented in following Table 9.

Table 9 : Correlation between protein content and grain yield

Sn	Crop	Correlation	Reference
1	Bean	-ve	Dickson and Hackler, (1973)
2	Pigeonpea, Mungbean Frenchbean.	-ve	Dahia *et al.*, (1977), Malhotra (1970), Laji *et al.*, (1972)
3	Mungbean	-ve	Sandhu *et al.*, (1978)
4	Pigeonpea	-ve	Ganeshmurthy and Dhanraj, (1990)
5	Peas	+ve	Know, (1964), George *et al.*, (1968), Malalyder and Autones, (1973)
6	*Fababean*	-ve NS	De Vries, (1979)
7	*Pea*	-ve NS	Blixt, (1978)
8	Soybean	-ve NS	Johnson *et al.*, (1955)
9	*Phaseolus vulgaris*	-ve NS	Conti and Zuffrano, (1979)
10	Peas	-ve S	Slinkard, (1980)
11	Lupine	-ve S	Green *et al.*, (1977)
12	Dry beans	-ve S	Kelly and Bliss, (1975)

A negative correlation between grain yield and protein content has been found in most of the legume species, but this correlation was significant only in a limited number of experiments. Negative but not significant correlations were found in *Vicia faba, Pisum sativum, Glycine max* and in *Phaseolus vulgaris* (Monti and Grillo, 1983). To the contrary, negative and significant correlations were found only by few researchers in peas, lupine and dry beans (Monti and Grillo, 1983). These differences can be due to the different materials and to the different environmental conditions.

Breeding implication of negative correlation of protein content with grain yield

The reported negative correlation between the protein content and grain yield in legumes seems to be small enough to allow selection of plants high in protein without reducing yield. Under this situation selection for higher protein content should be performed only among lines yielding at least at the same level as the original populations; the production of proteins per hectare must be taken as a selection criterion. The alternative possibility is to select for increased yield while trying to maintain the protein level constants. The choice between the two alternatives depends on the correlations found between the two traits in the experimental material. To summarize, even in species in which a negative correlation has been found between yield and crude protein content, progress in breeding work can be obtained because enough variation is present in each species to allow the selection of plants with higher yields and a constant or even higher protein content. Unlike in the cereals, in these species it is still possible to select at the same time for both yield and protein level, because the physiologic competition between the carbohydrate and protein production has probably not reached its limit.

Breeding approaches

Restricted index selection

Restricted index selection has been successfully utilized in soybean for improvement of protein concentration while holding oil content (Openshaw and Hadley, 1984) and for improvement in yield by holding protein concentration constant (Holdbrook *et al.*, 1989).

Recurrent selection

Recurrent selection was also used to increase seed protein content and seed yield (Brim and Burton, 1978) and for improvement of protein content (Xu and Wilcox, 1992 and Carter *et al.*, 1982) in soybean. Recurrent mass selection for 2 cycles, based on a desired grain index (Sullivan and Bliss, 1983), used for simultaneous increase of seed yield and seed protein percentage, could increase seed protein content from 21.9 to 24.6 %. Selection on the basis of single plant protein content in segregating generations resulted in increase of protein content with little change in yield (Meng *et al.*, 1990).

Selection strategy for improvement of protein content

Selection strategy depends on factors such as the crop species involved, the breeding system of crop species, the selection objectives, the influence of the environmental factors on the expression of the character under selection. Traits which are governed by major genes and have high heritability estimates should be selected in early segregating generations e.g., seed colour and seed texture. Traits which are governed by polygenes and have low heritability estimates should be selected in F_4 and F_5. For the trait like protein content between family selection should be practiced in later generations among the superior families already selected for higher yield. When the negative correlation between protein content and grain yield is low, selection should be made first for high yield and then within high yielding families, plants having the highest percentage protein should be identified for further selection and inter-mating.

Possibility of improvement in amino acid composition through selection

In conclusion, a possibility seems to exist to obtain, by genetic techniques, an improvement of protein quality in legume species by increasing their sulphur amino acid content. As noted before, the main difficulties for a direct selection are due to the low variability which exists for each amino acid, to the relationships among the amino acids and to the negative correlations between the protein content and the sulphur amino acid level. Variation in the content of an amino acid is related to the levels of the others (Lafiandra *et al.*, 1979), since they are bound inside the polypeptide chains. The influence of the protein content (%) on the levels of each amino acid (g/16 g N). The breeding possibilities are better for increasing the protein or the globulin fraction with a better amino acid composition similar to methods already employed in cereal breeding. Whenever such studies have been performed, negative but not always significant correlations were found between protein content and sulphur amino acid content. This indicated that a possibility is there for increasing the sulphur containing amino acid content without decreasing the protein content. Lines with both high protein and sulphur amino acid contents were obtained in *Vicia faba*, and in other species (Sjodin *et al.*, 1981). Particular attention in the selection work with legumes should, however, be given to the tryptophan content, which is the third limiting amino acid in the legume species and which is also present only in limited amounts in the cereals.

Relation of grain protein content and amino acid content

Like the belief of negative correlation between grain yield and protein content there is also a belief of negative association between protein concentration in grain and content of amino acids in protein. But there are also examples of positive correlations. In *Phaseolus vulgaris* seed protein content was positively correlated with leucine content (Kosson, 1989). Genotypes with high protein content and high lysine content were developed in soybean (Tymchuk *et al.*, 1990). Similarly mutants with high protein, lysine and seed yiled were identified in *Vigna mungo* and *Vigna radiata* (Ignacimuthu and Babu, 1989).

Breeding for Anti-nutritional factors

Legumes often contain anti-metabolites and even toxic substances. This problem, except in specific instances, is of less significance because, either the varieties commonly consumed have been selected to contain relatively small amounts of these substances, or, the food is cooked or processed so as to eliminate or inactivate them (this doesn't apply if used as an animal feed) *Boulter* (1982). Furthermore, there is an increasing body of evidence that some of these substances play a protective role (Gatehouse *et al.*, 1979), e.g. in insect or fungal resistance, and if this is the case their removal could cause problems in the field.

Genetic variability for antinutritional factors

Large variation in the trypsin and chymotrypsin inhibitors has been reported among the cultivated and wild species of pigeonpea (Singh and Jambunathan, 1981). The high levels of protease inhibitors are found in some of the wild species (Kumar *et al.*, 2011). It is pointed out that the use of such wild species as a source of high protein in breeding program to develop high protein cultivars should be made very carefully and all inter-generic lines obtained from crosses of cultivated species with wild species should be tested for the levels of protease inhibitors. It is evident that trypsin inhibitory activity of pigeonpea is much lower than in common bean, lima bean and soybean when compared under similar assay conditions. Although the improved protein efficiency ratio of the heat treated samples of pigeonpea meal could suggest the role of some heat labile factors, the reasons for low protein digestibility of heated pigeonpea meal samples could not be explained (Liener, 1979).

Recent studies indicated that pigeonpea condensed tannins ranged between 0 to 0.2% (Price *et al.*, 1980). But pigeonpea seeds contain considerable amount of polyphenolic compounds which may or may not be tannins and most of these compounds are located in seed coat.

Further perspectives

Opportunities to exploit legumes for nutrition can be observed in a growing appreciation for the potential for agriculture to contribute to public health. Biofortification is now recognized as a viable tool to be employed together with increasing the diversity of diets (Mayer *et al.*, 2008). Several studies are in progress with seed legume species and remarkable successes have been obtained already with traditional genetic techniques to make these plants more useful for human beings. By using special techniques such as *in vitro* culture and some chromosome induced mutants (Monti, 1983), greater chances of genetic improvement of protein structure are offered. Also, mutagenesis should be stressed (Monti and Saccardo, 1979), as an important tool in order to eliminate anti-nutritional factors. It is also true that by increasing our knowledge of nitrogen metabolism in these plants (Porceddu *et al.*, 1979) and of the genetic basis of the variation in the seed protein fraction and components, increased genetic possibilities can be obtained for improving protein content and quality. These studies are particularly advanced in peas, where five loci have been identified, determining five variant systems in the legumin and vicilin fractions of storage proteins (Thomson and Schroeder, 1978), and where some of the loci responsible for vicilin and convicilin fractions have been localized on chromosomes 7 and 2 (Monti and Grillo, 1983).

The other targets for legume seed protein improvement include removal of anti-nutritional factors and activities that generate undesirable flavour, removal of potential allergens, improved digestibility and improved functional behaviour for processing (Wang *et al.*, 2003). The molecular basis of many of these has not been sufficiently well defined to enable directed improvement, either by breeding using genetic variation or by genetic manipulation.. The great potential of several attempts in past to alter amino acid composition of globulins through use of natural variation, and through genetic manipulation is still unrealized but should yield seed protein with enhanced quality in future. Advances in genomics can make contribution by focusing a specific gene for proteins and micronutrient accumulation and this can be employed across genome species

REFERENCES

Adams MW (1973) On the quest for quality in the field bean. In: Nutritional Improvement of Food Legumes by Breeding. Milner M. (Ed.), Protein Advisory Group of the United Nations System, NY, 143-149

Adarsh-Bala, Satija DR, Gupta VP (1994) Variation and association analysis among protein quality traits in chickpea. *Indian J Genet Pl Breed* 54: 105-109

Anonymous (2011) Economic Survey of India, Ministry of Agriculture, Govt of India

Apostolatos G (1984) A survey of protein and methionine content of edible dry bean (*Phaseolus vulgaris* L.). *Genetica Agraria* 34: 91-99

Baldi G and Salamini F (1973) Variability of essential aminoacid content in seeds of 22 *Phaseolus* species. *Theoret Appl Genetics* 43: 75-78

Battisti DS and Naylor RL (2009) Historical warnings of Future food insecurity with unprecedented Seasonal heat. *Science* 323: 240— 244

Beldar DR, Chavan JK, Deshmukh RB and Kadam SS (1989) Nutritional analysis of promosing germplasm of blackgram and green gram. *J. Maharastra Agric. Univ.* 14: 47-51

Bhatty RS and Christison G I (1984). Composition and nutritional quality of pea (*Pisum sativum* L.), fababean (Vicia *faba* L. spp. minor) and lentil (*Lens culinaris* Medik.) meals, protein concentrates and isolates. *Qual Plant Plant Foods Hum Nutr* 34: 41-51

Birader BD, Goud JV and Patil SS (1994) Genetic studies on seed size, protein content and grain yield of cowpea. *Crop Res. Hisar* 7: 263-68

Blixt S (1978) Natural and induced variability for seed protein in temperate legumes. FAO/ IAEA int syrup seed protein improvement in cereals and grain legumes, Neuherberg, Sept 4-8

Bond DA (1976) In vitro digestibility of the testa in tannin-free field beans. *J. Agric. Sci. Cambridge* 86: 561-66

Boulter D (1977) In *Genetic Diversity in Plants,* p. 387 [A. Muhammed, R. Aksel and R. C. von Borstel, editors]. New York: Plenum Publ. COT

Boulter D (1980) In *Advances in Legume Science,* vol. I, p. 127 [R. J. Summerfield and A. H. Bunting, editors]. London: Ministry of Agriculture, Fisheries and Food

Boulter D and Gatehouse JA (1978) *FEBS Fedmatir,n of European Biochemical Societies* 44, Syp. A3, I 8 I. Oxford & New York: Pergamon Press

Brim CA and Burton JW (1977) Recurrent selection for increased seed protein percentage in two soyabean varieties (Abstr). In: Agron Abstr. Am. Soc. Agron.USA, pp.50

Burr HK (1975) Pulse protein. In: Friedman M (ed). Protein nutritional quality of foods and feeds. Dekker Inc, New York, p 119

Burton JW, Purcell EA and Walter WM Jr (1982) Methionine concentration in soybean protein from population selected from increased per cent protein. *Crop Sci.*22: 430-32

Carter JE (Jr), Burton JW and Brim CA (1982) Recurrent selection for percent protein in soybean seed-indirect effects on plant N-accumulation and distribution. *Crop Sci.* 22: 513-19

Chatterjee SR and Abrol YP (1975) Amino acid composition of new varieties of cereals and pulses and nutritional potential of cereal-pulse combinations. *J Food Sci & Technol* 12: 221-227

Chauhan MP and Singh IS (1995) Inheritance of protein content in lentil (Lens culinaris Medik). *Legume Res.* 18: 5-8

Chen W, Sharma HC and Muehlbauer F (2010) Chickpea and Lentil Crop Protection Compendium. American Phytopathological Society, St Paul, Minnesota, USA. 255 pp

Conti L and Zuffrano M (1979) Variabilith intra interspecificadella quantitie qualitt deue proteine del seine di fagiolo. CNR Meeting on 'prospettive delle proteaginose in Italia', Perugia (Italy), May 30-June 2, 1979

De Vries APH (1979) In search of characters to be used for indirect selection on grain and protein yield in *Vicia faba* L. EEC seminar on 'some current research on *Vicia faba* in Western Europe', Bari (Italy), April 1978, 324-341

Dickson MH And Hackler LR (1973) Protein quantity and quality in high yielding beans. In: Nutritional Improvement of food Legumes by Breeding. Rome, Italy, 185-92

Din van Luen and Lysinko VN (1991) Use of experimental mutagenesis in breedind soybean. Izvestiya Akademii Nauki SSR Moldova, Biolocheshe hmicheshie Nauki No. 1: 28-30

Dudits D (1970) Use of the mutation method for protein improvement in pea. *Hungarian Acad. Agric. Dept. Proc.* 29: 377-88

Eggum BO (1973) Biological availability of amino acid constituents in grain protein. In Nuclear techniques for seed protein improvement, IAEA, Vienna, p 422

Eggum BO (1977) Nutritive value of food crops and factors affecting the utilization of dietary protein and energy. Second FAO/SIDA Seminar on Field Food Crops in Africa and the Near East, Lahore, Pakistan

Ershine W, Williams PC and Nakkoul H (1985) Genetic and environmental variation in the seed size, protein yield and cooking quality of lentils. *Field Crops Res.* 12: 153-61

FAO/WHO (Food and Agriculture Organization/World Health Organization) (1973). Energy and protein requirements. Report of a Joint FAO/WHO Ad Hoc Expert Committee. WHO technical report series 522, Geneva, Switzerland, pp 63

FAO/WHO (1973) Expert group of energy and protein requirements. World Health Org Techn Rep Ser No 522

Gottschalk W, Muller HP and Wolff G (1976) Further investigations on the genetic control of seed production in *Pisum* mutants. Proc Symp, Vienna, IAEA, 157-177

Hanelt P, Rudolph A, Hammer K, Jank H-W, Muntz K and Scholz F (1978) *Kulturpflanze pp 234-247*

Hartwig EE and Hilen TC (1991) Yield and composition of soybean seeds from parents with different protein, similar yield. *Crop Sci.* 31: 290-92

Holdbrook CC, Burton JW and Carter (Jr.) TE (1989) Evaluation of recurrent restricted index selection for increasing yield while holding seed protein constant in soybean. *Crop Sci.* 29: 324-29

Holt NW and Sosulski FW (1979) Aminoacid composition and protein quality of field peas. *Can J Plant Sci* 59: 653-660

Hu MX, Yu DY, Meng XX and Zhou SC (1984) Genetic studies on the seed protein content of hybrid progenies of soybean. *Scientia Agricultura Sinica* No. 6: 40-44

Ignacimuthu S and Babu CR (1989) Induced variation in protein quantity and quality in the wild cultivate Urd and Mung beans. *Indian J. Genet Pl. Breed.* 49: 173-81

Jambunathan R and Singh U (1981) Relationship between total sulphur and sulphur amino acids in chickpea *(Cicer arietinum* L.) and pigeonpea *(Ca~anus ca~an* L.). *Qual Plant Plant Foods Hum Nutr* 31 : 109-117

Johnson HW, Robinson HF and Comstock RE (1955) Genotypic and phenotypic correlations in soybean and their implications in selections. *Agronomy J* 47: 477-83

Lafiandra D, Polignano GB and Colaprico G (1979) Protein content and aminoacid composition of seed in varieties of *Vicia faba* L. *Z Pflanzenzuchtg* 83: 308-314

Lelji OI, Diickson M H, Crowder LV and Bourke JB (1972) Inheritance of crude protein percentage and its correlation with seed yield in Beans(*Phaseolus vulgaris* L.) *Crop Science* 12: 168-71

Kasthuri R, Jayapragasam M, Rangasamy S, Owri G and Sadasivam S (1987) Protein and methionine contents of green gram and black gram genotypes. *TVIS News. Taiwan,* 2(2): 7-8

Kelly JD and Bliss FA (1975) Heritability estimates of percentage seed protein and available methionine and correlations with yield in dry beans. *Crop Sci.* 15: 753-57

Kelly JF (1971). Genetic variation in the methionine levels of mature seeds of common beans(*Phaseolus vulgaris* L.). *J. Am. Soc. Hort. Sci.* 96: 561-63

Kharkwal MC (1998) Induced mutation for improvement of protein in chickpea (*Cicer arietinum* L.). *Indian J. Gen. Pl. Breed.* 58: 61-68

Koroleva OS, Pelageichik TY, Uralskaya ER and Mironenko AV (1991) (Amino acid composition of the globulins in different varieties and forms of pea). *Vestsi Akademii Navuk BSSR, Seryya Biyalagichnykh Navuk* No.6: 103-106

Kosson R (1989) Correlation of prtein content ot flatulence galacto-oligosaccharides and exogenous amino acids in seeds of *Phaseolus vulgaris* L. *Acta Societatis Botanicorum Polaniae* 58: 65-70

Kumar P, Joshi PK and Birthal PS (2009) Demand projections for food grains in India. *Agricultural Economics Research Review* 22: 237— 243

Kumar S, Bejiga G, Ahmed S, NakkoulH and Sarker A (2011) Genetic improvement of grasspea for low neurotoxin (ODAP) content. *Food and Chemical Toxicology* 49: 589—600. Doi 10.1016/j.fct2010.06.051

Larkins BA (1983) Genetic engineering of seed storage proteins. In: Genetic Engineering of Plants: An Agricultural Perspective. T. Kosuge, Meredith, C.P. and Hollaender, A. (Eds). Pleunum Press, New York and London, 93-118

Liener IE (1979) Protease inhibitors and lectins. In: Neuberger, A, Jukes, T.H. (eds).

Biochemistry of nutrition. University Park Press, Baltimore, 27: 97-122

Malhotra RS, Gupta PK and Arora ND (1979) Combining ability and inheritance studies for protein content in mungbean. Qual.Ptant. - P1.Fds.hum.Nutr. XXVIII, 4: 323-331, 1979

Marentes E and Grusak MA (1998) Iron transport and storage within seed coat and embryo of developing seeds of pea (*Pisum sativum* L.). *Seed Sci. Res* 8: 367-375

Martensson P (1980) Variation in legumin vicilin ratio between and within cultivars of *Vicia faba L var minor.* Proc EEC seminar, ambridge, June 27-29, 1979

Mayer JE, Pfeiffer WH, Beyer P (2008) Biofortified crops to alleviate micronutrient malnutrition. *Current Opinion in Plant Biology* 11: 166— 170

Meng XX, Hu MX, Li AP, Wang SM and Yi CN (1990) Selection efficiency for protein content and its effect on seed yield and oil content in the early generation from three crosses of soybean). *Acta Agronomica Sinica* 16: 377-80

Miller JE and Fehr WR (1979) Direct and indirect recurrent selection for protein in soybean. *Crop Sci.* 101-106

Monti ML (1983) Natural and induced variability in peas for protein production. EEC seminar on 'Perspectives for peas and lupins as protein crop', Sant' Agnello- Sorrento (Italy), October, 19-22, 1981, pp 23-39

Monti LM and Grillo S (1983). Legume seed improvement for protein content and quality. *Qual. Plant Plant Foods Hum. Nutr.* 32: 253-266

Monti ML and Saccardo F (1979) The gamete irradiation technique in pea genetics and breeding. *Monografia IV Genetica Agraria* 141-152

Munk L, Holmberg E, Kardsson KE, Lofquist B and Sjodin J (1973) Breeding for protein quantity and quality in field bean(Vicia faba) and barley(Hordeum vulgare). In: Nutritional Improvement of Food Legumes by Breeding. M. Milner (Ed.), Protein Advisory Group of the United Nations System, N.Y., 197-204

Off JH and Hymowitz T (1979) Inheritance of the absence of the Kunitz Trypsin inhibitor in seed protein of Soybeans. *Crop Science* 99: 107-109

Olabisi F, Adekola and Funsho Oluleye (2007) Influence of mutation induction on the chemical composition of cowpea *Vigna unguiculata* (L.) Walp. *African Journal of Biotechnology* 6 (18): pp. 210-11

Openshaw SJ and Hadley HH (1984) Selection indexes to modify protein concentration of soybean seeds. *Crop Sci.* 24: 1-4

Picard J (1979) Some reflections on problems and prospects in *Vicia faba* breeding. Proc EEC seminar, Baxi (Italy), April 27-29, 1978

Porceddu E, Scarascia MG, Monti LM (1979) Genetica e miglioramento genetico della produzione proteiea delte leguminose da granella. CNR Meeting on 'prospettive delle proteaginose in Italia', Perugla (Italy), May 30-June 2, 1979

Porter WM, Maner JH, Axtell JD and Keim WF (1974) Evaluation of the nutritive quality of grain legumes by an analysis of total sulphur. *Crop Sci.*14: 652-54

Price ML, Hagerman AE, Butler LG (1980) Tannin content of cowpeas, chickpeas, pigeonpea and mung beans. *J. Agri. Food Chem* 28: 459-461

Rangasamy P, Kasthuri R, Gomathinayagam P, Khan AKF, Murugarajendran C and Nagarajan P (1993) Study on relationshipbetween methionine and total protein content in grain legumes. *Madras Agric. J.* 80: 709-10

Rech (1992) Particle bombardment-msdiated transient expression of abrazil nut methionine rich albumin in bean(*Phaseolus vulgaris* L.). *Plant Molecular Biology* 20 : 357-59

Sabbouh MY, Edwards LH and Keim KR (1998) Heterisis and combining ability for protein and oil concentration in the seeds of soybean (*Glycine max* (L.) Merr.). *SABRAO J. Breed. Genet.* 30: 7-17

Salimath PM; Bahl PN and Kumar J (1988) Genetic analysis for seed protein content in chickpea (*Cicer arietinum* L.). *Legume Res.* 11 (2): 71-74

Sandhu SS, Keim WF, Hodges HF, Nyquiest WF (1968) Inheritance of protein and sulphur content in chick peas, *Cicer arietinum* I. *Agronomy Abst* 20

Sandhu TS, Barar JS, Gumber RK, Sharma AK, Singh B and Singh B (1994). Estimation of combining ability effects for quality traits in mungbean. *Indian J. Pulses Res.* 7: 70-71

Sandhu TS, Rang A, Sharma AK, Singh H (1984) Genetics of protein and amino acids in chickpea. *Crop Improvement* 11: 47-51

Saxena KB, Kumar RV and Sultana R (2010) Quality nutrition through pigeonpea—a review. *Health* 2(11): 1335— 1344

Sengbush Ryon (1948) Susslupinen und Ollupinen. *Landw J* 1: 720-880

Shevlyakova NF (1987) Breeding significance of genotypic differences in high protein lines of pea. *Nauchnotech Cheskii Byulletin Vsesoyunogo* No. 35: 14-18

Sichkar VL and Levitskii AP (1984) Obtaining high-yielding soybean mutants with increased content and improved quality of protein). In: Khim. Mutagenez v povysh. Produktiv. S-kh. Rast.Moscow, USSR(1984); 128-34

Silbernagel MJ (1971) Bean protein improvement work by USDA, Bean and Pea Investigations. In: Report on the 10th Dry Bean Res. Conf. ARS. 74-56. Agric. Res. Service, USDA

Singh U and Jambunathan R (1981) Methods of protein estimation in pigeonpea and relationship between whole seed and dhal protein. *J Sci Food Agric* 32: 705-710

Singh U and Jambunathan R (1982) Distribution of seed protein fractions and amino acids in different anatomical parts of chickpea *(Cicer arietinurn L.)* and pigeonpea *(Cajanus cajan* L.). *Qual Plant Plant Foods Hum Nutr* 31: 347-354

Singh U And Jambunathan R (1981) Protease inhibitors and *in vitro* protein digestibility of pigeonpea *(Cajanus cajan L.)* and its wild relatives. *J Food Sci Technol* 18: 246-247

Singh KN and Santoshi US (1989) Heterosis for yield and protein content in pea. *Legume Res.*12: 196-98

Singh SB and Singh JP (1996) Inheritance of yield, protein content and recovery of dal in field pea (*Pisum sativum*). *Indian J. Agric. Sci.* 66: 366-69

Sjodin J (1982) Protein quantity and quality in *Vicia faba*. Faba bean improvement. G. Hawtin and C. Webb (Eds.), pp. 319-31

Sjodin J, Martensson P and Magyarosi T (1977) The variation in crude protein content and seed weight within field bean plants *(Vicia faba L.)* analyzed with the halfseed technique. *J Swedish Seed An* 87: 33-42

Sjodin J, Martensson P and Magyarosi T (1981). Selection for antinutritional substances in field bean (Vicia *faba* L.) *Z Pflanzenzuchtg* 86: 231-247

Slinkard AE (1980) Breeding field peas for increased yield and protein. In: Production, Utilization and Marketing of Field Peas. Final report

Sullivan JG and Bliss FA (1983) Recurrent mass selection for increased seed and seed protein percentage in the common bean, *Phaseolus vulgaris* L. Using a selection index. *J. Am. Soc. Hort. Sci.* 108: 42-46

Takagi Y and Kishikawa H (1988) Inheritance of protein content in a cross of soybean varieties. *Bulletin of the Faculty of Agriculture, Sega Univ.,* No. 64: 1-10

Thomson JA and Schroeder HE (1978) Cotilodonary storage proteins in *Pisum sativum.* II Hereditary variation in components of the legumin and vicilin fractions. *Aust J Plant Physiol* 5: 281-294

Tiwari AS and Rmanujam (1976) Combining ability and heterosis for protein and methionine contents in mungbean. *Indian J. Genet and Pl. Breed.* 36: 353-57

Tymchuk NF, Bondarenko VI and Matushkin VA (1990) Biochemical value of source material of soybean for use in breeding. In: Rezenvey povysheniya produktivnosti soi, Nvosibirsk, Russia, 59-61

Wang TL, Domoney C, Hedley CL, Casey R and Grusak MA (2003) Can we improve the nutritional quality of legume seeds? *Plant Physiol.*13: 886-891

Weber CR (1950) Inheritance and interrelation of some agronomic and chemical characters in an inter-specific cross in Soybean *Glycine max G ussuriensis. Iowa State Agr Exp Sta Bull* 374

Xu HJ and Wilcox JR (1992) Recurrent selection for maturity and per cent seed protein in *Glycine max* based on S0 plant evaluation. *Euphytica* 62: 51-57

Yau J L, Li AZ and Zhao BY (1991) Correlation between seed and qualitative characters in mung bean. *Acta Agriculturae Boreali Sinica* 6: 96-98

Yohe JM, Swindell RE, Watt EE, Bashandi MM, Seahler DT and Poehelman JM (1972) Evaluation of mung bean (*Phaseolus vulgaris* (Roxb.)) strains. *Dept. Agron. Misc. Pub. No.* 72-9

Yohe JM, Watt EE, Bashandi MM, Sechle DT and Poehlman JM (1971) Evaluation of mungbean strains at Columbia, Missouri in 1970. Dept. of agron. Misc. pub. No. 71-4. Mo. Agric.Exp.Sta

Yu MA and Bliss FA (1978) Seed proteins of common bean. *Crop Science* 18: 431-437

5

Integrated Approach in Garlic Improvement

Abhilash Nair, Anil Khar, Amandeep Hora and *C.P. Malik*

Garlic (*Allium sativum*) is valued as a valuable vegetable, spice with divergent functions. It is an extraordinary crop being completely propagated asexually for the past several centuries. It has several diverse phenotypes amongst different clones. This makes garlic a model species for investigating heritage and diversity. Using molecular techniques like AFLP, RAPD, ISSR molecular markers diversity between strains has been studied and some workers have developed a phylogenetic tree. Micropropagation through organogenesis and somatic embryogenesis has been carried out. Mostly stem disc (consisting in the apical meristem and the lateral buds of the clove) has been used to regenerate plants. Medicinal properties and active principles of garlic have been extensively described. Genetic improvement of garlics through genetic transformation has also been done.

Introduction

India shares about 12% of the world output of vegetables from about 2.0% of cropped area in the country (Sidhu, 1998). The major objectives of research on vegetables in India is improving production per unit area by solving chronic problems of production through breeding high yielding, disease and pest resistant varieties, developing F_1 hybrids, standardization of agro-techniques for different agro-ecological situations, disease and insect pest management and post-harvest studies with a view to reduce post-harvest losses. Amongst vegetables, *Alliums* are perhaps the most important species consumed throughout the world. The genus *Allium* comprises approximately 600 known species distributed over Europe, North America, North Africa and Asia. The number of species belonging to the genus has grown steadily over the time as botanist has discovered new species throughout the world. The general characteristics of the *Allium* species are that they are almost exclusively herbaceous, perennials and usually form bulbs. Some species, however, often forms thickened rhizomes. In genus *Allium*, Garlic (*Allium sativum*) and Onion (*Allium cepa*) are mostly consumed throughout the world. In India especially, no culinary preparation is complete without onion or garlic. Apart from being used as a prime vegetable, garlic also has numerous medicinal properties. Use of these two alliums dated back to 3200 – 2780 B.C. According to Tackholm and Drar (1954), onion and garlic were popular foods in Egypt. It is believed that pyramid builders consumed these crops in great quantities. One of the most important testimonials to the use of onion and garlic in ancient Egypt is from the Bible, where, reference about consumption of these has been made (Numbers xi, 5). Garlic and onion, among other crops are also mentioned in Koran, Part I: 61 (Jones and Mann, 1963).

In India too, references about cultivation of onion and garlic are found from ancient times onwards as is evidenced by mention in *Charaka-Samhita*, a famous early medical treatise of our country (Jones and Mann, 1963). The varied medicinal uses and pharmacological actions like the anti-inflammatory, antihelmintic and heart friendly nature of garlic were known to ancient Hindus many centuries ago and seem to be supported by modern research (Kirtikar and Basu, 1975).

Garlic enjoys a wide distribution worldwide and has long been known as a cultivated species. Its native home and existence as a truly wild species is still not clear and has been discussed by Regel (1875, 1887) and Candolle (1885). According to Vavilov (1951) and Vvedensky (1946), *Allium longicuspis* Rgl. is an endemic species of central Asia and wild ancestor of garlic. The work of Maab and Klaas (1995) too suggests west to middle Asia to be the primary centre of origin and the Mediterranean area as the secondary centre of origin of garlic. However, their work suggests that the garlic of the subtropical regions is different from the other varieties of the crop and probably originated independently a long time ago from *A. longicuspis* group, perhaps in northern India. Jones and Rees (1968) have suggested that the wild ancestor of garlic was flowering and produced seeds or aerial bulbils under different soil and climatic conditions in the ancient civilization leading to the origin of different varieties gradually. The present day non-flowering varieties are believed to have originated as a result of continued human interference with the natural life cycle caused by storage.

Table 1 Production trends of major garlic producing countries

Country	Production (Tonne)
China	1,20,88,000
India	6,45000
South Korea	3,25000
Egypt	2,58,608
Russia	2,54000
USA	2,21,810
Spain	1,42,400
Argentina	1,40,000

The latest global review of production shows that garlic ranks second in production (Table 1). In India, garlic is grown in the Northern parts especially in Gujarat, Haryana, Punjab, Maharashtra, etc. and in a few southern localities, especially the cold hilly tracts of Ooty and Kodaikanal, Tamil Nadu. The major garlic growing states and production for the year 2010-11 are given in Table 2.

Table 2 Area, production and productivity of garlic in different states of India

State	Area ('000 ha)	Production('000 MT)	Yield (Ton/ha)
Andhra Pradesh	0.40	4.00	10.00
Bihar	4.25	29.75	7.00
Chhattisgarh	3.90	21.10	5.41
Gujarat	40.00	275.00	6.88
Himachal Pradesh	3.60	44.70	12.42
Jammu and Kashmir	2.30	32.00	13.91
Karnataka	4.20	30.20	7.19
Madhya Pradesh	54.00	228.00	4.22
Maharashtra	3.50	34.10	9.74
Orissa	11.00	35.80	3.25
Punjab	3.70	40.50	10.95
Rajasthan	25.00	150.00	6.00
Tamil Nadu	0.40	3.10	7.75
U.P	35.10	190.50	5.43
Uttaranchal	1.20	7.30	6.08
West Bengal	3.50	33.90	9.69
Others	13.29	104.74	7.88
Total	**209.34**	**1,264.69**	

1) *DIRECTORATE OF ECONOMICS AND STATISTICS, Krishi Bhawan, New Delhi. (Till 2007-08)*
2) *HORTICULTURE DIVISION, MINISTRY OF AGRICULTURE, GOVT. OF INDIA (From 2008-09 onwards)*

Taxonomy

Allium sativum, commonly known as garlic, is an important vegetable valued mainly for its strong and pungent characteristic odor and an unavoidable ingredient of almost all Indian cuisine, besides being an important medicinal plant belonging to the family *Alliaceae*. Bentham and Hooker (1880) had initially classified Alliums under the tribe *Alliaceae* of the family *Liliaceae*, on account of its superior ovary. Later, Takhtajan (1967) included this under the order *Amaryllidales*, family *Alliaceae*, subfamily *Alloideae* and tribe *Allieae*. However, Hutchinson (1934, 1973) and Traub (1968) placed it under the tribe *Allieae* of the family *Amaryllidaceae*, due to the bulbous nature and umbellate inflorescence. It was Dahlgren *et al.* (1985) and Walter *et al.* (1999) who gave Alliums its present status under the family *Alliaceae* and order *Asparagales*. Molecular analysis of the genus *Allium* within the *Alliaceae* was initially done by Klass (1998) through phylogenetic positioning of a plastid DNA sequence coding for the large subunit of ribulose-1-5-bi phosphate carboxylase. At least two distinct botanical varieties of garlic are recognized, *ophioscorodon* and *sativum* (Jones and Mann, 1963). Under ideal growing conditions, variety *ophioscorodon* is characterized by an initially coiled, tall woody scape and relatively few brownish-purple cloves per bulb. Variety *sativum*, produces a weak flower stalk, if it bolts at all, and is characterized by a bulb with many pure white or pink-blushed cloves (England, 1991).

According to some workers (Fritsch and Friesen, 2002) *Allium sativum* is a complex comprising 3 main groups: (a) Sativum, (b) Longicuspis and (c) Ophioscorodon, and two subgroups: Subtropical and Pekinense.

Some workers (Heredia-García, 2000) divide the the commercial types into: (1) violet or Asian, being cultivated in subtropical regions, (2) pink, which needs long photoperiods (3) white, requiring long photoperiods, and (4) purple, which needs long photoperiods and periods of cold .

They are also classified into hard-neck and soft-neck garlic. The former has a floral scape whose flowers normally abort and whose end produces topsets, while soft-necked garlic is devoid of a scape. Most commercially cultivated garlics are soft-neck being easy to cultivate and have long shelf life (Kamenestsky, 2007).

Most cultivated alliums have the basic chromosome count of $x = 8$. *Allium sativum* is a diploid with $2n = 16$ (Jones and Mann, 1963; Mc Collum, 1987; Figliuolo *et al.*, 2001). However, some species show a basic chromosome count of 7 and a few species even have 9 as their basic chromosome number (Jones and Mann, 1963).

Morphology and development

Allium sativum L. is a monocot having flat leaves, slender scape, long beaked spathes and heads bearing bulbils (Kirtikar and Basu, 1975). The small bulbs are always mixed with the flowers of the inflorescence and the flowers usually abort at the bud stage. Though there are reports on the seed production in the literature (Konokov, 1953), confirmation is needed that the garlic ever produces seed (Jones and Mann, 1963).

Commonly, garlic is vegetative propagated through cloves (Jones and Mann, 1963; Figliuolo *et al.*, 2001). Cloves are the sole organs of storage and these are the modified axillary buds of the foliage leaves. At maturity, the main stem of the bulb, roots and leaves attached to it all die. Only the cloves remain to carry the plant on to the next season (Jones and Mann, 1963).The clove consists of two mature leaves and a vegetative bud. The outer most leaf is the *protective leaf*, which becomes thin, dry and dead at maturity. Within the protective leaf is the single storage organ, the *storage leaf*. Within and at the base of the storage leaf are several very small leaves, which form the bud for next year's growth. The outer most leaf of the bud, the *sprout leaf*, lacks a blade and when the clove sprout, pushes up through the soil, but grows no further. From within the sprout leaf the foliage leaves arises, which constitutes the green top of the plant (Jones and Mann, 1963).

A wide range of adaptability to soil types, temperatures and day length makes garlic farming possible from tropics to temperate latitude (Figliuolo *et al.*, 2001). It is grown worldwide in nearly all cool regions (Novak *et al.*, 1986). Garlic is usually grown as a winter crop in India, *i.e.* during *rabi* season. It grows vigorously and matures in the spring or mid summer, depending upon the locality and cultivar. Temperature and day length play a very critical role in bulb induction in garlic. Exposure of dormant cloves or young plants to temperatures of 0^0 to 10^0 C, for one or two months and long days are a pre requirement for bulb induction. Plants that are never exposed to temperatures below 20^0 C fail to produce any bulbs, upon maturity, even under long days (Jones and Mann, 1963).

Chemical composition and medicinal values of garlic

In addition to food flavoring agent, this herb also has several medicinal uses for various ailments e.g. those caused due to aging, arthritis, cancer, artheroesclerosis, immune deficiencies, diabetes, respiratory diseases, etc. (Keusgen, 2002). Likewise, it has been observed that garlic has antioxidant properties, and reduces blood cholesterol and triglycerides levels, lowers blood pressure and minimizes the possibility of blood clot formation by improving arterial oxygenation (Bordia *et al.*, 1998). Garlic bulb contains sulphur compounds and hence has antibacterial properties and inhibits several pathogenic (Keusgen, 2002).

Garlic bulbs are also rich sources of carbohydrates and proteins. Analysis of garlic indicates that it contains 61 to 64% moisture, 31% carbohydrate, 5 to 6% protein and only 0.2% fat. Significant levels of phosphorous (3.9 to 4.6 mg/g), potassium (1.0 to 1.2 mg/g) and calcium (0.5 to 0.9 mg/g) are present (Kaufmann *et al.*, 1999). All *Allium* species produce volatile chemicals which act as repellants to many insects. Garlic has some other chemicals which deter other insects by its volatile metabolites. Sequestered in vacuoles within the plant's cells is an odorless, sulfur based compound, (+) S-allyl-1 cysteine sulfoxide or alliin.This comprises about 0.24% of the fresh weight of the bulb. In the cytosol, normally separate from alliin, is an enzyme, allinase. When the cell is disrupted, the cell's vacuoles are broken and alliin mixes with allinase resulting finally in the formation of allicin, a strong smelling fiery tasting chemical that repels almost every animal.

Garlic has several medicinal properties; the bulb is pungent; hot, oleaginous; tonic, aphrodisiac, fattening, digestive, antihelmintic; improves appetite, voice complexion; useful in the diseases of the heart, low fevers, bronchitis, inflammation, piles, leucoderma, asthma, "vata", lumbago, tumors, epileptic fits, thirst and earache, according to Ayurveda. It is a diuretic, carminative, alexipharmac, aphrodisiac; useful in inflammation, paralysis, pain in the body and the joints, troubles of the spleen, liver and lungs; clears the voice; good for lumbago, chronic fevers, caries of the teeth, leucoderma and thins the blood, according to Yunani (Kirtikar and Basu, 1975).

Louis Pasteur (1858) showed its antibacterial activity (Kaufmann *et al.*, 1999). Garlic is a stimulant, diaphoretic, expectorant, diuretic and tonic. It is rubefacient when applied externally. It has a foul smell due to the presence of unsaturated sulfides. The garlic juice is traditionally used for various ailments of the stomach including amoebic dysentery, anti - tubercular, anti - epileptic, anti - fertility drug and in reducing the blood sugar levels. It also lowers tissue lipids, antibiotic / antibacterial and insulin like effect on the human body (Augusti, 1990). Garlic juice is also reported to be antibiotic, anticancerous, antithrombotic and lipid - lowering cardiovascular effects (Agarwal, 1996).

Breeding approaches in garlic

Commercial garlic cultivars are propagated vegetatively; hence the increase of genetic variation through conventional crossing is extremely low, or even absent. Hence clonal selection, induced mutations, somaclonal variation or genetic engineering are the only options for breeding improved cultivars (Robinson, 2007). Among these, clonal selection is most widely used method for generating new garlic material. It is based on the variability existing in populations due to cross pollination between various garlic types. It is desirable to breed garlics having tolerance to biotic and abiotic stress, earliness, yield and quality, though lack of of sexuality limits the increase in variability which is essential for its improvement (Kamenetsky, 2007). Contrarily, vegetative propagation has many disadvantages for this crop: (a) a low multiplication rate (b) expensive and short-term storage requiring extensive space (c) transmission of phytopathogens from one production area to another, decreasing yield up to 70%, and (d) loss in quality of produce(Kamenetsky, 2007).

Biotechnological tools for garlic improvement

Some of the biotechnological techniques which have been variably applied for breeding garlics include micropropagation, especially meristems culture to obtain virus-free plants, somaclonal variation, and genetic transformation and are discussed below:

Micropropagation

For micropropagation, tissue culture techniques for garlic improvement were initiated as early as 1970. This technique has several advantages over clove reproduction, as it requires a few cells or small tissue fragments to generate large number of plants. It is possible to carry out micropropagation by two ways:

(1) Organogenesis, which leads to the formation of organs (shoots or roots), and
(2) Somatic embryogenesis, which yields the formation of structures having a similar morphology to that of a zygotic embryo.

Both processes may involve initial callus phase. It may be stated that morphogenetic ability in garlic declines in the old callus and often leads to the production of abnormal plants. Hence regeneration which does not involve callus phase is prefered. Moreover, embryogenesis has many advantages over organogenesis, e.g. higher potential for elevated plant output, lower labour costs etc. It may be mentioned that several micropropagation protocols have been established using both types of morphogenesis. Moreover different explant types have been tried though most developed protocols have followed the organogenetic way.

Organogenesis

Meristem culture is used to secure virus-free plants, and also for micropropagation. Messiaen *et al.* (1970) were pioneers in regenerating garlic plants from meristems. Shoot or bud formation from callus was achieved using a combination of 6- furfurylaminopurine (kinetin), indol-3-acetic acid (IAA) and 2,4-diclorophenoxiacetic acid (2,4-D) (9.28 µM, 11.4 µM and 4.5 µM, respectively). Similarly, Havránek and Novak (1973) obtained numerous growth areas on calli produced from meristems on a culture medium with 2,4-D. The subculture of calli to a medium containing IAA (11.4 µM) and kinetin (46.5 µM) induced formation of adventitious shoots.

In a different work, calli obtained from meristems of three garlic varieties (*Rose de Lautrec, California Early* and *California Late*) cultivated in a medium with 2,4-D (4.5 µM) and IAA (5.7 µM) produced adventitious shoots when transferred to a medium with IAA (5.7 µM) and kinetin (4.6 µM) (Kehr and Shaeffer, 1976). For his part, Abo-El-Nil (1977) started cultures from meristems, stems and leaf discs of the variety *Extra Early White,* on a medium with p-chlorophenoxyacetic acid (p-CPA) (10 µM), 2, 4-D (2 µM) and kinetin (0.5 µM), from which callus formation was achieved, which in turn resulted in the formation of adventitious shoots with kinetin (10 µM) and IAA (10 µM). In summary meristems were cultivated on B5 medium (Gamborg *et al.,* 1968) with 2.5 µM 2-isopentenyladenine (2ip) and 5.5 µM α-naphtalenacetic acid (NAA) (Bhojwani, 1980), or in Linsmaier and Skoog (LS) (Linsmaier and Skoog, 1965) medium with 9 µM N6-benzyladenine (BA) alone or with 11.1 µM and multiple shoots weresecured (Osawa *et al.,* 1981).

There are only a few reports on suspension cell culture in garlic. For instance, Nagasawa and Finer (1988) were the first to establish suspension cell cultures obtained from calli derived from meristems of the cultivar *Howaito-Roppen* grown in NAA (5.5 µM) and BA (9 µM). Adventitious shoots with leaf primordia started differentiating only the calli were transferred to agar-solidified medium. Kim *et al.* (2003) obtained cells in suspension after cultivating shoots of the cultivar *Danyang* differentiated *in vitro* in Murashige and Skoog (MS) (1962) medium with 2.5 µM 2iP. These cultures regenerated on an average of 21.5 shoots per explant when they were exposed to a light intensity of 50 µmol m-2 s 1. Thirty bulbs developed per explant in a medium containing 11% sucrose and 135 bulbs in the presence of 10 µM jasmonic acids.

Subsequently, Nagakubo *et al.,* (1993) evolved a micropropagation protocol for six varieties (*Isshuwase, Isshu-gokuwase, Shanhai, Santo, Furano* and *Howaito-roppen*) starting from shoot-tips which were cultivated in a medium supplemented with NAA (1 µM) and BA (1µM). Regenerated adventitious shoots were subcultured for four generations in NAA (5 µM) and BA (10 µM) for their multiplication. With this protocol it was possible to produce of 256 plants from one shoot-tip in 10 months. A novel regeneration protocol was developed by dissecting and sectioning longitudinally the shoots developed from cloves of the cultivar *Extra Select Sets.* These shoots were cultivated on a medium with BA (8 µM) and NAA (0.1 µM), and after five weeks they produced eight more shoots compared to the ones that had been kept intact (Mohamed-Yassen *et al.,* 1994).

Interestingly the roots produced by cloves have proved to be a good explant for plant regeneration. When the root tips are cultivated in a medium with NAA (1 μM) and BA (10 μM) the shoot formation is achieved for 75% of the explants, without callus phase. By using this method up to 380 shoots could be produced starting from a single clove (Haque *et al.*, 1997). Other protocols have been developed, which involve callus formation from this type of explants upon their cultivation in MS or N6 culture media (Chu *et al.*, 1975) supplemented with 2,4-D (4.5 μM) alone or with kinetin (4.7 μM). When the calli were transferred to a medium with 4.4 μM BA permitted the regeneration of 169 plants per gram of callus, which have the ability of forming microbulbs (Robledo-Paz *et al.*, 2000). Khan *et al.* (2004) also regenerated adventitious shoots from calli developed from root tips of two garlic varieties. The highest callus formation frequency was observed when a combination of 2, 4-D (6.8μM) and kinetin (23.8μM) was used. Shoot differentiation rose exponentially with increasing BA concentration, reaching the highest value at 45μM BA, while shoot rooting occurred in the absence of growth regulators. A variation in the number of shoots and their regeneration time was observed depending on the genotype used. Approximately 75% of the regenerated plants established successfully when transferred to greenhouse.

The roots developed from adventitious shoots obtained *in vitro* also allowed garlic micropropagation when cultivated on a medium that induced callus formation and then transferred to a medium with BA (13.3μM) and 4-amino-3,5,6-trichloro-picolinic acid (picloram) (1.4 μM). This method enables the regeneration of 5.4 shoots per explant (Myers and Simon, 1998a). A protocol named one-step was developed when the same type of explants was cultivated on a modified B5 medium supplemented with 0.1 μM 2, 4-D, 11.1 μM NAA and 13.6 μM BA, under two light conditions (16 hours photoperiod and complete darkness). In general, the root tips cultivated under low light conditions displayed the highest percentage of organogenic calli. The application of this protocol allowed the formation of callus and the regeneration of 250 shoots per gram of callus in the same culture medium and under the same light conditions (Martín-Urdíroz *et al.*, 2004). Zheng *et al.*, (2003) also obtained adventitious shoots by using apical and non-apical root fragments, originating from plants generated *in vitro* of four cultivars grown in Europe (*Messidrome, Morado de Cuenca, Morasol* and *Printanor*). The explants were cultivated on MS medium with 4.5μM 2,4- D and 0.5μM 2iP in order to induce calli formation, which were then transferred to a medium containing 4.7μM kinetin to promote shoot differentiation. The highest regeneration rate was obtained when non-apical fragments were used, although the difference was not significant.

In another study Ayabe and Sumi (1998) used the stem disc (consisting in the apical meristem and the lateral buds of the clove) to regenerate plants of the cultivar *Fukuchi-howaito*. When this was cut into various fragments and then cultivated on a medium with BA (0.4 μM), 20-25 adventitious shoots were obtained. The same result was observed when protoplasts isolated from shoot primordia were cultured in NAA (0.5 μM) and 2iP (2.4 μM), adenine and coconut milk (Ayabe *et al.*, 1995). Barandiaran *et al.* (1999) used immature bulbs of 23 accessions as a source of axillary buds, which were cultivated during six weeks on B5 medium with 2.5μM 2iP and 0.55μM NAA (establishment phase). Multiplication of regenerated shoots was done on the same culture medium and 20 weeks later shoot clusters were separated in order to cultivate them individually and to induce bulb formation at a low temperature (4°C). Although plants and bulbs were obtained for all accessions under tested conditions, response depended on genotype (accession). Three months later, 60% of bulbs that were transferred to soil survived and produced shoots. This protocol allowed the use of the same culture medium for all phases of micropropagation (establishment, multiplication and bulb formation) and for all accessions, which enabled the handling of all materials tested at the same time, as only three subcultures were required over a period of seven months. Primordial leaf obtained from cloves are also able to produce adventitious shoots when cultivated on a medium with 2,4-D (4.5µM), and develop into plants when transferred onto a medium containing picloram (1.4µM) and BA (13.3µM) (Myers and Simon, 1999). Haque *et al.* (2003) developed a protocol for plant regeneration and bulb formation from shoot and root meristems of the cultivar *Bangladesh Local*. Meristems were cultivated on MS medium without growth regulators or containing various concentrations of BA (1-10µM) and NAA (1-5µM). None of the combinations of regulators tested produced a higher response than the one observed in their absence (95.5%). In fact, the presence of these compounds suppressed shoot formation in a directly proportional manner to concentration; 45% of root explants formed adventitious shoots, 60% of which produced bulbs. Although a higher number of buds resulted in shoot formation, the root meristems produced more shoots per explant. Bulbs derived from root meristems were smaller than the ones derived from bud meristems.

On the other hand, Luciani *et al.*, (2006) tested different explants for micropropagation of variety *069*, which were cultivated on BDS medium (Dustan and Short, 1977), supplemented with picloram, 2,4-D and BA. The basal plates and meristems resulted in the highest values of shoot regeneration, and 2, 4-D proved to be better than picloram for inducing callus and shoot formation. By using a combination of 0.25µM 2, 4-D and 4.43µM BA, 100% of explants were able to produce calli, which differentiated into both embryos and shoots. It is worth mentioning that *in vitro* propagation is frequently associated with a process known as hyperhydricity or vitrification, which is a physiological disorder caused by the *in vitro* culture conditions that affects the behavior of regenerated plants. This disorder promotes abnormalities at physiological, anatomical and morphological level, which limit the successful establishment of differentiated plants upon their transfer to greenhouse. Hyperhydric plants have a slow growth rate, thick and deformed stems. Their leaves are translucent, thick and wet (Olmos and Hellin, 1998; Kevers *et al.*, 2004).

A study of biochemical and ultrastructural traits of hyperhydric garlic shoots regenerated *in vitro* was carried out by Wu *et al.* (2009), who observed that organelles such as mitochondria and chloroplasts were compressed against cell wall, in these shoots. In addition, protein content decreased significantly and O_2 and H_2O_2 generation rate increased 45.3% and 63.9%, respectively. Activity of oxidative stress-related enzymes (lipoxygenase, superoxide dismutase, peroxidase, catalase, ascorbate peroxidase) also increased. Moreover, a rise in the level of electrolytes lixiviation was observed, indicating a damage of membrane lipids. Authors concluded that hyperhydric condition of tissues is closely linked to oxidative stress.

Somatic embryogenesis

Formation of structures called embryoids was reported for the first time in 1977. They differentiated from calli obtained from stem tips, bulb leaf discs cultivated in the presence of kinetin (20µM) and IAA (10µM) (Abo-El-Nil, 1977). This response was observed again after a long time when basal plates and floral receptacles were cultivated on a medium containing NAA (1µM) and BA (10µM) (Xue *et al.*, 1991; Al-Zahim *et al.*, 1999). Likewise, Ali and Metwally (1992) induced embryo formation from calli generated from root segments; however, regeneration rate was low. In a different work, Barrueto-Cid *et al.* (1994) established cultures in suspension of the variety *Chonan* starting from calli initiated on MS medium with 5µM 2,4-D, 5µM picloram and 10µM kinetin. Calli were subcultured onto a medium with 4.5µM 2,4-D and hydrolyzed casein before using them for cell suspension cultures. Plant regeneration occurred after transferring cells to a medium containing 77-153µM adenine.

Later, Sata *et al.* (2001) obtained somatic embryos directly from basal sections of cloves of the cultivar *Malepur* grown on White medium (White, 1963) supplemented with 4.5µM 2, 4-D and 2.3µM kinetin. Under these conditions, each explant formed 20 to 25 embryos, which in the presence of higher concentrations of 2,4-D and kinetin turned into masses of hyperhydric tissue. In the same way, Fereol *et al.* (2002) produced somatic embryos and plants of the variety *Rouge de la Réunion* after cultivating calli obtained from root tips on a modified B5 medium supplemented with 2,4-D (0.4µM) and kinetin (2.3µM). Thirty percent of the somatic embryos developed into plants which acclimated successfully to greenhouse conditions. Later, Fereol *et al.*, (2005b) established a protocol for embryo regeneration through suspension cultures by using young leaf sections from cloves of the variety *Morasol*. Embryogenic calli were obtained when explants were grown on B5 medium with 4.5µM 2,4- D and 0.47µM kinetin, then transferred to a modified B5 medium with 2.2, 1.1, 1.1, 0.4µM 2,4-D, IAA, NAA and kinetin, respectively, plus 175mM sucrose and 2mM proline. Calli were maintained on this medium for five months and were later used to initiate suspension cultures in a modified N6 medium supplemented with 1.3 µM 2, 4-D, 0.4 µM BA and 131mM sucrose. Embryo production was induced on a medium with 2.3µM kinetin and 0.4 µM 2, 4- D. Embryos developed into plants, which could produce microbulbs *in vitro*. By using the same explants type and the same culture conditions described above, induction of embryogenic suspension cultures of four garlic cultivars (*Rouge de la Réunion, Morasol, Messidrome* and *Printanor*) was achieved. Ninety percent of calli differentiated into embryos at globular stage after two months of culture. Out of the regenerated embryos, 50% developed into plants that were successfully established in greenhouse. The histological analysis of the culture revealed that regenerated somatic embryos had a unicellular origin (Fereol *et al.*, 2005a).

Meristem culture

Meristem culture technique has been widely used for the production of virus-free clones. Virus elimination through meristem culture is based on the fact that these meristematic cells are free or almost free of virus and therefore plants regenerated from them will also be virus-free (Salomon, 2002). For this purpose, it is recommended to isolate explants of maximum 5mm, although sometimes their size may limit their establishment *in vitro*. Meristem culture enabled virus-free plants to be produced in various regions in the world (Walkey, 1987; Bhojwani *et al.*, 1982; Peña-Iglesias and Ayuso, 1982; Bertacinni *et al.*, 1986; Conci and Nome, 1991). In Slovenia, Eastern Europe, a trial was conducted to eliminate the OYDV in plants of the cultivar *Ptujksi-spomladanski* through thermotherapy and meristem culture. Meristems of 0.3-0.6mm were first cultivated on B5 medium with 1μM IAA and 1μM BA, then transferred to a multiplication medium containing 5μM jasmonic acid and 5μM 2iP. Meristems obtained from plants that had undergone thermotherapy regenerated a lower number of shoots (1.0-2.2) than the non-treated plants (9.3); 90 to 100% plants were found to be free of the OYDV (Ucman *et al.*, 1998).

Sidaros *et al.* (2004) attempted to produce plants of three garlic cultivars (*Chinese, Italian* and *Balady*) through meristem culture and chemotherapy. Chemotherapy was carried out by using virazole [or ribavirin (1-%-D-ribofuranosil-1, 2,4-triazole-3-carboxamide)] in culture medium.

The highest percentage (100%) of virus-free plants was obtained when meristems of 3mm were cultivated on MS medium containing 50mg L-1 virazole. In a different study, thermotherapy, chemotherapy and meristem culture were combined in order to obtain plants of the varieties *Taiwan* and *Chileno* free of potyvirus.Thermotherapy consisted in maintaining regenerated plants from embryos dissected from cloves that showed negative results on an ELISA (Enzyme Linked Immuno Absorbent Assay) for potyvirus during one week at 32°C, followed by two weeks at 36°C, and three weeks at 38°C. Embryos were removed from cloves of these plants and cultivated in presence of 205μM ribavirin. Meristems (0.1-0.5mm) of regenerated plants that showed negative results by ELISA were used to generate new plants. Thermotherapy had a more negative effect on plant survival than meristem culture and chemotherapy. However, thermotherapy proved to be more efficient for virus elimination (60.0 to 70.9%) than meristem culture (64.0%), while chemotherapy was not efficient for potyvirus elimination. On the other hand, 10.7% of plants of the cultivar Taiwan grown in field became reinfected, while the Chileno cultivar showed an 8.9% of reinfection after three consecutive cycles of the crop (Ramírez-Malagón *et al.*, 2006). The use of stems and scape tips of the variety *Red Six Cloves* allows formation of adventitious shoots when cultivated on a medium with NAA (2.6μM) and kinetin (2.3μM). These shoots developed into plants free of the garlic mosaic virus (GMV) 65 days after starting the culture (Ma *et al.*, 1994). Alternative protocols have been developed for generating virus-free plants starting from inflorescence meristems, bulbils or roots, as apart from being virus free they are available in higher numbers than the apical meristems (Appiano and D'Agostino, 1983).

In this way, Verbeek *et al.* (1995) cultivated meristems obtained from cloves and bulbils (0.15-1.00mm), 71-71% of which regenerated plants; 38% of explants obtained from cloves and 25% of the explants obtained from bulbils were found to be virus-free. In addition, it was observed that meristems smaller than 0.4mm failed to produce shoots. Similarly, Ebi *et al.* (2000) established a system for elimination of mite-borne mosaic virus using meristems (0.2-0.4mm) obtained from bulbils. These meristems produced plants after being cultivated on MS medium with 5.4μM NAA. The immunoblot assay indicated that several of regenerated plants were virus-free. Senula *et al.*, (2000) obtained plants of 87 accessions free of the viruses OYDV, LYSV, GCLV, SLV and MbFV by cultivating meristems of 0.3-0.8mm originated from bulbils. OYDV and LYSV were eliminated in 85-95% of the regenerated plants. Addition of ribavirin to culture medium reduced regeneration potential, but increased virus elimination. Later, Xu *et al.* (2001) regenerated virus-free plants from meristems obtained from inflorescences of nine lines. Explants were cultivated on B5 medium containing 0.22µM BA and 0.3mM adenine. By using this protocol it was possible to obtain 50-90% plants free of the OYDV, 70-100% plants free of the LYSV and 60-80% free of the SLV. Production of bulbs from virus-free plants was higher than from infected plants. Four to five years would be necessary to obtain virus-free elite seeds that can be established in the field. The economical analysis indicated a net profit of 50.3 to 244.5% (depending on the genotype) for garlic seed producers.

Somaclonal variants

Tissue culture tools such as *in vitro* selection, embryo rescue, somatic hybridization, genetic transformation and somaclonal variation can be used to generate crop variation. Larkin and Scrowcroft (1981) defined somaclonal variation as the phenotypic variation seen in plants regenerated *in vitro* with respect to the original plant. At genetic level, somaclonal variation can be brought about by various DNA changes that include: (a) chromosomal rearrangements, (b) aneuploidy, (c) poliploidy, (d) modification of gene expression by methylation, amplification, inactivation or reactivation, (e) genetic conversion, (f) somatic recombination, (g) transposons movement, (h) genes mutations, etc. (Scowcroft, 1984; Peschke and Phillips, 1991). Various experiments have been undertaken in garlic in order to generate somaclonal variants that could be used in its improvement. For instance, Novak (1983) treated meristems 0.5-0.7mm with a solution of colchicine (3g L-1) to induce polyploidy. Meristems were treated in two different ways: (1) cultured for 7 days on a solid medium with colchicine, and (2) cultured for 2 days in a liquid medium with colchicine. The latter treatment proved to be more effective. By using this experimental strategy, 35% of regenerated plants were found to be tetraploid, and 14% chimaeras with diploid and tetraploid cells. Dolezel *et al.* (1986) pointed out that in garlic probability of generating somaclonal variation is higher when disorganized growth occurs, specially for longer periods of time. Similarly, plants generated from old explants or with a high level of differentiation have an increased possibility of suffering this type of variation. The cultivar *Frankon* is resistant to populations of the nematode *Ditylenchus dispsaci* found in Israel, but produces bulbs and cloves of small size and has low yield. A protocol for *in vitro* regeneration was developed to generate somaclonal variants that can produce bulbs with commercial traits. This protocol consisted in cultivating basal plates on BDS medium supplemented with 2.2µM 2,4-D and 2.3µM kinetin to induce callus formation, which were then transferred to a medium containing 9.5µM kinetin and 11.7µM IAA to develop adventitious shoots. Once shoots produced roots, they were transplanted to soil in greenhouse. Assessment of plant characteristics and bulb development revealed that there was variation in bulb size and color, and also in number and size of cloves per bulb compared to the original plant type, which indicates that this could represent a promising material for generation of improved somaclonal variants (Koch and Salomon, 1994).

In a different study, Madhavi *et al.*, (1991) compared ability of calli (organized and disorganized) and bulbs to produce sulphur compounds like alliin. Apart from finding differences in sulphur level compounds, they also observed changes related to proteins, aminoacids, carbohydrates and enzymes. For instance, specific activity of the enzyme alliin lyase in callus was 50% lower than in bulb, while enzymes amino transferase, malate dehydrogenase, polyphenol oxidase, peroxidase and alkaline phosphatase displayed a higher activity than in bulb. Incorporation of precursors to volatile fraction was also higher in organized calli. The somaclone *118.15*, derived from variety *Rosado*, which is grown commercially in Argentina, possesses agronomically desirable traits. A cytological and phenotypical study revealed that both somaclone and original type had the same chromosome number (2n = 16). In addition, some individuals of this clone contained polyploid, aneuploid and haploid cells, probably derived from processes such as endomitosis, nuclear fusion or homologue chromosome pairing in somatic cells. Binucleate cells and differences in length of chromosome pairs were also observed. Plants of clone *118.15* were taller, had a higher diameter of pseudostem and produced bulbs with less, but bigger cloves than the original type (Ordoñez *et al.*, 2002).

El-Aref (2002) regenerated plants of cultivar *Balady* from leaves and roots cultivated first on BDS medium with 2,4-D (4.5μM) and kinetin (9.5μM) for callus initiation, then in presence of 9μM BA and 5.5μM NAA for plant regeneration. These plants formed bulbs which were planted in soil to generate new plants. Isoenzyme analysis showed that 9 out of 29 regenerated plants were different from original plant with regard to some of the enzymes analyzed (phosphatase acid, alcohol dehydrogenase, malate dehydrogenase, esterase). Parental bands were found in all plants, and 5 new bands were observed in 31% of them. Esterase and acid phosphatase displayed a higher polymorphism than alcohol dehydrogenase and malate dehydrogenase. Later, Mukhopadhyay *et al.*, (2005) studied chromosome stability in plants of cultivar *Rossete* generated from callus. It was observed that plants generated from calli iniciated on solid MS medium with 2,4-D (9μM) and kinetin (0.93μM) and sub-cultured in liquid medium with NAA and kinetin exhibited chromosome stability, while the ones grown solely on the initiation solid medium contained hypo- or hyperdiploid cells along with the diploid ones. Frequency of aneuploid cells (2.2-48.9%) increased with callus age.

Badria and Ali (1999) identified somaclonal variants regenerated from calli obtained from root meristems cultivated on MS medium with kinetin, 2,4-D and IAA. The somaclonal variants formed bulbs without division in the first generation, and displayed normal phenotype in the following generation. After four cycles in field, somaclonal variants that exhibited significant differences of bulb characteristics were found. Cytogenetical analysis revealed that these somaclonal variants had the same chromosome number as original plants. Quantification of alliicin production showed that some somaclones contained three times more of this compound (14.5mg g-1) than control plants (3.8mg g-1). Authors suggest that this technique could be useful for improving alliicin content in garlic.

In vitro germplasm conservation of garlic

Conservation of valuable garlic accessions involves their yearly cultivation, as bulbs cannot be stored for long periods of time (6 months at -3°C). This practice is expensive as it requires land use and manpower. Moreover, germplasm grown in field is exposed to environmental changes, pests and diseases that reduce its quality (Panis and Lambardi, 2006). The majority of plant germplasm is stored in seed repositories at temperatures between -15 and -20°C. However, for species whose seeds are recalcitrant (they cannot be dried to humidity levels low enough for storage) or for species that do not produce seeds, like garlic, slow growth storage and cryopreservation are the only tools for conserving them. Slow growth storage involves a condition that maintains tissue growth at a minimum and it allows the medium term storage of material (Botau *et al.*, 2005). It is based on using organs cultured *in vitro* at 4°C and 10 to 15°C for plants growing in temperate and tropical areas, respectively (Keller *et al.*, 2006).

Cryopreservation is one of the most commonly used tools for germplasm conservation because it requires minimum amount of space and maintenance. In addition, it reduces loss of accessions by contamination, human errors and somaclonal variation which may occur during slow growth storage (Panis and Lambardi, 2006; Sakai and Engelman, 2007). This technique involves the use of liquid nitrogen (which has a freezing temperature of -196°C) for long-term storage of plant material. At this temperature, the majority of biochemical and physical processes are effectively stopped. Cryopreservation is only useful if formation of intracellular ice crystals does not take place, as they may cause irreversible damage to the cell membrane (Panis and Lambardi, 2006). Ice formation without an extreme reduction of the cell water content can only be avoided by a process known as vitrification, in which an aqueous solution turns into an amorphus and glassy state (Sakai, 2000). This procedure substitutes cellular dehydration that occurs during freezing by a reduction of cell water content that is achieved by treating tissues prior to the cooling process with highly concentrated solutions (PVS2, PVS3) containing glycerol, ethylen glycol, dimethylsulfoxid and sucrose, or by air drying (Sakai and Engelman, 2007).

Some countries are already making use of the previously mentioned techniques to conserve their valuable germplasm. For example, China possesses *in vitro* virus-free germplasm banks and their respective databases. These banks have been established taking into account factors such as genotype, culture medium components, light conditions, temperature of incubation rooms, etc. Moreover, studies have been carried out in order to optimize conservation conditions. In this respect, Xu *et al.*, (2005) studied the behavior of six genotypes during their *in vitro* storage. They found out differences in conservation period, depending on genotype. Two genotypes, namely *Cangshan Zaotai* and *Tianjin Baodi* could be stored for 25 months and had a survival percentage of 100%. They also observed that shoots grown at low temperatures on B5 medium with 1.3-2.2μM BA, 0.5-1.6μM NAA and 38-115μM of abscisic acid could be conserved for a longer time. Evaluation of stored material indicated that it was genetically stable and 0.1-0.2% of it became infected with virus. Similarly, the Institute of Plant Genetics and Crop Plant Research at Gatersleben, Germany, one of the biggest gene banks in Europe, possesses a collection of 3039 accessions of species of the genus *Allium*, including the European garlic core collection. Before storage of germplasm, virus elimination is undertaken through meristem culture, and then either slow growth storage is carried out for 12 months at 2 and 19°C or cryopreservation-vitrification using a mix of glycerol and sucrose 1:1 as cryoprotectant inside the aluminum foil (Keller *et al.*, 2006).

In the United States, investigations have been carried out for cryopreservation of garlic accessions at the Western Regional Plant Introduction Station, Pullman, Washington. In this respect, Ellis *et al.*, (2005) tested two vitrification solutions to cryopreserve 12 accessions. Shoot tips excised from cloves were treated with the vitrification solutions 2 (PVS2; 15% DMSO, 15%

ethylene glycol, 30% glycerol, 0.4M sucrose) and 3 (PVS3; 50% sucrose, 50% glycerol). Eleven out of the 12 accessions could be successfully cryopreserved by using vitrification solutions 2 and 3 as cryoprotectants. Cryopreservation resulted in better regrowth of 7 and 3 accessions when PVS2 and PVS3, respectively, was used. Only one genotype displayed good recovery in both solutions, which indicated that response depended on genotype.

In Romania, investigations were carried out to establish conditions for slow growth of Romanian garlic landraces. Shoots regenerated from leaf discs were cultured under a slow growth condition that consisted in a medium lacking sucrose at 16°C and under normal conditions on a medium with 3% of sucrose at 24°C. After four weeks, shoots cultivated under slow growth conditions reduced their growth rate by 42%. Shoots grown on sucrose-free medium displayed shorter internodes, but had higher foliar surface, and shorter roots. Some landraces proved to be more sensitive to lack of sucrose than to low temperature. Differences in response to growth conditions tested were also observed among the different genotypes under study. The developed protocol allowed medium-term preservation of landraces under study (Botau *et al.*, 2005).

Other investigation groups have focused their efforts on defining the best conditions for slow growth storage or cryopreservation. For example, Makowska *et al.* (1999) studied the response of apexes from different sources (cloves and bulbils) to cryopreservation after being treated with vitrification solutions. They found out that after freezing, apexes treated with PVS2 solution (30% glycerol, 15% ethylene glycol, 15% DMSO, 0.4M sucrose) restored their growth in a higher proportion compared to the ones that had been submerged in solution PVS3 (40% glycerol and 40% sucrose). A higher number of apexes excised from big bulbils restored their growth, while the ones dissected from small bulbils failed to regrow. On the other hand, apexes obtained from cloves had higher survival rate than the ones obtained from bulbils.

Later, Sudarmonowati (2001) tested different vitrification solutions in an attempt to define the most suitable method for cryopreservation of embryogenic calli of cultivar *Lumbu Hijau*. Calli with embryos at globular stage were precultured on MS medium with 9μM BA and 0.4M sucrose for 1 to 7 days. Next, they were submerged in MS medium with 2M glycerol and 0.4M sucrose (loading solution) in order to be later exposed to different vitrification solutions for various periods of time (5 to 60 min.). Calli were placed in cryotubes with one drop of vitrification solution, then plunged into liquid nitrogen for 30 minutes. After freezing, calli were first plated on MS medium with 5μM 2iP, 2.3μM kinetin and 0.4M sucrose, and later on the same medium containing a lower amount of sucrose. Of the three vitrification solutions tested, the one that contained a mix of glycerol (22%), ethylene glycol (17%), propylene glycol (17%) and DMSO (7%) proved to be better for calli conservation. The highest calli survival percentage was 30%, which indicated that it was necessary to improve the tested methodology.

Hassan *et al.* (2007) established slow growth cultures of two garlic varieties grown in Egypt (*Balady* and *Seds 10*). They cultivated bulblets on MS medium with 0.35M sucrose, 5g L-1 charcoal and 0.04μM BA, then on a medium with different concentrations of sorbitol and sucrose (0.1, 0.2 and 0.4M). Cultures were incubated in darkness at 4°C for their conservation.

Bulblets did not develop neither shoots nor roots under these conditions during the first three months. Addition of 0.1M sorbitol to culture medium delayed growth of shoots and roots of cultivar *Balady* to 6, 12 and 18 months, while sucrose (0.1 or 0.2M) had the same effect on bulblets of cultivar *Seeds*. The survival rate was of 100% after 18 months of maintaining cultures under these conditions.

Genetic engineering for garlic improvement

Although several methods are available to introduce DNA into plant cells, most of them have been developed using the bacterium *Agrobacterium tumefaciens* (direct method) or biolistic (indirect method) as a vehicle. *A. tumefaciens* is a soil bacterium having natural ability to transfer part of its DNA to plant cells of various species causing formation of crown gall tumors (Hooykaas and Schilperoot, 1992). This ability is conferred by the Ti plasmid (tumor-inducing), which contains a region called T-DNA, that is transferred to the host cell with helping of virulence genes, also present in Ti plasmid. The T-DNA contains genes that are involved in production of cytokinins (2-isopentyl-AMP) and auxins (IAA), which are responsible for tumor formation (Leemans *et al.*, 1982; Barry *et al.*, 1984). Genetic manipulation of this plasmid has resulted in the replacement of genes contained in the T-DNA of the wild strain by genes that confer desirable traits to transformed plants (Christou, 1996).

Starting from the successful transformation of monocotyledonous plants, such as maize by using the biolistic method (McCabe *et al.*, 1988; Fromm *et al.*, 1990), this has become one of the most used systems for gene transfer. This method consists in bombarding target cells with DNA-coated gold or tungsten microparticles accelerated to very high speeds by a gene gun, which allows them to cross the cell walls. Although there are various types of gene guns, the PDS1000 helium designed by Dupont has been the most widely used, especially for transformation of monocotyledonous plants (Vain *et al.*, 1993; Christou, 1995). A great variety of genetic transformation protocols have been developed, but this technology has not been applied with the same efficiency in every species. The species of the genus *Allium* represent an example in this respect, particularly garlic, for which only a small number of publications are available.

Biolistic approach

In 1998 a protocol for garlic transformation was reported for the first time by Barandiaran *et al.* (1998) who bombarded leaf tissue, immature bulbs, cloves and callus of the cultivar *Morado de Cuenca* with four constructs (pDE4, pCW101, pActl-D and pAHC25). Out of these vectors, the one carrying the reporter *uidA* gene (*gusA*) (coding for %-glucuronidase) under control of the promoter 35S from cauliflower mosaic virus (CaMV35S) and the terminator of the nopaline synthase gene (NOS), allowed expression of the *uidA* gene in 43.3% of leaf explants, 76.7% of bulbs, 23.3% of clove tissue and 13% of calli. Transitory expression of the *uidA* gene could only be detected after treating tissues with a nuclease inhibitor (aurintricarboxylic acid). However, regeneration of transgenic plants could not be achieved by using this protocol. Similarly, Myers and Simon (1998) bombarded cell suspensions of *RAL27* clone with *gusA* and *nptII* (conferring resistance to kanamycin) genes which were under control of CaMV35S and NOS promoters, respectively. After 14-16 weeks on selection medium, shoots were regenerated on calli. Incorporation of *gus* and *nptII* genes into garlic transgenic plants was confirmed by PCR assays.

Later on Ferrer *et al.* (2000) used this approach to introduce the reporting gene *uidA* and the selection gene *bar*, which codes for N-acetyl-transpherase, into leaf tissue, basal plate discs and embryogenic calli of cultivar *Moraluz*. The *uidA* and *bar* genes were under control of CaMV35S and maize ubiquitin (Ubi) promoters, respectively. Maximum expression of *uidA* was observed in calli and leaves. In a different investigation, Sawehel (2002) developed a transformation system using calli derived from immature cloves of cultivar *Giza 3*. Calli were bombarded with the plasmid pBI22.23 containing the *hpt* gene (coding for the enzyme hygromycin phosphotranspherase that confers resistance to antibiotic hygromycin), and the reporting gene *gusA*. Calli had been previously treated with aurintricarboxilic acid to inhibit activity of endogenous nucleases. Southern blot assays and histochemical analysis proved that this system allowed the transfer, expression and stable integration of transgenes into the garlic genomic DNA.

At the same time, Park *et al.* (2002) obtained transgenic plants resistant to herbicide chlorsulfuron after bombarding calli of cultivar *Danyang* with the plasmid pC1301-ALS, which contains *gus, hpt* and *als* (coding for acetolactate synthase) genes, under control of the promoter CaMV35S. Out of 1900 calli, 12 grew and regenerated plants resistant to chlorsulphuron (3mg L-1), which formed bulbs and reached maturity. PCR, Southern blot and Northern blot assays confirmed the expression and integration of transgenes into the genome.

In a different work, Robledo-Paz *et al.* (2004) established a transformation protocol using embryogenic calli derived from root tips of cultivar *GT96-1*. Calli were bombarded with the plasmid pWRG1515 containing *hpt* and *gusA* genes, both under the control of the promoter CaMV35S, and the 3′ region of the *nos* gene. Putative transgenic calli were identified after four months of culturing them on a medium containing hygromycin (20mg L-1), and later developed into plants. Molecular (Southern blot) and (GUS) analysis confirmed transgenic nature of regenerated plants. Transformation efficiency was of 2.2 clones per fresh weight gram of bombarded callus.

Agrobacterium tumefaciens mediated approach

Kondo *et al.* (2000) were the first to achieve the establishment of a transformation protocol in garlic using *A. tumefaciens*. They infected morphogenetic calli with the strain EHA101 carrying the plasmid pIG121, which in turn contained *nptII, hph* and *uidA* genes under control of the promoter CaMV35S. By using this protocol it was possible to regenerate

15 transgenic plants from 1000 inoculated calli grown on a selective culture medium for five months. Zheng *et al.* (2004) presented a transformation system that apart from producing plants resistant to antibiotics or herbicides, also enabled introduction of genes for resistance to insects. Inoculation of calli of three European cultivars was undertaken using the strain AGLO carrying four different plasmids containing *gusA* and *hpt* genes, and also *cry1Ca* and *HO4* genes from *Bacillus thuringiensis,* which confers resistance to the insect *Spodoptera exigua.* The highest transformation frequency (1.47%) was achieved with the cultivar *Printanor* and the plasmid pPB34. Of regenerated plants, only the ones that integrated the *cry1Ca* gene had

a good growth under greenhouse conditions and had the ability to form bulbs. These plants were totally resistant to *Spodoptera exigua* in bioassays carried out *in vitro*. Later on, Eady *et al.* (2005) inoculated immature embryos with the strain LBA4404 carrying the vector pBIN *m-gfp-ER* containing the gen *gfp* (encoding for the green flourescent protein) and the gene *nptII*. Out of the 3200 infected embryos, only two transgenic plants (0.06%) were regenerated. Khar *et al.* (2005) studied the transitory expression of the reporter gene *gusA* in two garlic cultivars after infecting them with *A. tumefaciens* strain carrying two plasmids. Plasmid pCAMBIA 1301 induced a higher transformation frequency (7.4%) than plasmid pTOK233 (4.1%). Genes conferring resistance to fungi are still not being commercially used for fighting diseases caused by these phytopathogens. In garlic, some workers have incorporated chitinase and glucanase genes in an attempt to confer resistance to the fungus *Sclerotium cepivorum*. These experiments revealed that regenerated transgenic plants were not totally resistant to the fungus, but displayed a delay in the infection speed.

Molecular markers deployed in garlic improvement

Genetic diversity plays an important role in plant breeding either to exploit heterosis or to generate productive recombinants. The choice of parents is of paramount importance in breeding program. So, the knowledge of genetic diversity and relatedness in the germplasm is a pre-requisite for crop improvement programs. The precise information on the nature and degree of genetic diversity present in garlic collections from its principal areas of cultivation would help to select for evolving superior varieties. For the genetic amelioration of this crop, diverse genotypes from the existing germplasm should be selected and used in further breeding program. Assessment of genetic diversity at the molecular level is more meaningful than at the phenotypic level as the later involves data on morphological traits which are environmental dependent. Though, they significantly contribute towards phenotypic variation but cannot be accurately phenotyped. So the study of polymorphism is best done at the level of arrangement of nucleotide bases in DNA, the primary source of all biological information. At this level, even seemingly identical accessions could display enormous differences, if only we could employ appropriate DNA profiling techniques.

Molecular tools provide valuable data on diversity through their ability to detect variation at the DNA level. Identification is of fundamental importance in diversity studies in a variety of different ways. For evaluation of species diversity, it is essential that individuals can be classified accurately.

AFLP (Amplified fragment length polymorphism) technique has also been used to characterize garlic (Lampasona *et al.*, 2003; Volk *et al.*, 2004). Ipek *et al.* (2003) reported that polymorphic AFLPs are abundant in garlic and demonstrated genetic diversity among closely related clones which would not be differentiated with RAPD markers and isozymes. They concluded that AFLP is an additional tool for fingerprinting and detailed assessment of genetic relationship in garlic.

According to Volk *et al.* (2004) 64% of the U.S. National Plant Germplasm System's garlic collection held at the Western Regional Plant Introduction Station in Pullman, Washington, USA, and 41% of commercial garlic collections were duplicates. Detection and elimination of those duplicated accessions in germplasm collection can reduce maintenance costs significantly.

RAPD analysis has been successfully applied to interspecific classification of number of crop species, in the genus *Allium*. Close relations of *A.cepa* have already been analyzed with RAPD markers by Wilkie *et al.* (1993).

Maab and Klaas (1995) used isozymes and RAPD markers to clarify classification scheme, to see the extent of genetic diversity in different groups and to gauge the effect of domestication, adaptation and the spread to different climatic condition on the loss in diversity. Based on the result they proposed an interspecific classification of garlic into four major groups and *A.longicuspis* to be the progenitor of *A.sativum*. RAPD technique is employed to assess genetic variation in garlic and examine relationship between cultivated garlic and the wild progenitor *A. longicuspis*.

Molecular marker studies in garlic have not been very exhaustive because of its asexual nature though RAPD technique has been widely used for characterization of Australian garlic by Bradley *et al.* (1996).

Shasany *et al.* (2000) classified 21 Indian garlic genotypes along with Argentinean clones and reported a diversity of 60% by RAPD analysis whereas a diversity of 20% only was assessed by morphological features. Xu *et al.* (2001) confirmed that the RAPD technique is useful and efficient tool in classification and identification of garlic genetic resources.

Etoh *et al.* (2001) studied RAPD variation of garlic clones from the primary center of origin(Central Asian)as compared with that of 30 garlic clones from the western most area of distribution(Iberian Peninsula) and concluded that the genetic similarity among the Iberian clones were high and among the Central Asian clones were comparatively low.

Kamenetsky *et al.* (2005) worked on the analysis of diversity in fertility potential and organo-sulphur compounds among garlics from central Asia. Panthee *et al.* (2006) worked on diversity analysis of garlic germplasm available in Nepal based on morphological characters. They used 179 garlic accessions for this study.

Low genetic diversity was detected among garlic (*Allium sativum* L.) accessions using Random Amplified Polymorphic DNA (Mario Paredes *et al.* 2007). The results showed that the clones analyzed had a genetic similarity rate of 94%. In addition, 70% of them were clustered in one major group. However, in spite of that situation several clones have different agronomic characteristics.

Genetic diversity studies of Brazilian garlic cultivars and quality control of garlic-clover production was carried out by Buso *et al.* (2008). Results revealed that the number of markers was efficient and sufficient to obtain a coefficient of variation of 10%. Similarity varied between 16 and 98% and cluster analysis showed that, in general, genetic similarities correlate with morphological characters of the cultivars and production cycle variation.

RAPD technique is useful and efficient tool in classification and identification of garlic genetical resources. Molecular markers associated with the trait of pollen fertility using RAPD have been identified by Etoh *et al.* (2001). Genetic diversity of eight selected Argentinean garlic clones was investigated at DNA level using AFLP procedure by *S. Garcia, Lampasona* and *Burba* (2003).

Maintenance of redundant garlic (*Allium sativum* L.) accession is expensive due to the necessity of yearly regenerating garlic accessions in germplasm centers. Therefore, rapid characterization of garlic accession is important for avoiding duplicate genotypes. For this purpose Ipek *et al.* (2008) developed several locus-specific polymerase chain reaction(PCR) based DNA markers and tested them for the characterization of garlic clones that were previously analyzed using amplified fragment length polymorphism (AFLP) and randomly amplified polymorphic DNA (RAPD) markers. Among 25 garlic clones, 4 locus-specific markers generated 24 polymorphic alleles.

In Indian context, except for one or two reports on molecular characterization of Indian garlic germplasm by RAPD technique, not much work has been done. A lot of efforts towards diversity assessment of Indian garlic are important. Beside this with the reports of flowering garlic, it has become imperative on our part to import those lines and start breeding work on introduction, adaptation and development of segregating families for further breeding work towards yield improvement and resistance breeding. Study of genetic diversity among garlic clones collected from Czech Republic done by Kucera *et al.* (2007) using AFLP, phenotypic descriptors and S-amino acid levels.

Analysis of genetic diversity among Indian garlic cultivars and breeding lines using RAPD markers was undertaken by Khar *et al.* (2008). They used 23 commercially released varietics and breeding lines of garlic along with onion as an out- group. The UPGMA dendrogram was constructed using Jaccard's similarity matrix discriminated all the genotypes into four clusters. Cluster 1 comprised of 14 genotypes with a similarity index of about 89%. Cluster II comprised of only one line, SKAUG-151 that had a similarity index of 62% with cluster I. Cluster III comprised of 8 garlic genotypes at a distance of about 52%. Variation amongst the garlic accessions was about 48%, whereas, with onion it showed a dissimilarity of 90%.They concluded that the clustering pattern of garlic genotypes in the present study discriminated them as per the variation requirement for bulb formation.

Singh *et al.* (2011) carried out RAPD profile based grouping of garlic germplasm with respect to photoperiodism. Forty garlic genotypes were evaluated using RAPD markers to determine the genetic relationships among genotypes showing morphological variations.

Inter simple sequence repeats (ISSR) primers target simple sequence repeats (microsatellites) that are abundant in eukaryotic genome and evolve rapidly, but they do not require any prior knowledge of DNA sequence for primer design reported by Fang and Roose (1997). Zietkiewicz *et al.* (1994) added that ISSR can rapidly differentiate closely related individuals, Fang and Roose (1997), Dagani *et al.* (2003), Salhi Hannachi *et al.* (2004), Salhi Hannachi *et al.* (2005) and Okpul *et al.* (2005) have reported that ISSR markers have been successfully used to assess genetic diversity among closely related cultivars which were difficult to distinguish with other molecular markers.

Very less work has been carried out in Alliums genetic diversity using ISSR (Inter Simple Sequence Repeat) markers. The genetic diversity of 32 onion (*Allium cepa* L.) germplasm resources were analyzed by ISSR(inter-simple sequence repeat) Primers by Xu *et al.* (2007) which indicated a rich genetic diversity of onion (*Allium cepa* L.) germplasm resources. According the result of UPGMA cluster analysis indicated that ISSR technique is useful in revealing the genetic diversity and genetic relationship among onion germplasm materials, providing a scientific basis for genetic breeding and patents selection. Gantait *et al.* (2010) worked on the determination of genetic integrity of long-term micropropogated plantlet of *Allium ampeloprasum* L. using ISSR markers. Detection of genetic integrity for *in vitro* regenerated clones was carried out using 10 ISSR primers among which 4 primers produced a total number of 336 distinct and scorable bands with an average of 7 bands per primers where the other 6 primers were not reproducible.

ISSR Fingerprinting was used to assess genetic diversity of Tunisian garlic population was undertaken by Jabbes *et al.* (2011). They were the first one to use ISSR primers to screen diversity in garlic and concluded that Factor analysis of distance table (AFTD) did not classify accessions on the base of geographical origin or morpho-physiological characters but confirmed the appurtenance of analyzed accessions to *sativum* botanical subspecies.

Very recently Cunha *et al.* (2012) have developed new microsatellite markers for garlic. A new set of SSR markers for garlic, an important medicinal spice, was developed to aid studies of genetic diversity and to define efficient strategies for germplasm conservation. The new SSR markers have the potential to be informative tools for genetic diversity, allele mining, mapping and associative studies, and in the management and conservation of garlic co.

Two most recent studies have appeared, the one deals with Indian garlics (Nair, 2012). This author has used 160 RAPD primers of which only 40 primers gave amplification and 20 primers gave good amplification, 189 bands were generated and 65 were polymorphic (34.39% polymorphism). OPE 08 showed maximum polymorphism (63.63%), OPM 09 showed minimum polymorphism (12.5%). Further, of the 100 ISSR primers screened only 27 gave amplification; of which 14 were good. In all 123 bands were generated and 45 were polymorphic (36.58% polymorphism). Primer 825 (62.5%), 808 (60%) showed maximum polymorphism whereas primer 834 showed least polymorphism (20%). The UPGMA dendrogram constructed using Jaccard's similarity matrix of RAPD data and discriminated germplasm in two major groups: Group I had 4 Sub Clusters (Ia, Ib, Ic and Id), Ia and Ib showing 100% similarity, while Ic and Id showed 85% similarity. Group II comprised 1 Sub Cluster and showed 89% similarity. Group 1 and 2 showed 74% similarity when compared. ISSR data discriminated all the germplasm into 2 major groups. Group 1 comprising 5 Sub Clusters (Ia, Ib, Ic, Id and Ie). Ia, Ib and Ic showed 100% similarity while Id and Ie showed 91% similarity. Group 2 consisted of 2 clusters and showed 90% similarity among them. Group 1 and 2 showed 79% similarity between them. When RAPD and ISSR similarity matrix compared they showed genetic similarity (GS) value ranging from 0.57 to 0.33.

In summary, of the 96 morph types, 16 types were identified as genetically divergent from the rest. Total number of bands varied from 186 with RAPD and 112 with ISSR .There was 100% similarity with both the primers. Banding pattern was clear, consistent and excellent. The studied morphotypes displayed enormous morphological diversity for several characters, but the molecular markers exhibited low genetic diversity in garlic germplasm .Total lack of sexual reproduction could be the reason for low genetic diversity. Garlic crop was domesticated nearly 5000-6000 years ago, the extent of genetic or chromosomal mutations did not appear to have contributed significantly for creating genetic variability. From amongst 96 types, 16 types are genetically distinct for various agronomic traits: Purple color - 8;

Single clone and red color -4; Red bulbs -4. It is inferred that most morphs are genetically related/ there are duplications in the germplasm bank at Pune.

High numbers of bands were detected indicating polymorphism but had low genetic diversity. Most of the polymorphic bands showed low frequency. It is proposed that two selection pressures – one for morphology and second for genetic traits worked independently in evolution of garlic's morphotypes.

Jo *et al.*, (2012) have classified the genetic variations of garlic. Seven selected simple sequence repeats (SSRs) revealed a total of 37 alleles across 120 garlic accessions, with an average of seven alleles per locus. The values for observed heterozygosity ranged from 0 to 0.99 (mean = 0.71). The average genetic diversity and polymorphic information content values were 0.586 and 0.518, respectively. Based on the phylogram constructed the garlic accessions were clustered into four main groups (G1–G4) in the phylogram. Group1 consisted of accessions of 'Aomori', Group 2 consisted of 64 accessions, Group 3 consisted of 25 accessions, and Group 4 consisted of 20 accessions. These workers correlated genetic diversity with geographical region. There may have been local selection pressure and differences in adaptability of the garlic to different geographical conditions. All of the tested loci deviated significantly ($P < 0.01$) from Hardy-Weinberg equilibrium. Thus, a number of disturbances occurred in the garlic population tested, including natural selection. These investigations help in explaining the genetic relationships and population structure of garlic accessions.

Although plant regeneration has been achieved from different explants types, use of root tips has advantages over other explants due to their virus-free condition and to their availability in a relatively high number (30 or more per clove). Moreover, roots developed from bulbs obtained *in vitro* can also be used for tissue culture (Robledo-Paz *et al.*, 2000). On the other hand, production of virus-free plants via meristem culture combined with thermotherapy and chemotherapy can reduce losses caused by phytopathogens, even when propagation of virus- free material is relatively expensive (Salomon, 2002). Tissue culture has also been applied to the establishment of germplasm banks in various parts of the world where valuable garlic collections are maintained for medium (slow growth) and long term (cryopreservation).

Techniques such as somaclonal variation and genetic engineering could play an important role in the genetic improvement of garlic because they generate genetic variability. However, the somaclonal variants with commercial potential are scarce, and further experiments are necessary to identify the optimal explant type and the culture conditions that enable formation of somaclones (Novak, 1990). In addition, although there have been advances in the field of genetic transformation in garlic, more investigations are required to establish reproducible and efficient protocols. This task will require selection of suitable target cells for inoculation with *Agrobacterium* or biolistics (e.g. embryogenic calli) (Myers and Simon, 1998b; Robledo-Paz *et al.*, 2004), strategies for transgene expression, a suitable selection method and efficient protocols for plant regeneration (McEloy and Brettell, 1994; Hansen and Wright, 1999). Furthermore, molecular markers will be key pieces in phylogenetic and taxonomic studies (Maab and Klaas, 1995) and germplasm conservation (Ipek *et al.*, 2008). Moreover, they will be used for detection of somaclonal variants (Al-Zahim *et al.*, 1999), fertile genotypes (Etoh and Hong, 2001), disease resistant genotypes (Nabulsi *et al.*, 2001) and clones producing compounds of economical importance (Ovesná *et al.*, 2007) which can be used for improving this important crop plant.

REFERENCES

Abdoli M, Habibi-Khaniani B, Baghalian KS, Shahnazi RH, Naghdi BH (2009) Classification of Iranian garlic (*Allium sativum* L.) ecotypes using RAPD markers. *J Med Plants* 8: 45-51

Abo-El-Nil MM (1977) Organogenesis and embryogenesis in callus culture of garlic. *Plant Sci Lett* 9: 259-264

Abrams GA, Fallon MB (1998) Treatment of hepatopulmonary syndrome with *Allium sativum* L. (garlic): a pilot trial. *J Clin Gastroenterol* 27: 232-235

Ali A, Metwally EE (1992) Somatic embryogenesis and plant regeneration as a tool for garlic improvement. *Egypt J. Appl Sci* 7: 727-735

Al-Zahim MA, Ford-Lloyd BV, Newbury HJ (1999) Detection of Somaclonal Variation in Garlic (*Allium sativum* L.) Using RAPD and Cytological Analysis. *Plant Cell. Rep* 18: 473-477

Al-Zahim MA, Newbury HJ, Ford-Lloyd BV (1997) Classification of Genetic Variation in Garlic (*Allium sativum* L) Revelated by RAPD. *HortSci* 32: 1102-1104

Appiano A, D'Agostino G (1983) Distribution of Tomato Bushy Stunt Virus in Root Tips of Systemically Infected *Gomphrena globosa. J. Ultrastructural Re.,*85: 239-248

Augusti KT (1990) Therapeutic and Medicinal Values of Onion and Garlic. In: *Onions and Allied Crops*, Rabinowitch, H. D.,Brewster, J. L., Eds., CRC Press, Boca Raton, Fl., U.S.A. III: 94-108

Ayabe M, Sumi S (1998) Establishment of a Novel Tissue Culture Method, Stem-Disc Culture, and Its Practical Application to Micropropagation of Garlic (*Allium sativum* L.). *Plant Cell. Rep* 17: 773-779

Ayabe M, Taniguchi K, Sumi S (1995) Regeneration of Whole Plants from Protoplasts Isolated from Tissue-Cultured Shoot Primordial of Garlic (*Allium sativum* L.). *Plant Cell. Rep* 15: 17-21

Badria FA, Ali AA (1999) Chemical and Genetic Evaluation of Somaclonal Variants of Egyptian Garlic (*Allium sativum* L.). *J. Med. Food* 2: 39-43 **Barandiaran, X, Di Pietro, A, Martin, J** (1998) Biolistic Transfer and Expression of a *uidA* Reporter Gene in Different Tissue of *Allium sativum* L. *Plant Cell. Rep* 17: 737-741

Barandiaran X, Martín N, Alba C, Rodríguez-Conde MF, Di Pietro A Martin J (1999) An efficient Method for the *in vitro* Management of Multiple Garlic Accessions. *In Vitro Cell. Dev. Biol.- Plant* 35: 466-469

Barrueto-Cid L, Piedrabuena AE (1994) Regeneration of Garlic Plants All*ium sativum* L. cv. "Chonan" Via Cell Culture in Liquid Medium. *In Vitro Cell. Dev. Biol.-Plant* 30: 150-155

Barry GF, Rogers SG, Fraley RT, Brand L (1984) Identification of a Cloned Cytokinin Biosynthetic Gene. *Proc. Natl. Acad. Sci* 81: 4776-4780

Bertacinni A, Marani F, Borgia M (1986) Shoot Tip Culture of Different Garlic Lines for Virus Elimination. *Rivista-della Ortoflorofrutticoltura* 70: 97-105

Bhojwani SS (1980) *In vitro* Propagation of Garlic by Shoot Proliferation. *Sci. Hort* 13: 47-52

Bhojwani SS, Cohen D, Fry PR (1982) Production of Virus-Free Garlic and Field Performance of Micropropagated Plants. *Sci. Hort* 18: 39-43

Block E (1985) The Chemistry of Garlic and Onion. *Sci. Am* 252: 114-119

Bordia T, Verma SK, Srivastava KC (1998) Effect of Garlic (*Allium sativum*) on Blood Lipids, Blood Sugar, Fibrinogen and Fibrinolytic Activity in Patients with Coronary Artery Disease. *Prostaglandins Leukot. Essent. Fatty Acids* 58: 257-263

Botau D. Danci M, Danci O (2005) *In vitro* Medium Term Preservation of Different Romanian Landraces. *Acta Biol. Szegediensis* 49: 41-42

Bradley KF, Rieger MA, Collins GG (1996) Classification of Australian Garlic Cultivars by DNA Fingerprinting. *Australian J. Exp. Agric* 36: 613-618

Brewster J L (1994) *Onions and Other Vegetable Alliums.* CAB International, ISBN 1845933990, Wallingford, U.K.

Bujanos-Muñiz R, Marín-Jarillo A (2000). Plagas: Descripción, Daños y Control. In: *El Ajo en México.* Heredia-García, E., Delgadillo-Sánchez, F., Eds., pp. 64-67, SAGAR- INIFAP, Campo Experimental Bajío, ISBN 968-800-486-3, Celaya, Gto., México.

Buso GSC, Paiva MR, Torres AC, Resende FV, Ferreira MA, Buso JA, Dusi AN (2008) Genetic Diversity Studies of Brazilian Garlic Cultivars and Quality Control of Garlic-Clover Production. *Gen. Mol. Res* 7: 534-541

Christou P (1995) Strategies for Variety Independent Genetic Transformation of Important Cereals, Legumes and Woody Species Utilizing Particle Bombardment. *Euphytica* 85: 13-27

Christou P (1996) Transformation Technology. *Trends Plant Sci* 1: 423-431

Chu CC, Wang C, Sun SC, Hsú C, Yin KC, Chu CY, Bi FY (1975) Establishment of an Efficient Medium for Anther Culture of Rice Through Comparative Experiments on the Nitrogen Sources. *Sci. Sinica* 16: 659-688

Conci V, Nome S (1991) Virus Free Garlic (*Allium sativum* L.) Plants Obtained by Thermotherapy and Meristem-Tip Culture. *J. Phytopath* 132: 186-192

Cunha CP, Eulalia S, Hoogerheide S, Zucchi Maria I, Monteiro M, Pinheiro J B (2012) New microsatellite markers for Garlic *Allium sativum* Alliaceae *American Journal of Botany* 17-19

Delgadillo-Sánchez F (2000) Enfermedades: Descripción y Tratamiento. In: *El Ajo en México.* Heredia-García, E, Delgadillo-Sánchez, F., Eds., SAGAR- INIFAP, Campo Experimental Bajío 968-800-486-3, Celaya, Gto., México 68-77

Dolezel J, Novak FJ, Havel L (1986) *Cytogenetics of Garlic (Allium sativum L.) in vitro Culture.* IAAE, ISBN 92-0-010086-4, Vienna, Austria

Dustan DI, Short KC (1977) Improved Growth of Tissue Cultures of The Onion, *Allium cepa. Physiol. Plant* 41: 70-72

Eady C, Davis S, Catanach A, Kenel F, Hunger S (2005) *Agrobacterium tumefaciens*- Mediated Transformation of Leek (*Allium porrum*) and Garlic (*Allium sativum*). *Plant Cell. Rep* 24: 209-215

Ebi M., Kasai N, Masuda K (2000) Small Inflorescence Bulbils are Best for Micropropagation and Virus Elimination in Garlic. *HortSci* 35 : 735-737

El-Aref HM (2002) An Effective Method for Generating Somaclonal Variability in Egyptian Garlic (*Allium sativum* L.). Faculty of Agriculture, Assiut University, Assiut, Egypt

Ellis D, Skogerboe D, Andre C, Hellier B, Volk G (2005) Cryopreservation of 12 *Allium sativum* (garlic) Accessions: a Comparison of Plant Vitrification Solutions (PVS2 and PVS3). *In Vitro-Cellular and Developmental Biology – Plants*11-12, In Vitro Biology Meeting, June 2005, Baltimore, Ma., U.S.A.

Etoh T (1985) Studies on Sterility in Garlic, *Allium sativum* L. *Memoirs of the Faculty of Agriculture, Kagoshima University, Japan* 21: 77-132

Etoh T (1986) Fertility of Garlic Clones Collected in Soviet Central Asia. *J. Japan Soc. Hort. Sci* 55: 312-319

Etoh T, Hong CJ (2001) RAPD Markers for Fertile Garlic. *Acta Hort* 555: 209- 212

Etoh T, Simon PW (2002) Diversity, Fertility and Seed Production of Garlic. In: *Allium Crop Sciences: Recent Advances*, Rabinowitch HD, Currah L, Eds., CAB International Wallingford, U. K 101-117

FAOSTAT data. http: //apps.fao.org/default.htm (Access June 12th, 2011).

Fereol L, Chovelon V, Causse S, Michaux-Ferriere N, Kahane R (2002) Evidence of a Somatic Embryogenesis Process for Plant Regeneration in Garlic (*Allium sativum* L.). *Plant Cell Rep* 21: 197-203

Fereol L, Chovelon V, Causse S, Kalumbueziko ML, Kahane R (2005a). Embryogenic Cell Suspension Cultures of Garlic (*Allium sativum* L.) as Method for Mass Propagation and Potential Material for Genetic Improvement. *Acta Hort* 688: 65-74

Fereol L, Chovelon V, Causse S, Triaire D, Arnault I, Auger J, Kahane R (2005b). Establishment of Embryogenic Cell Suspension Cultures of Garlic (*Allium sativum* L.), Plant Regeneration and Biochemical Analyses. *Plant Cell Rep* 24: 319- 325

Ferrer E, Linares C, Gonzalez JM (2000) Efficient Transient Expression of the %-Glucuronidase Reporter Gene in Garlic (*Allium sativum* L.). *Agronomie* 20: 869-874

Fritsch RM, Friesen N (2002) Evolution, Domestication and Taxonomy. In: *Allium Crop Sciences: Recent Advances*. Rabinowitch HD, Currah L, Eds., CAB International Wallingford, U. K. 5-30

Fromm ME, Morrish F, Armstrong C, Williams R, Thomas J, Klein TM (1990) Inheritance and Expression of Chimeric Genes in the Progeny of Transgenic Maize Plants. *Nature Biotechnology* 8: 833-839

Gamborg OL, Miller RA, Ojima K (1968) Nutrient Requirements of Suspension Culture of Soybean (*Glycine max* Merril) Root Cells. *Exp. Cell Res* 50: 151- 158

García-Lampasona S, Martínez L, Burba JL (2003) Genetic Diversity Among Selected Argentinean Garlic Clones (*Allium sativum* L.) Using AFLP (Amplified Fragment Length

Polymorphism). *Euphytica* 132: 115-119

Hansen G, Wright MS (1999) Recent Advances in the Transformation of Plants. *Trends Plant Sci* 4: 226-231

Haque MS, Wada T, Hattori K (1997). High Frequency Shoot Regeneration and Plantlet Formation from Root Tip of Garlic. *Plant Cell Tiss. Org. Cult* 50: 83-89

Haque MS, Wada T, Hattori K (2003) Shoot Regeneration and Bulblet Formation from Shoot and Root Meristem of Garlic cv Bangladesh Local. *Asian J. Plant Sci* 2: 23-27

Hassan NA, El-Halwagi AA, Gaber A, El-Awady M, Klalaf A (2007) Slow-Growth *in vitro* Conservation of Garlic Cultivars Grown in Egypt: Chemical Characterization and Molecular Evaluation. *G. J. Mol. Sci* 2: 65-75

Havránek P, Novak FJ (1973) The Bud Formation in Callus Cultures of *Allium sativum* L. Z. *Pflanzen* 68: 308-318

Heredia-García E (2000) Clasificación Taxonómica. In: *El Ajo en México*. Heredia-García E, Delgadillo-Sánchez F, Eds., SAGAR-INIFAP, Campo Experimental Bajío, ISBN 968-800-486-3, Celaya, Gto., México 18-22

Hong CJ, Watanabe H, Etoh T, Iwai S (2000) A Search of Pollen Fertile Clones in the Iberian Garlic by RAPD Markers. *Memoirs of the Faculty of Agriculture, Kagoshima University, Japan* 36: 11-16

Hooykaas PJJ, Schilperoot RA (1992) .*Agrobacterium* and Plant Genetic Engineering. *Plant Mol. Biol*19: 15-38

Ipek M, Ipek A, Simon PW (2003) Comparison of AFLPS, RAPD Markers, and Isozymes for Diversity Assessment of Garlic and Detection of Putative Duplicate in Germplasm Collections. *J. Amer. Soc. Hort. Sci* 128: 246-252

Ipek M, Ipek A, Simon PW (2008) Rapid Characterization of Garlic Clones with Locus-Specific DNA Markers. *Turk. J. Agric. For* 32: 357-362

Ipek M, Simon PW (1998) Genetic Diversity in Garlic (*Allium sativum* L.) as Assessed by Amplified Fragment Length Polymorphism (AFLP). *1998 National Onion (and other Allium) Research Conference*. Sacramento, Cal., U.S.A, December, 1998.

Kamenetsky, R. (2007). Garlic: Botany and Horticulture. *Hort. Rev* 33: 123-171

Jo Man Hyun, In Ki Ham, Kyaw Thu Moe, Soon-Wook Kwon, Fu-Hao Lu, Yong-Jin Park, Woon Seob Kim, Mi Kyoung Won, Tae Il Kim, Eun Mo Lee (2012) Classification of genetic variation in garlic (*Allium sativum* L.) using SSR markers. *AJCS* 6(4): 625-631

Kamenetsky R, Rabinowitch HD (2001) Floral Development in Bolting Garlic. *Sex. Plant Reprod* 13: 235-241

Kazakova AA (1971) Most Common Onion Species, Their Origin and Intraspecific Classification. *Trudy po Prikladnoi Botanike, Genetike I Seleksii* 72: 135-136 (In Russian)

Kehr AE, Schaeffer T (1976) Tissue Culture and Differentiation of Garlic. *HortSci* 11: 422-423

Keller ERJ, Senula A, Leunufna S, Grúbe M (2006). Slow Growth Storage and Cryopreservation-Tools to Facilitate Germplasm Maintenance of Vegetatively Propagated Crops in Living Plant Collections. *Int. J. Refrigeration* 29: 411- 417

Keusgen M (2002) Health and *Alliums*. In: *Allium Crop Sciences: Recent Advances*. Rabinowitch HD, Currah L, Eds., CAB International, ISBN 0-85199-510-1, Wallingford, U.K. 357-378

Kevers C, Franck T, Strasser RJ, Dommes J, Gaspar T (2004) Hyperhydricity of Micropropagated Shoots: a Typically Stress-Induced Change of Physiological State. *Plant Cell Tiss. Org. Cult* 77: 181-191

Khan N, Alam MS, Nath UK (2004) *In vitro* Regeneration of Garlic Through Callus Culture. *J. Biol. Sci* 4: 189-191

Khar A, Yadav RC, Yadav N, Bhutáni RD (2005) Transient *gus* Expression Studies in Onion (*Allium cepa* L.) and Garlic (*Allium sativum* L.). Akdeniz Universitesi Ziraat Fakultesi Dergisi 18: 301-304

Kim EK, Hahn EJ, Murthy HN, Paek K Y (2003) High Frequency of Shoot Multiplication and Bulblet Formation of Garlic in Liquid Cultures. *Plant Cell Tiss.Org Cult* 73: 231-236

Klaas M, Friesen N (2002) Molecular Markers in *Allium*. In: *Allium Crop Sciences: Recent Advances*. Rabinowitch HD, Currah L, Eds., CAB International, ISBN 0-85199-510-1, Wallingford, U.K. 159-185

Koch HP, Lawson LD (1996). *Garlic: The Science and Therapeutic Application of Allium sativum L. and related species*. Williams and Wilkins Press, ISBN 0683181475, Baltimore, Ma., U.S.A.

Koch M, Salomon R (1994) Improvement of Garlic Via Somaclonal Variation and Virus Elimination. *Acta Hort* 358: 211-214

Kondo T, Hasegawa H, Suzuki M (2000) Transformation and Regeneration of Garlic (*Allium sativum* L.) by *Agrobacterium*-Mediated Gene Transfer. *Plant Cell Rep* 19: 989-993

Konvicka O, Nienhaus F, Fischbeck G (1978) Untersuchungen über die Ursachen der Pollensterilität bei *Allium sativum* L. Z. *Pflanzen* 80: 265-276

Kotlinska T, Havranek P, Navratill M, Gerasimova L, Pimakov A, Neikov S (1991) Collecting Onion, Garlic and Wild Species of *Allium* in Central Asia. *Plant Gen. Res. Newslett* 83/84: 31-32

Koul AK, Gohil RN, Langer A (1979) Prospects of Breeding Improved Garlic in the Light of its Genetic and Breeding Systems. *Euphytica* 28: 457-464

Lallemand J, Messian CM, Briand F, Etoh T (1997) Delimitation of Varietal Groups in Garlic (*Allium sativum* L.) by Morphological, Physiological and Biochemical Characters. *Acta Hort*. 433: 123-132

Larkin PJ, Scowcroft WR (1981. Somaclonal Variation-a Novel Source of Variability from Cell Cultures for Plant Improvement. *Theor. Appl. Genet.*, 60: 197-214

Lawson LD (1996) The Composition and Chemistry of Garlic Cloves and Processed Garlic. In: *Garlic: The Science and Therapeutic Application of Allium sativum L. and related species*. Koch HP, Lawson LD, Eds., Williams and Wilkins Press, ISBN 0683181475, Baltimore, Ma., U.S.A 37-107

Leemans J, Deblaere R, Willmitzer L, De Greve H, Hernalsteens JP, Van Montagu M,

Schell J (1982) Genetic Identification of Functions of TL-DNA Transcripts in Octopine Crown Galls. *EMBO J.*, Vol. 1: 147-152

Linsmaier EM, Skoog F (1965) Organic Growth Factor Requirements of Tobacco Tissue Cultures. *Physiol. Plant* 18: 100-127

Luciani GF, Mary AK, Pellegrini C, Curvetto NR (2006) Effects of Explants and Growth Regulators in Garlic Callus Formation and Plant Regeneration. *Plant Cell Tiss. Org. Cult.* 87: 139-143

Ma Y, Wang HL, Cun-Jin Z, Zhang CJ, Kang YQ (1994) High Rate of Virus-Free Plantlet Regeneration via Garlic Scape-Tip Culture. *Plant Cell Rep* 14: 65- 68

Maab HI, Klaas M (1995) Intraspecific Differentiation of Garlic (*Allium sativum* L.) by Isozyme and RAPD Markers. *Theor. Appl. Genet* 91: 89-97

Madhavi DL, Prabha TN, Singh NS, Patwarhan MV (1991) Biochemical Studies with Garlic (*Allium sativum*) Cell Cultures Showing Different Flavour Levels. *J. Sci. Food Agric* 56: 15-24

Makowska Z, Keller ERJ, Engelman F (1999). Cryopreservation of Apices Isolated from Garlic (*Allium sativum* L.) Bulbils and Cloves. *Cryoletters* 20: 175-182

Mann L K (1952) Anatomy of the Garlic Bulb and Factors Affecting Bulb Development. *Hilgardia* 21: 195-249

Martín-Urdíroz N, Garrido-Gala J, Martín J, Barandiaran X (2004) Effect of Light on the Organogenic Ability of Garlic Roots Using a One-Step *in vitro* System. *Plant Cell Rep* 22: 721-724

McCabe DE, Martinelli BJ, Christou P (1988) Stable Transformation of Soybean (*Glycine max*) by Particle Acceleration. *Biotechnology* 6: 923-926

McEloy D, Brettell RSI (1994) Foreign Gene Expression in Transgenic Cereals. *Trends Biotechnol* 12: 62-68

Messiaen CM, Lot H, Delecolle B (1994) Thirty Years of France's Experience in the Production of Disease-Free Garlic and Shallot Mother Bulbs. *Acta Hort* 358: 275-279

Messiaen CM, Marrov J, Quiot JB, Leclant F, Leroux JP (1970) Etude dans le Sud- est de la France d' un Schéma de Sélection Sanitaire de l' ail et de l' échalote. *Comptes Rendus de la 7 Conf. de Pathologie des Plantes*. pp. 101-103. C.N.R.A. Montfavet, France.

Mohamed-Yassen Y, Splittstoesser WE, Litz R E (1994) *In vitro* Shoot Proliferation and Production Sets from Garlic and Shallot. *Plant Cell Tiss. Org. Cult* 36: 243-247

Mukhopadhyay MJ, Sengupta P, Mukhopadhyay S, Sen S (2005) *In vitro* Stable Regeneration of Onion and Garlic from Suspension Culture and Chromosomal Instability in Solid Callus Culture. *Sci Hort* 104(1): 1-9

Murashige T, Skoog F (1962) A Revised Medium for Rapid Growth and Bioassays with Tobacco Tissue Cultures. *Physiol. Plant* 15: 473-497

Myers JM, Simon PW (1998a) Continuous Callus Production and Regeneration of Garlic (*Allium sativum* L.) Using Root Segments from Shoot Tip-Derived Plantlets. *Plant Cell Rep* 17: 726-730

Myers JM, Simon PW (1998b) Microprojectile Bombardment of Garlic, *Allium sativum* L. *1998 National Onion (and other Allium) Research Conference*. Sacramento, Ca., U.S.A. December, 1998.

Myers JM, Simon PW (1999) Regeneration of Garlic Callus as Affected by Clonal Variation, Plant Growth Regulators and Culture Conditions Over Time. *Plant Cell Rep* 19: 32-36

Nabulsi I, Al-Safadi B, Mir-Ali N, Arabi MIE (2001) Evaluation of Some Garlic (*Allium sativum* L.) Mutants Resistant to White Rot Disease by RAPD Analysis. *Ann. Appl. Biol* 138: 197-202

Nagakubo T, Nagasawa A, Ohkawa H (1993) Micropropagation of Garlic Through *in vitro* Bulblet Formation. *Plant Cell, Tiss. Org. Cult* 32: 175-183

Nagasawa A, Finer JJ (1988) Development of Morphogenic Suspension Cultures of Garlic (*Allium sativum* L.). *Plant Cell. Tiss. Org. Cult* 15: 183-187

Novak FJ (1972) Tapetal Development the Anthers of *Allium sativum* L. and *Allium longicuspis. Regel Experientia* 28: 363-364

Novak FJ (1983) Production of Garlic (*Allium sativum* L.) Tetraploids in Shoot-Tip *in vitro* Culture. *Z. Pflanzen* 91: 329-333

Novak FJ (1990) *Allium* Tissue Culture. In: *Onions and Allied Crops*. Rabinowitch HD, Brewster, J. L., Eds., pp. 233-250, Vol. II, CRC Press, ISBN 0849363012, Boca Raton, Fl., U.S.A.

Olmos E, Hellin E (1998) Ultrastructural differences of Hyperhydric and Normal Leaves from Regenerated Carnation Plants. *Sci. Hort* 75: 91-101

Ordás A, Malvar RA, Ron AM (1994) Relationships among American and Spanish Populations of Maize. Euphytica 39: 149-161

Ordonez A, Torres LE, Hidalgo MG, Munoz JO (2002) Análisis Citológico de una Variante Genética Somática de Ajo (*Allium sativum* L.) Tipo Rosado. *Agriscientia* 19: 37-43

Osawa K, Kuriyama T, Sugawara Y (1981) Clonal Multiplication of Vegetatively Propagated Crops Through Tissue Culture. I. Effective Balance of Auxin and Cytokinin in the Medium and Suitable Explants Part for Mass Propagation of Plantlets in Strawberry, Garlic, Scallion, Welsh Onion, Yam and Taro. *Bull. Veg. Ornamental Crops Res. Stn. Ano Mie (Japan)* 9: 1-46

Ovesná J, Kucera L, Králová J, Leisová L, Staveliková H, Velisek J (2007) Genetic Diversity Among Garlic Clones as Revealed by AFLP, Phenotypic Descriptors and S-Amino Acid Level. *Veg. Crops Res. Bull* 66: 105-116

Panis B, Lambardi M (2006) Status of Cryopreservation Technologies in Plants (Crops and Forest Trees). In: *The Role of Biotechnology in Exploring and Protecting Agricultural Genetic Resources*. Ruane J, Sonnino A, Eds., 68-78, FAO, ISBN, Rome, Italy.

Park MY, Yi NR, Lee HY, Kim ZT, Kim M, Park JH, Kim JK, Lee JS, Cheong JJ, Choi YD (2002) Generation of Chlorsulfuron Resistant Transgenic Garlic Plant (*Allium sativum* L.) by Particle Bombardment. *Mol. Breed* 9: 171-181

Peña-Iglesias A, Ayuso P (1982) Characterization of Spanish Garlic Viruses and Their Elimination by *in vitro* Shoot Apex Culture. *Acta Hort* 127: 183-193

Peschke VM, Phillips RL (1991) Activation of the Maize Transposable Element Suppressor-Mutator (Spm) in Tissue Culture. *Theor. Appl. Genet* 81: 90-97

Pooler MR, Simon PW (1993) Characterization and Classification of Isozyme and Morphological Variation in a Diverse Collection of Garlic Clones. *Euphytica* 68: 121-130

Pooler MR, Simon PW (1994). True Seed Production in Garlic. *Sex. Plant Reprod* 7: 282-286

Purseglove JW (1988) *Tropical Crops. Monocotyledons.* I. Longman, ISBN 0470205687, New York, N.Y., U.S.A.

Rabinkov A, Zhu XZ, Grafi G, Mirelman D (1994) Allin Lyase (Alliinase) from Garlic (*Allium sativum*), Biochemical Characterization and cDNA Cloning. *Appl. Biochem. Biotechnol* 48: 149-171

Ram K (2001) Historical Perspective on Garlic and Cardiovascular Disease. *J. Nutr* 131: 977-979

Ramírez-Malagón R, Pérez-Moreno L, Borodanenko A, Salinas-González GJ, Ochoa- Alejo N (2006) Differential Organ Infection Studies, Potyvirus Elimination, and Field Performance of Virus-Free Garlic Plants Produced by Tissue Culture. *Plant Cell. Tiss. Org. Cult* 86: 103-110

Robinson RA (2007) *Self-Organizing Agro-ecosystems.* Sharebooks Publishing, ISBN 698-0-9783634-1-3

Robledo-Paz A, Villalobos-Arámbula VM, Jofre-Garfias AE (2000). Efficient Plant Regeneration of Garlic (*Allium sativum* L.) by Root Tip Culture. *In Vitro Cell Dev. Biol. Plant* 36: 416-419

Robledo-Paz A, Cabrera-Ponce JL, Villalobos-Arámbula VM, Herrera-Estrella L, Jofre-Garfias AE (2004) Genetic Transformation of Garlic (*Allium sati*vum L.) by Particle Bombardment. *Hort Sci* 39: 1208-1211

Sakai A (2000) Development of Cryopreservation Techniques. In: *Cryopreservation of Tropical Plant Germplasm – Current Research Progress and Applications.* Engelman F, Takagi H. Eds. 1-7 International Plant Genetic Resources Institute, ISBN 9290434287, Rome, Italy.

Sakai A, Engelman F (2007) Vitrification, Encapsulation-Vitrification and Droplet-Vitrification: a Review. *Cryoletters* 25: 219-226

Saker MS, Sawahel WA (1998) Cultivar Identification and Detection of Somaclonal Variations Using RAPD Fingerprinting in Garlic. *Arab. J. Biotech* 1: 69-75

Salomon R (2002) Virus Diseases in Garlic and Propagation of Virus-Free Plants. In: *Allium Crop Sciences: Recent Advances.* Rabinowitch HD, Currah L, Eds., pp. 311-328, CAB Internacional, ISBN 0-85199-510-1, Wallingford, U.K.

Sánchez-Chiang N, Jiménez VM (2009) Técnicas Moleculares para la Detección de Variantes Somaclonales. *Agricultura Mesoamericana* 20: 135-151

Sata SJ, Bagatharia SB, Thaker VS (2001) Induction of Direct Somatic Embryogenesis in Garlic (*Allium sativum*). *Meth. Cell. Sci* 22: 299-304

Sawehel WA (2002) Stable Genetic Transformation of Garlic Planting Using Particle Bombardment. *Cell. Mol. Biol. Lett* 7: 49-59

Scowcroft WR (1984) *Genetic Variability in Tissue Culture: Impact on Germplasm Conservation and Utilization*, IBPGR, Bars Code: 0161793, Rome, Italy.

Senula A, Keller ERJ, Leseman DE (2000) Elimination of Viruses Through Meristem Culture and Thermotherapy for the Establishment of an *in vitro* Collection of Garlic (*Allium sativum*). *Acta Hort* 530: 121-128

Sidaros SA, Omar RA, El-Kewey SA, El-Khalik SA (2004) Virus Elimination from Infected Garlic Plants Using Different Techniques. *Egyptian J. Virol* 1: 333-341

Songstad DD, Somers DA, Griesbach RJ (1995) Advances in Alternative DNA Delivery Techniques. *Plant Cell Tiss. Org. Cult* 40: 1-15

Sudarmonowati E (2001) Cryopreservation of Garlic (*Allium sativum*) cv Lumbu Hijau Using Vitrification Techniques. *Annales Bogorienses* 8: 39-46

Ucman R, Zel J, Ravnikar M (1998) Thermotherapy in Virus Elimination from Garlic: Influences on Shoot Multiplication from Meristems and Bulb Formation *in vitro*. *Sci. Hort* 73: 193-202

Vain P, McMullen MD, Finer JJ (1993) Osmotic Treatment Enhances Particle Bombardment–Mediated Transient and Stable Transformation of Maize. *Plant Cell Rep* 12: 84-88

Vavilov NI (1926) Studies on the Origin of Cultivated Plants. *Bull. Appl. Bot* 16: 1-248

Verbeek M, Van Dijk P, Van Well PMA (1995) Efficiency of Four Viruses From Garlic (*Allium sativum*) by Meristem-Tip Culture. *Eur. J. Plant Pathol* 101: 231-239

Vinson JA, Hao Y, Su X, Zubik L (1998) Phenol Antioxidant Quality in Foods: Vegetables. *J. Agric. Food Chem* 46: 3630-3634

Volk GM, Henk AD, Richards CM (2004) Genetic Diversity Among U.S. Garlic Clones as Detected Using AFLP Methods. *J. Amer. Soc. Hort. Sc* 129: 559-569

Walkey DGA (1987) Production of Virus-Free Garlic (*Allium sativum* L.) and Shallot (*A. ascalonicum* L.) by Meristem Tip Culture. *J. Hort. Sci* 62: 211-220

Walkey DGA (1990) Virus Diseases. In: *Onions and Allied Crops*. Rabinowitch HD, Brewster JL, Eds., pp. 191-212, Vol. II, CRC Press, ISBN 0849363012, Boca Raton, Fl., U.S.A.

White PR (1963) *The Cultivation of Animal and Plant Cells*. The Ronald Press Company, New York, N.Y., U.S.A.

Wu Z, Chen LJ, Long YJ (2009) Analysis of Ultrastructure and Reactive Oxygen Species of Hyperhydric Garlic (*Allium sativum* L.) Shoots. *In Vitro Cell Dev. Biol.- Plant* 45: 483-490

Xu P, Yan C, Qu S, Yan C, Srinives P (2001) Inflorescence Meristem Culture and Economic Analysis of Virus–Free Garlic (*Allium sativum* L.) in Commercial Production. *Acta Hort* 555: 283-288

Xu P, Yan C, Yan CY (2005) Biotechnology Applied to Garlic and Onion. *Acta Hort* 688: 59-75

Xue HE, Araki H, Shi L, Yakuwa T (1991) Somatic Embryogenesis and Plant Regeneration in Basal Plate and Receptacle Derived-Callus Culture of Garlic (*Allium sativum* L.). *J. Japan Soc. Hort. Sci* 60: 627-634.

Zheng SJ, Henken B, Ahn YK, Krens FA, Kik C (2004) The Development of a Reproducible *Agrobacterium tumefaciens* Transformation System for Garlic (*Allium sativum* L.) and the

Production of Transgenic Resistant to Beet Armyworm (*Spodoptera exigua* Hübner) *Mol. Breed* 14: 293-307

Zheng SJ, Henken B, Krens FA, Kik C (2003) The Development of an Efficient Cultivar Independent Plant Regeneration System From Callus Derived From Both Apical and Non-Apical Root Segments of Garlic (*Allium sativum* L.). *In Vitro Cell Dev. Biol.- Plant* 39: 288-292

6

Vermicompost and Sustainable Agriculture

A. Haribhushan, S. H. Wani, D. Kumar, Y. K. Singh, N. Jyotsna, D. Kamei and *Kh. Nodiachand*

There is a grave problem of disposing bio-waste in a safer way. Huge quantities of biodegradable waste are being generated in rural and urban areas that threaten the environment with pollution and contamination. These wastes can be converted into valuable composts which have long been recognized in agriculture as beneficial for plant growth and yield and the maintenance of soil fertility rather than when they are directly applied. Vermicompost is stable, fine granular organic manure, which enriches soil quality by improving its physicochemical and biological properties. It is highly useful in raising seedlings and for crop production. It is becoming popular as a major component of organic farming system. In this review the details about the process of vermicompost production will be described with its effect on plant growth, soil and environment.

Introduction

A massive quantity of biodegradable organic waste is being generated in cities and towns in the country, creating dumping problems. This waste may be of animal manures, human waste, food wastes, yard wastes and sewage sludge's. Even in rural areas, several wastes like agriculture, animal husbandry and related activities generate in a large quantities. Considerable quantities of tender twigs, dry leaves, grass, weeds, and other agricultural wastes are also available. These organic wastes contain organic carbon and plant nutrients in substantial amounts. Modern approaches to the use of organic amendments in farming have provided a priceless surrogate for chemical fertilizers and proven to be valuable means of improving soil structure, enhancing soil fertility and increasing crop yields and also it reduces environmental pollution. Organic matter are excellent source of plant-available nutrients and their addition to soil could maintain high microbial populations and activities with increased values of biomass C, basal respiration, biomass C:total organic C ratio, and metabolic quotient (qCO_2). Crop yields have increased with the corresponding improvements in soil quality from additions of organic matter (Haribhushan *et al.*, 2008; Chadha *et al.*, 2012; Garai *et al.*, 2013).

In India, the incorporation of crops and livestock waste and use of manure as fertilizer were customarily the basis of farming systems. But progress of chemical fertilizer industry during the green revolution period perverted opportunities for low-cost supply of plant nutrients in inorganic forms which lead to rapid disarticulation of organic manures derived from livestock excreta. The deterioration of soil fertility through loss of nutrients and organic matter, erosion and salinity, and pollution of environment are the negative consequences of modern agricultural practices.

Composting has been documented as a low cost and environmentally sound process for treatment of many organic wastes (Hoitink, 1993). Furthermore, the rapid decomposition and raised temperatures during composting produce a relatively homogeneous, odor-free, pathogen-free and easy-to-handle product. Bevacqua and Mellano (1993) reported that compost-treated soils had lower pH and increased levels of organic matter, primary nutrients, and soluble salts. In crop studies, Bryan and Lance (1991) found that tomatoes grown in compost-amended soils yielded more. Maynard (1993) also reported increases in fruit yield of compost-amended plants compared with those growing in soil alone. Other benefits from the use of compost include the possible reduction of hazards from nitrate leaching into groundwater compared to those from inorganically fertilized controls (Maynard, 1989). Furthermore, composting and composts have been reported to suppress plant pathogens. Hoitink and Fahy (1986) reported that composts were suppressive to soil borne plant diseases such as *Pythium* damping-off and *Rhizoctonia* damping-off. Population development of *Helicotylenchus spp.* and *Pratelynchus dianthus* (D'Errico and Di Maio, 1980) were suppressed by composts made from municipal refuse. Composts have also been reported to enhance population development of beneficial nematodes such as *cephaloids* and *rhabditids* (saprophagous nematodes).

A method associated to composting which can advance the advantageous deployment of organic wastes is vermicomposting. It is a non-thermophilic process by which organic materials are converted by earthworms and microorganisms into rich soil amendments with greatly increased microbial activity and nutrient availability. Vermicompost is highly rich in nutrient and the percentages of these nutrients in the compost vary as the process proceeded until their ideal concentrations for the promotion of plant growth is accomplished (Ansari and Rajpersaud, 2012).

Vermicompost

Aristotle called the earthworms as "intestines of the earth" and stated that there may not be any other creature that has played so important a role in the history of life on earth. Earthworms constitute a large part of biomass (living bodies) inhabiting soil. In recent years efforts have been made to use the potential of earthworms in recycling of nutrients, waste management and development of vermicomposting systems at commercial scale. These are also called as "Ecosystem engineers" as they increase the numbers and types of microbes in the soil by creating conditions under which these creatures can thrive and multiply.

Organic wastes can be ingested by earthworms and egested as a peat-like material termed "vermicompost". Recycling of wastes through vermicomposting reduces the problem of non-utilization of livestock excreta. During vermicomposting, the important plant nutrients such as N, P, K, and Ca, present in the organic waste are released and converted into forms that are more soluble and available to plants. Vermicompost also contains biologically active substances such as plant growth regulators. Moreover, the worms themselves provide a protein source for animal feed.

Vermicompost is rich in organic carbon, which plays a key role in soil fertility, and contains all essential plant nutrients in appropriate proportions. Thus it is a complete and balanced plant food. It also contains biochemical substances that promote plant growth and fight plant diseases. Vermi composting is a promising biotechnology for many waste management applications (Piper *et al.*, 2005). Vermicomposting is a method of preparing enriched compost with the use of earthworms. It is one of the easiest methods to recycle agricultural wastes and to produce quality compost. Earthworms consume biomass and excrete it in digested form called worm casts. Worm casts are popularly called as Black gold. The casts are rich in nutrients, growth promoting substances, beneficial soil micro flora and having properties of inhibiting pathogenic microbes. It is stable, fine granular organic manure, which enriches soil quality by improving its physicochemical and biological properties. It is highly useful in raising seedlings and for crop production. It is becoming popular as a major component of organic farming system.

Prerequisite for vermicomposting

Earthworms, organic wastes and water are the three major necessities for vermicomposting.

1. Earthworms: Non burrowing, organic debris consuming species of earthworms, such as *Eisenia foetida, Eudrilus eugeniae, Pheretima elongata, Perionyx excavatus* are ideal for vermicomposting. These are available with many krishi vigyan kendra's, agricultural research centres, non-government organizations and private companies. They can also be obtained from villages where vermicomosting is already being done.

2. Organic wastes: The following organic wastes can be utilized for vermicomposting.

i) Crop based waste materials such as straw, stubble, chaff, husk, etc., tree leaves and twigs, weeds

ii) Animal based wastes such as cattle dung, sheep and goat dung, poultry wastes

iii) Household wastes such as vegetable and fruit wastes, leftover and spoilt foods

3. Water: Adequate water is essential for maintaining moisture in the compost heap.

Types of vermicomposting

The types of vermicomposting depend upon the amount of producton and composting structures. Small-scale vermicomposting is done to meet the personal requirement and farmer can harvest 5-10 tonnes of vermicompost annually. While, large-scale vermicomposting is done at commercial scale by recycling large quantity of organic waste with the production of more than 50 – 100 tonnes annually.

Techniques of vermicomposting

Vermicomposting is done by various methods, among them bed and pit methods are more common:

Bed method: Composting is done on the pucca / kachcha floor by making bed (6x2x2 feet size) of organic mixture. This method is easy to maintain and to practice.

Pit method: Composting is done in the cemented pits of size 10x3x2 feet. The unit is covered with thatch grass or any other locally available materials. This method is not preferred due to poor aeration, water logging at bottom, and more cost of production. We can also go for a low cost silpauline and bamboo made pit for homestead purpose.

Process of vermicomposting

Following steps are followed for vermicompost preparation

i) Vermicomposting unit should be in a cool, moist and shady site.

ii) Cow dung and chopped dried leafy materials are mixed in the proportion of 3:1 and are kept for partial decomposition for 15 – 20 days.

iii) A layer of 15-20cm of chopped dried leaves/grasses should be kept as bedding material at the bottom of the bed.

iv) Beds of partially decomposed material of size 6x2x2 feet should be made.

v) Each bed should contain 1.5-2.0q of raw material and the number of beds can be increased as per raw material availability and requirement.

vi) Earthworm (1500-2000) should be released on the upper layer of bed.

vii) Water should be sprinkled with can immediately after the release of worms.

viii) Beds should be kept moist by sprinkling of water (daily) and by covering with gunny bags/polythene.

ix) Bed should be turned once after 30 days for maintaining aeration and for proper decomposition.

x) Compost gets ready in 45-50 days.

xi) The finished product is 3/4th of the raw materials used.

Advantages of using vermicompost:

a) The use of vermicompost not only increases the rate of water intake into soil but also improves the soil's ability to hold water.

b) Its use enhances colour, smell, taste, flavour and keeping quality of flowers, fruits, vegetables and food grains and helps the growers to sell their products at a higher price in the market.

c) The making of vermicompost provides livelihood support to the unemployed in rural areas. Enterprising villagers can take up commercial production of vermicompost and earn good profits by selling the compost in the market, which is constantly growing.

Fig.1 : Bed method of vermicomposting

Fig. 2 : Vermicomposting unit at KVK- Senapati, Manipur

d) Earthworms convert waste materials that would otherwise pose disposal problems, into valuable compost. Thus vermicomposting prevents environmental pollution and helps in keeping the surroundings clean and free of garbage.

Table 1 Nutrient composition of vermicompost

Nutrient	Content (%)
Organic carbon	20-25
Nitrogen	1.5-2.0
Phosphorus	0.5-1.5
Potassium	0.5-1.0
Calcium	0.4-0.8
Magnesium	0.3-0.6
Sulphur	100-500 ppm*
Iron	6.7-9.3 ppm
Copper	2.0-9.5 ppm
Zinc	5.7-11.5 ppm

*ppm - parts per million (**Source:** CRIDA, Hyderabad)

Agronomic Importance of Vermicompost

a) High levels of bio-available nutrients for plants: Vermicompost contains most nutrients in plant-available forms such as 'nitrates' (N), 'phosphates' (P), 'soluble' potassium (K), & magnesium (Mg) and 'exchangeable' phosphorus (P) & calcium' (Ca) (Edward *et al.*, 1988; 2004). Vermicomposts have a large particulate surface area that provides many micro-sites for microbial activities and for the strong retention of nutrients (Arancon *et al.*, 2004; 2006).

b) High level of beneficial soil microorganisms promoting plant growth: Vermicomposts are rich in 'microbial populations & diversity', particularly 'fungi', 'bacteria' and 'actinomycetes' (Brown *et al.*, 1995; Chaoui *et al.*, 2003; Scheu, 1993; Singh *et al.*, 2009; Tiwari *et al.*, 1989). Teotia *et al.*, 1950 and Parle, 1963 reported bacterial count of 32 million per gram in fresh vermicompost compared to 6-9 million per gram in the surrounding soil. Scheu (1987) reported an increase of 90% in respiration rate in fresh vermicast indicating corresponding increase in the microbial population. Suhane (2007) found that the total bacterial count was more than 1010 per gram of vermicompost. It included *Actinomycetes, Azotobacter, Rhizobium, Nitrobacter* and phosphate solubilizing bacteria which ranged from 102-106 per gm of vermicompost. The PSB has very significant role in making the essential nutrient phosphorus (P) 'bio available' for plant growth promotion (Rodriguez and Fraga, 1999). Although phosphates are available in soils in rock forms but are not available to plant roots unless solubilized. Pramanik *et al.*, (2007) studied the microbial population in vermicompost prepared from cow dung and municipal solid wastes (MSW) as substrates (raw materials) and found that it was in highest abundance in cow dung vermicompost. The total bacterial count was 73 x 10^8, the cellulolytic fungi was 59 x 10^6 and the nitrogen-fixing bacteria was 18 x 10^3. It was least in vermicompost obtained from MSW. The total bacterial count was 16 x 10^8, the cellulolytic fungi were 21 x 10^6 and the nitrogen-fixing bacteria were 5 x 10^3. Application of lime in the substrate enhanced the population of all above mentioned microbes irrespective of the substrates used for vermicomposting. Plant growth promoting bacteria (PGPB) directly stimulates growth by nitrogen (N) fixation, solubilization of nutrients, production of growth hormones such as 1-aminocyclopropane-1-carboxylate (ACC) deaminase and indirectly by antagonising pathogenic fungi by production of siderophores, chitinase, ß-1,3-glucanase, antibiotics, fluorescent pigments and cyanide (Han *et al.*, 2005). There is also substantial body of evidence to demonstrate that microbes, including bacteria, fungi, actinomycetes, yeasts and algae, also produce 'plant growth regulators' (PGRs) such as 'auxins', 'gibberellins', 'cytokinins', 'ethylene' and 'ascorbic acids' in appreciable quantities and as their population is significantly boosted by earthworms large quantities of PGRs are available in vermicompost (Frankerberger, Jr. *et al.*, 1995).

c) Rich in growth hormones: Biochemical stimulating total plant growth: Researches show that vermicompost further stimulates plant growth even when plants are already receiving 'optimal nutrition'. Vermicompost has consistently improved seed germination, enhanced seedling growth and development and increased plant productivity much more than would be possible from the mere conversion of mineral nutrients into plant-available forms. (Arancon *et al.*, 2004) found that maximum benefit from vermicompost is obtained when it constitutes between 10 to 40% of the growing medium. Neilson (Neilson, 1951; 1965 and Tomati *et al.*, 1988) have also reported that vermicompost contained growth promoting hormone 'auxins', 'cytokinins' and flowering hormone 'gibberellins' secreted by earthworms. It was demonstrated by (Grappelli *et al.*, 1985; Tomati *et al.*, 1983; 1987; 1988) that the growth of ornamental plants after adding aqueous extracts from vermicompost showed similar growth patterns as with the addition of auxins, gibberellins and cytokinins through the soil.

d) Rich in humic acids: Biochemical promoting root growth & nutrient uptake: Atiyeh *et al.* (2000a; 2000b; 2000c) speculates that the growth responses of plants from vermicompost appears more like 'hormone-induced activity' associated with the high levels of humic acids and humates in vermicompost rather than boosted by high levels of plant-available nutrients. This was also indicated by Canellas *et al.*, (2000) who found that humic acids isolated from vermicompost enhanced root elongation and formation of lateral roots in maize roots. Pramanik *et al.* (2007) also reported that humic acids enhanced 'nutrient uptake' by the plants by increasing the permeability of root cell membrane, stimulating root growth and increasing proliferation of 'root hairs'.

e) Vermicompost is free of pathogens: Nair *et al.* (2007) studied that 21 days of a combination of thermocomposting and vermicomposting produced compost with acceptable C:N ratio and good homogenous consistency of a fertilizer. The study also indicated that vermicomposting leads to greater reduction of pathogens after 3 months upon storage. Whereas, the samples which were subjected to only thermophilic composting, retained higher levels of pathogens even after 3 months.

f) Vermicompost is free of toxic chemicals: Several studies have found that earthworms effectively bioaccumulate or biodegrade several organic and inorganic chemicals including 'heavy metals', 'organochlorine pesticide' and 'polycyclic aromatic hydrocarbons' (PAHs) residues in the medium in which it inhabits.

g) Vermicompost protects plants against various pests and diseases: There has been considerable evidence in recent years regarding the ability of vermicompost to protect plants against various pests and diseases either by suppressing or repelling them or by inducing biological resistance in plants to fight them or by killing them through pesticidal action (Anonymous, 2001; Al-Dahmani *et al.*, 2003).

i) Induce biological resistance in plants: Vermicompost contains some antibiotics and actinomycetes which help in increasing the 'power of biological resistance' among the crop plants against pest and diseases. Pesticide spray was significantly reduced where earthworms and vermicompost were used in agriculture (Singh, 1993; Suhane, 2007).

ii) Repel crop pests: There seems to be strong evidence that worms varmicastings sometimes repel hard-bodied pests (Arancon *et al.*, 2004). Edwards and Arancon (2004) reports statistically significant decrease in arthropods (aphids, buds, mealy bug, spider mite) populations and subsequent reduction in plant damage, in tomato, pepper and cabbage trials with 20% and 40% vermicompost additions. George Hahn, doing commercial vermicomposting in California, U.S., claims that his product repels many different insect pests. His explanation is that this is due to production of enzymes 'chitinase' by worms which breaks down the chitin in the insect's exoskeleton (Munroe and Glenn, 2007).

iii) Suppress plant disease: Edwards and Arancon (2004) have found that use of vermicompost in crops inhibited the soil-born fungal diseases. They also found statistically significant suppression of plant-parasitic nematodes in field trials with pepper, tomatoes, strawberries and grapes. The scientific explanation behind this concept is that high levels of agronomically beneficial microbial population in vermicompost protects plants by out-competing plant pathogens for available food resources i.e. by starving them and also by blocking their excess to plant roots by occupying all the available sites. This concept is based on 'soil-foodweb' studies pioneered by Dr. Elaine Ingham of Corvallis, Oregon, U.S. (http://www.soilfoodweb.com). Edwards and Arancon (2004) reported the agronomic effects of small applications of commercially produced vermicompost, on attacks by fungus *Pythium* on cucumber, *Rhizoctonia* on radishes in the greenhouse, by *Verticillium* on strawberries and by *Phomposis* and *Sphaerotheca fulginae* on grapes in the field. In all these experiments vermicompost applications suppressed the incidence of the disease significantly. They also found that the ability of pathogen suppression disappeared when the vermicompost was sterilized, convincingly indicating that the biological mechanism of disease suppression involved was 'microbial antagonism It has been found that the aqueous extracts of vermicomposts depress soil-borne pathogens and pests. (Szczech *et al.*, 1993; Orlikowski, 1999; Rodriguez *et al.*, 2000 and Zaller, 2006). They found in their field experiment that only half as many plants of tomatoes sprayed with aqueous extract of vermicompost were infected with *Phytopthora infestans* (that cause 'late-blight' disease) as those of control ones.

Vermicompost and Plant Growth

Vermicompost significantly stimulates the growth of a wide range of plant species including several horticultural crops such as tomato (Atiyeh *et al.*, 1999; Atiyeh *et al.*, 2000a, Atiyeh *et al.*, 2000b; Atiyeh *et al.*, 2001; Hashemimajd, *et al.*, 2004; Gutiérrez-Miceli *et al.*, 2007), pepper (Arancon *et al.*, 2004a), garlic (Argüello *et al.*, 2006), aubergine (Gajalakshmi and Abbasi, 2004), strawberry (Arancon *et al.*, 2004b, Arancon *et al.*, 2005b), sweet corn (Lazcano *et al.*, 2011) and green gram (Karmegam *et al.*, 1999). Vermicompost has also been found to have positive effects on some aromatic and medicinal plants (Anwar *et al.*, 2005), cereals such as sorghum and rice (Bhattacharjee *et al.*, 2001; Reddy and Ohkura, 2004; Sunil *et al.*, 2005), fruit crops such as banana and papaya (Cabanas-Echevarria *et al.*, 2005; Acevedo and Pire, 2004) and ornamentals such as geranium (Chand *et al.*, 2007), marigolds (Atiyeh *et al.*, 2002), petunia (Arancon *et al.*, 2008), chrysanthemum (Hidalgo and Harkess, 2002a) and poinsettia (Hidalgo and Harkess, 2002b). Positive effects of vermicompost have also been observed in forestry species such as acacia, eucalyptus and pine tree (Donald and Visser, 1989; Lazcano *et al.*, 2010a, 2010b). Vermicompost has been found to have beneficial effects when used as a total or partial substitute for mineral fertilizer in peat-based artificial greenhouse potting media and as soil amendments in field studies. Likewise, some studies show that vermicomposting leachates or vermicompost water-extracts, used as substrate amendments or foliar sprays, also promote the growth of tomato plants (Tejada *et al.*, 2008), sorghum (Gutiérrez-Miceli *et al.*, 2008), and strawberries (Singh *et al.*, 2010). Positive effects of vermicompost include stimulated seed germination in several plant species such as green gram (Karmegam *et al.*, 1999), tomato plants (Atiyeh *et al.*, 2000b; Zaller, 2007), petunia (Arancon *et al.*, 2008) and pine trees (Lazcano *et al.*, 2010a). Vermicompost also has a positive effect on vegetative growth, stimulating shoot and root development (Edwards *et al.*, 2004). The effects include alterations in seedling morphology such as increased leaf area and root branching (Lazcano *et al.*, 2009). Vermicompost has also been shown to stimulate plant flowering, increasing the number and biomass of the flowers produced (Atiyeh *et al.*, 2002; Arancon *et al.*, 2008), as well as increasing fruit yield (Atiyeh *et al.*, 2000b; Arancon *et al.*, 2004a, 2004b; Singh *et al.*, 2008). In addition to increasing plant growth and productivity, vermicompost may also increase the nutritional quality of some vegetable crops such as tomatoes (Gutiérrez-Miceli *et al.*, 2007), Chinese cabbage (Wang *et al.*, 2010), spinach (Peyvast *et al.*, 2008), strawberries (Singh *et al.*, 2008), lettuce (Coria-Cayupán *et al.*, 2009), and sweet corn (Lazcano *et al.*, 2011). Nevertheless, despite the large body of scientific evidence showing the positive effects of vermicompost on plant growth and yield, there is also strong evidence that these effects are not general or constant, and that there is great variability in the magnitude of the effects reported in different studies. In fact, some studies report that vermicompost may decrease growth and even cause plant death (Roberts *et al.*, 2007; Lazcano *et al.*, 2010c). The variability in the effects of vermicompost may depend on the cultivation system into which it is incorporated, as well as on the physical, chemical and biological characteristics of vermicompost, which vary widely depending on the original feedstock, the earthworm species used, the production process, and the age of vermicompost (Rodda *et al.*, 2006; Roberts *et al.*, 2007; Warman and AngLopez, 2010). There is also a large variation in the effects of vermicompost depending on the plant species or even the variety considered. This was observed in tomato plants where the replacement of a fertilized commercial potting media with vermicompost had different effects on germination, seedling elongation, biomass allocation, fruit morphology and chemical properties of three tomato varieties (Zaller, 2007). Similar variation was observed in an experiment studying the effects of vermicompost and vermicompost extracts on the germination and early growth of six different progenies of maritime pine (Lazcano *et al.*, 2010a). In this experiment, the speed of maturation increased, relative to the control without vermicompost, in three out of the six pine progenies, decreased in two of the progenies and was unaffected in the other. It may be expected that different hybrids or plant genotypes will respond differently to vermicompost, considering that plant genotype determines important differences in nutrient uptake capacity, nutrient use efficiency and resource allocation within the plant. Different genotypes may therefore enhance root growth or modify root exudation patterns in order to increase nutrient uptake (Kabir *et al.*, 1998; Cavani and Mimmo, 2007), and all of these strategies will determine the establishment of different interactions with the microbial communities at the rhizosphere level. In fact, after the application of vermicompost to sweet corn crops, the different genotypes showed important differences in their rhizosphere microbial community (Aira *et al.*, 2010). In light of this evidence, it is clear that vermicompost constitutes a promising alternative to inorganic fertilizers in promoting plant growth. However, further research into the exact mechanisms and circumstances that stimulate plant growth by this organic substrate is necessary in order to maintain consumer confidence in this type of fertilizer.

Effect of Vermicompost on Plant Growth Mechanisms

Vermicompost is a microbiologically-active, nutrient-rich, peat-like material, which when added to plant growing media may influence plant growth directly or indirectly through different chemical, physical and biological mechanisms (Fig 3).

Direct Mechanisms

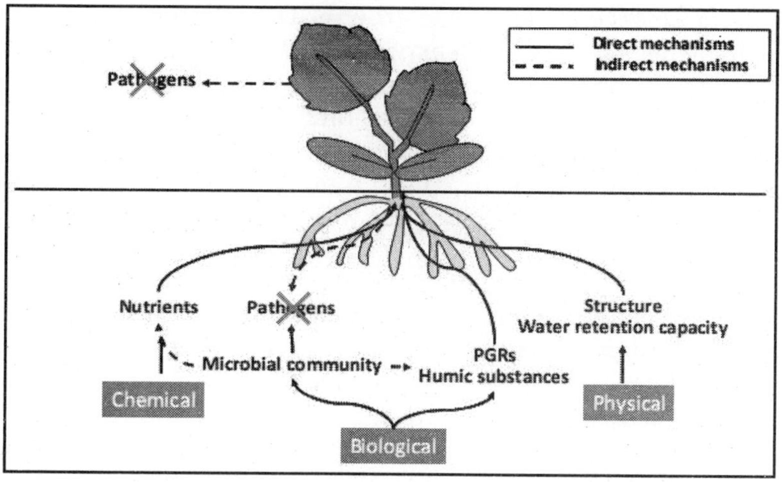

Fig 3. Some of the proposed chemical, biological and physical mechanisms by which vermicompost may directly or indirectly influence plant growth and development. (Source: Lazcano and Domínguez, 2011)

As regards the direct effects on plant growth, vermicompost constitutes a source of plant macro- and micronutrients. Although some of these nutrients are present in inorganic forms and are readily available to plants, most are released gradually through mineralization of the organic matter, thus constituting a slow-release fertilizer that supplies the plant with a gradual and constant source of nutrients (Chaoui *et al.*, 2003). However, in contrast to chemical fertilizers, the amount of nutrients provided may vary greatly depending on the original feedstock, processing time and maturity of the vermicompost (Campitelli and Ceppi, 2008).

Vermicompost is a finely-divided peat-like material and because of this fine structure, the addition of vermicompost to plant potting media causes significant changes in the physical properties, altering water and air availability in the substrates and conditioning root growth. Nonetheless, the type and magnitude of these effects depends on the physical characteristics of the original growing medium. For example, the bulk density of vermicompost is usually higher and the particle size lower than in some of the most commonly used peat-based substrates (Atiyeh *et al.*, 2001); mixing of these two substrates produces a significant increase in the bulk density and water holding capacity, while decreasing particle size and total porosity (Atiyeh *et al.*, 2001; Bachmann and Metzger, 2007). Hidalgo and Harkess (2002a) and Hidalgo *et al.* (2006) reported a significant increase in total porosity and water holding capacity after addition of vermicompost to a greenhouse potting medium comprising a mixture of sand, pine bark and peat. Vermicompost may also have significant effects on soil physical properties. Ferreras *et al.* (2006) observed that addition of 20 tn ha^{-1} of vermicompost to an agricultural soil in two consecutive years significantly improved soil porosity and aggregate stability. The number of large, elongated soil macropores increased significantly after a single application of a dose of vermicompost equivalent to 200 kg ha^{-1} of N to a corn field (Marinari *et al.*, 2000). Similarly, Gopinath *et al.* (2008) reported a significant decrease in soil bulk density and a significant increase in soil pH and total organic carbon after application of vermicompost in two consecutive growing seasons, at a rate equivalent to 60 kg ha^{-1} of N. Together these changes in soil properties improve the availability of air and water, thus encouraging seedling emergence and root growth. Despite the above-mentioned physical and chemical mechanisms, there is much experimental evidence showing that vermicompost enhances plant growth further than expected because of nutrient supply and improvements in the physical condition of substrates. This was first suggested by Scott (1988) and Edwards and Burrows (1988), who observed that small doses of vermicompost added to the potting media of several ornamental species, produced a much larger increase in plant growth than the equivalent dose of nutrients. These effects were maintained even when vermicompost was diluted 1:20 with other potting media, resulting in a dose of vermicompost would be expected to have negligible physical effects (Edwards and Burrows, 1988). This was also observed in field experiments carried out by Arancon *et al.* (2004b) and Singh *et al.* (2008). Both studies indicated a significant increase in the growth and productivity of strawberries cultivated with 5 and 7.5 t ha^{-1} of vermicompost respectively, in comparison with strawberries cultivated with equivalent doses of mineral fertilizers. Furthermore, many studies have shown that increases in growth and yield often involve changes in plant development and/or plant morphology such as increased leaf area, root volume and root branching (Singh *et al.*, 2008; Lazcano *et al.*, 2009). It has been postulated that such enhancement of plant growth may be attributed to the existence of biologically-mediated plant growth promoting mechanisms.

Vermicompost may influence plant growth directly via the supply of plant growth regulating substances (PGRsas proposed by (Tomati *et al.*, 1983; 1987; 1988; 1990; Grapelli *et al.*, 1987; Tomati and Galli, 1995).

Hormonal activity has also been associated with the humic substances present in vermicompost. Canellas *et al.* (2002) and Zandonadi *et al.* (2006) reported that the humic substances extracted from earthworm compost were capable of inducing lateral root growth in maize plants by stimulation of the plasma membrane H^+-ATPase activity, thus producing similar effects such as the exogenous application of indole-3-acetic acid (IAA). Furthermore, the induction of lateral root initiation by vermicompost-derived humic substances in arabidopsis has been related to the activation of the transcription of some auxin responsive genes (Trevisan *et al.*, 2010). The hypothesis of the auxin activity of vermicompost humic substances is reinforced by the presence of exchangeable auxin groups in their macrostructure (Canellas *et al.*, 2002). It has been suggested that earthworms may be important agents capable of influencing the production of PGRs by microorganisms through the stimulation and promotion of microbial activity in organic substrates (Nielson, 1965; Grappelli *et al.*, 1987; Tomati *et al.*, 1983; 1987; 1988; Tomati *et al.*, 1990; Tomati and Galli, 1995; Nardi *et al.*, 1988; 1980; Dell'Agnola and Nardi, 1987). Nevertheless, some authors suggest that earthworms, and not microorganisms, are responsible for the production of PGRs. Nielson (1965) reported the first evidence of the presence of indole compounds in the tissues of *Aporrectodea caliginosa, Lumbricus rubellus,* and *Eisenia fetida.* El Harti *et al.* (2001a, 2001b) showed that a crude extract of the earthworm *Lumbricus terrestris* was able to stimulate rooting in bean seeds due to the presence of indole compounds of endogenous origin.

Indirect Mechanisms

Vermicompost has also been found to have a wide range of indirect effects on plant growth such as the mitigation or suppression of plant diseases (Fig 3). Suppression of plant diseases has been extensively investigated in other organic amendments such as manure and compost (Noble and Coventry, 2005; Termorshuizen *et al.*, 2006; Trillas *et al.*, 2006). Likewise, some studies have shown that vermicompost can suppress a wide range of microbial diseases, insect pests and plant parasitic nematodes. As regards the suppression of fungal diseases, Orlikowski (1999) observed that the addition of vermicompost extracts to three ornamental plant species significantly reduced sporulation of the pathogen *Phytophthora cryptogea.* Similarly Nakasone *et al.* (1999) observed that aqueous extracts of vermicompost were capable of reducing the growth of pathogenic fungi such as *Botrytis cinerea, Sclerotinia sclerotiorum, Corticium rolfsii, Rhizoctonia solani* and *Fusarium oxysporum.* The addition of solid vermicompost to tomato seeds significantly reduced infection caused by *Fusarium lycopersici* (Szczech, 1999) and *Phytophthora nicotianae* (Szczech and Smolinska, 2001). Nevertheless, Szczech and Smolinska (2001) did not find any significant suppressive effects of a sewage sludge vermicompost on *Phytophthora nicotianae,* in comparison with peat. Edwards *et al.* (2006) observed that the suppressive effect exerted by several types of vermicompost on several plant pathogens such as *Pythium, Rhizoctonia, Verticillium,* and *Plectosporium,* disappeared after sterilization of the vermicompost, and concluded that disease suppression may be related to the presence of biological suppressive agents in vermicompost. As regards the effects of vermicompost on insect pests and mites, field studies have shown that the addition of vermicompost to soil significantly reduces the incidence of the psyllid *Heteropsylla cubana* (Biradar *et al.*, 1998), the sucking insect *Aproaerema modicella* (Ramesh, 2000), jassids, aphids, beetles, and spider mites (Rao, 2002). Furthermore, some studies show that vermicompost reduces not only the degree of plant infestation, but also the populations of plant parasitic insects in soils. This has been reported by Arancon *et al.*, (2005b) who showed that vermicompost produced from food waste significantly reduced the populations of two species of beetles in soil (*Acalymma vittatum*, *Diabotrica undecimpunctata*), in comparison with inorganic fertilization. Similarly, Yardim *et al.* (2006) observed a significant decrease in the larvae of the worm *Manduca quinquemaculata* after addition of vermicompost to a cucumber crop. Arancon *et al.* (2007) observed a significant reduction in the populations of spider mites (*Tetranychus urticae*), mealy bugs (*Pseudococcus* sp.) and aphids (*Myzus persicae*) after the addition of food waste vermicompost to several vegetable crops (tomatoes, cucumber, cabbage, bush beans and eggplants). Vermicompost may also have significant effects on both the incidence and abundance of plant-parasitic nematodes in soil. This was first reported by Swathi *et al.* (1998), who found that addition of vermicompost to soil at a rate of 1kg m-2 significantly inhibited the incidence of the parasite nematode *Meloidogyne incognita* in tobacco plants. Similar reductions in the degree of plant infestation by *Meloidogyne incognita* were observed by Morra *et al.* (1998), while Ribeiro *et al.* (1998) reported a significant decrease in the egg mass of *Meloidogyne javanica* after the application vermicompost to the growth medium. Arancon *et al.* (2003) indicated significant reductions in the abundance of plant-parasitic nematodes in soil plots amended with different doses of two types of vermicompost, in comparison with the effects of inorganic fertilizers. According to the mechanisms proposed for compost (Hointink and Bohem, 1999; Noble and Coventry, 2005), disease suppression by vermicompost may be attributed to either direct suppression of pathogens or to the induction of systemic resistance in the plant (Fig 3). Direct suppression of the pathogen by the vermicompost-associated microflora and/or microfauna may be general or specific, depending on the existence of a single suppressive agent or the joint action of several agents. In both cases, the proposed mechanisms are competition, antibiosis and parasitism. The mechanisms responsible for disease suppression by vermicompost are not yet well known, and the effects of biopesticides are usually attributed to general suppressive effects. Vermicompost increases microbial biomass in soil and changes the diversity and abundance of soil fauna (Gunadi *et al.*, 2002; Arancon *et al.*, 2006), and thus a broader range of organisms may act as biocontrol agents. Furthermore, there is recent evidence showing that the use of vermicompost extracts as foliar sprays in different crop plants effectively reduces the incidence of fungal diseases such as *Phytophthora infestans* (Zaller, 2006), *Erysiphe pisi* and *Erysiphe cichoracearum* (Singh *et al.*, 2003). Reductions in disease incidence are sometimes accompanied by an increase in the production of defence substances by the plant (Singh *et al.*, 2003), thus suggesting the induction of plant systemic resistance by vermicompost. Some of the indirect effects of vermicompost have been related to the change in the microbiological properties of the soil or the potting media. Processing by earthworms during vermicomposting has a strong effect on the microbial community of the initial waste (Dominguez *et al.*, 2010). Vermicompost therefore has a rather different microbial community structure than the parent waste, with lower biomass and activity but enhanced metabolic diversity (Lores *et al.*, 2006; Aira *et al.*, 2007, Gómez-Brandón *et al.*, 2010). Application of such a microbiologically active organic substrate may have important effects on the microbial properties of soil or greenhouse potting media thereby influencing plant growth. However, information regarding the impacts of vermicompost on soil microbial properties is still limited. A single application of vermicompost to a strawberry crop has been shown to produce a significantly higher increase in soil microbial biomass than application of an inorganic fertilizer, independently of the dose used (Arancon *et al.*, 2006). As well as increasing microbial biomass, vermicompost increases microbial activity (Ferreras *et al.*, 2006) promotes the establishment of a specific microbial community in the rhizosphere different from that of plants supplemented with mineral fertilizers or other type of organic fertilizers such as manure (Aira *et al.*, 2010). Kale *et al.*, (1992) observed that the application of vermicompost may produce significantly greater increases in the abundance of N-fixers, actinomycetes and spore formers than in soil supplemented with inorganic fertilizers. Soil enzyme activity is also significantly increased by vermicompost addition as compared to equivalent rates of mineral fertilizers (Marinari *et al.*, 2000; Arancon *et al.*, 2006; Saha *et al.*, 2008).

Application of vermicompost towards sustainable agriculture

The success of sustainable agriculture is very much dependent upon the availability of cheap and good quality organic manures. Among the sources of available organic manures, vermicompost is a potential source due to the presence of readily available plant nutrients, growth enhancing substances, and a number of beneficial microorganisms like nitrogen fixing, P solubilising and cellulose decomposing organisms (SailajaKumari *et al.*, 2002). Sustainable agriculture can be defined as a set of practices that conserve resources and the environment without compromising human needs, and the use of organic fertilizers such as animal manure has been indicated as one of its main pillars (Tilman *et al.*, 2002). Animal manure is a valuable resource as a soil fertilizer because it provides large amounts of macro- and micronutrients for crop growth and is a low-cost, environmentally-friendly alternative to mineral fertilizers. However, the use of manure in agriculture is being abandoned because of increasing transportation costs and environmental problems associated with the indiscriminate and inappropriately-timed application to agricultural fields (Hutchison *et al.*, 2005). Processing of this waste material through controlled bio-oxidation processes, such as composting, reduces the environmental risk by transforming the material into a safer and more stable product suitable for application to soil (Lazcano *et al.*, 2008), and also reduces the transportation costs because of the significant reduction in the water content of the raw organic matter. Composted materials are therefore gaining acceptance as organic fertilizers in sustainable agriculture, and there has been a considerable increase in research dedicated to the study of the effects of compost-like materials on soil properties and plant growth. The results of several long-term studies have shown that the addition of compost improves soil physical properties by decreasing bulk density and increasing the soil water holding capacity (Weber *et al.*, 2007). Moreover, in comparison with mineral fertilizers, compost produces significantly greater increases in soil organic carbon and some plant nutrients (García-Gil *et al.*, 2000; Bulluck *et al.*, 2002; Nardi *et al.*, 2004; Weber *et al.*, 2007). Long-term beneficial effects of composted materials are also observed in soil humic substances (due to an increase in the complexity of their molecular structure, which increases the humic/fulvic acid ratio), as well as in soil sorption properties (with increased cation exchange capacity and base saturation) (Weber *et al.*, 2007). In addition to the changes exerted on the chemical and physical properties, composted materials have a clear impact on soil biological properties, such as increases in microbial biomass and activity (Knapp *et al.*, 2010), as well as changes in the activity of soil enzymes (Garcia-Gil *et al.*; 2000, Ros *et al.*, 2006) and in the structure of the soil microbial community (Ros *et al.*, 2006).

Vermicompost under protected cultivation

Many greenhouse experiments have demonstrated that vermicomposts can have consistently positive effects on plant germination growth and yields. Edwards and Burrows (1988) reported that vermicomposts increased ornamental seedling emergence compared with those in control commercial plant growth media, using a wide range of test plants such as pea, lettuce, wheat, cabbage, tomato and radish. In similar investigations, ornamental shrubs such as *Eleagnus pungens, Cotoneaster conspicua, Pyracantha, Viburnum bodnantense, Chaemaecyparis lawsonia, Cupressocyparis leylandii and Juniperus communis* usually grew better in vermicompost-supplemented mixtures than in a commercial plant growth medium when transplanted into larger pots or grown outdoors. In greenhouse pot trials by Buckerfield *et al.* (1999), using 0% to 100% mixtures of vermicompost and sand, similar increased growth trends were reported. Although the germination of radishes decreased with increasing vermicompost concentrations radish harvest weights were increased proportionally to the application rates of vermicomposts, with the yields of plants in 100% vermicompost being up to ten times greater than those in 10% vermicompost. Buckerfield *et al.*,(1999) reported that vermicompost applications inhibited germination initially, but subsequently weekly applications of the diluted extracts improved plant growth and increased radish yields significantly by up to 20%. The growth of tomatoes, lettuces, and peppers were reported to be best at substitution into soils at rates of 8-10%, 8%, and 6%, respectively, using duck waste vermicompost and peat mixture (Wilson and Carlile, 1989). At higher vermicompost substitution rates, same inhibition of growth that occurred was attributed to higher electrical conductivity (salt content) and excessive nutrient levels.

The effects of a variety of vermicomposts produced from cow manure, sheep manure, poultry manure, goat manure (mixed with carpet underfelt, lawn clippings, cardboard, and domestic waste), kitchen scraps, cardboard (mixed with wheat, maize, meat, lucerne and linseed meals, rice pollard and oat hulls), and pig wastes on plant growth were investigated by Handreck (1986) after mixing *Pinus radiata* bark and quartz sand in the growth media to which 30% of each of the vermicompost was added. He reported that all the mixtures increased the dry weights of stocks (*Mathiola incana)* compared those in the control receiving no vermicompost. Similar trends in the germination of tomatoes and peppers grown in vermicomposts mixed with a commercial peat/sand planting medium have been reported. Chan and Griffiths (1988) reported stimulating effects of pig manure vermicomposts on the growth of soybeans (*Glycine max)*, particularly in terms of increased root lengths, lateral root numbers, and internode lengths of seedlings. In another rooting experiment, that used vermicomposts improved in the establishment of vanilla (*Vanilla planifolia)* cuttings better than other growth media such as mixtures of coir pith and sand (Siddagangaiah *et al.*, 1996). Similar responses in growth were observed from cloves (*Syzygium aromaticum*) and black peppers (*Piper nigrum*) sown into 1:1 mixtures of vermicompost and soil (Thankamani *et al.*, 1996). Black pepper cuttings raised in vermicomposts were significantly taller and had more leaves than those grown in commercial potting mixtures. Plant heights, numbers of branches, and the longest taproots were on cloves grown in the vermicompost mixtures. Vermicomposts produced from coir dust increased the yields of onions (*Allium cepa)* (Thanunathan, 1997).

In another study, substitution of Metro-Mix 360 by 30% or 40% pig manure vermicomposts produced more vegetative growth and flowers of marigolds in mixtures containing 40% pig manure vermicompost (Atiyeh *et al.*, 2002). The increased in flowering of marigolds were not associated with the amounts of available mineral-N, nor amounts of microbial biomass, during the later growth and fruiting stages of marigolds since the plants are provided with needed nutrients (Atiyeh *et al.*, 2002; Arancon *et al.*, 2005a).

All treatments in the greenhouse experiments were watered with nutrient solutions regularly, which eliminate nutrients as the major contributing factor in increasing yields of tomatoes or peppers. However, it is possible that other growth-enhancing factors, resulting from the introduction of smaller concentrations of food waste vermicompost into Metro-Mix 360 may have caused the growth change. These factors could include: improvements of in the physical structure of the container medium, increases in enzymatic activities, increased numbers of beneficial microorganisms or biologically-active plant growth-influencing substances such as plant growth regulators and humic acids.

Vermicompost under field conditions

A number of field experiments have reported positive effects of quite low application rates of vermicomposts to field crops. These applications were comparable with rates that improved growth on the same crops in greenhouse experiments. When cabbage was grown in compressed blocks made from pig waste vermicompost, after transplanting to the field they were larger and more mature at harvest compared to those grown in commercial blocking material (Edwards and Burrows, 1988). In a field experiment applying cassava peel mixed with guava leaves and vermicomposts from poultry droppings, Mba (1983) reported more shoot biomass and increased seed yields of cowpeas. Masciandro *et al.* (1997) investigated the effects of direct applications of vermicomposts produced from sewage sludge into the soil as well ferti-irrigation with humic extracts from vermicomposts. They reported a greater growth index of garden cress (*Lepidium sativum)* treated with vermicomposts than in control treatments with no vermicompost applications. Soil analyses after the vermicompost applications showed marked improvements in the overall physical and biochemical properties of the soil. A surface application of vermicompost derived from grape marc, spread under grape vines covered with a straw and paper mulch increased yields of a grape variety Pinot Noir by 55% (Buckerfield and Webster, 1998). The increases in yields included large increases in both bunch-weights and bunch numbers and no losses in flavor. In an experiment at a second site, vermicompost applications from animal manures, under straw mulch, increased Chardonnay grape yields by up to 35% and vermicompost applications tended to have greater effects on yields when applied under mulches than when applied directly to the soil surface possibly through degradation of vermicompost on exposure to sun and air. In later experiments, Webster reported that a single application of vermicompost to grapes still had positive effects on yields for 5 years. Venkatesh *et al.* (1999) reported that yields of Thompson Seedless grapes were significantly greater when vermicomposts were applied. Seyval grapes produced greater marketable yields, more fruit clusters per vine and bigger berry sizes after applications of food waste and paper waste vermicomposts at rates of 2.5 t/ha or 5 t/ha supplemented with inorganic fertilizers. Vadiraj *et al.* (1998) compared vermicomposts application rates of 5 t/ha up to 25 t/ha in 5 t/ha increments on growth of three varieties of coriander. The responses to the vermicompost applications differed for all three varieties tested. However, he reported RCr-41 produced the most herbage among the three. Maximum herbage yields from all three were occurred 60 days after sowing. The varieties RCr-41, Bulgarian, and Sakalespur Local attained largest yield at vermicompost applications rates of 15T/ha, 10-25T/ha, and 20T/ha, respectively.

Some field experiments have involved amending soils with vermicomposts in conjunction or combination with recommended inorganic fertilization programs. Vermicomposts applied at 12 t/ha to field soils together with 100% or 75% of the recommended application rate of inorganic fertilizers increased yields of okra (*Abelmoschus esculentus* Moench) significantly (Ushakumari *et al.*, 1999). Amending soils with vermicomposts, applied at rates of 2 kg/plant, together with 75% of the recommended rate of inorganic fertilizers, promoted shoot production of bananas (Athani *et al.*, 1999). Vermicompost applications to field soils combined with 50% of the recommended inorganic fertilizers increased the yields of tomatoes (Kolte *et al.*, 1999). Vasanthi and Kumaraswamy (1999) reported increased yields of rice after amending soils with vermicomposts at rates of 5t/ha or 10t/ha supplemented with recommended application rates of inorganic fertilizer. A lower application rate of 2t/ha vermicomposts plus recommended amounts of inorganic fertilizers, increased tomato yields to a level similar to those of tomatoes in soils treated with 4 t/ha vermicomposts and 50% of the recommended rates of inorganic fertilizers (Patil *et al.*, 1998). Tuberlets from a true potato seed line, produced the greatest marketable yields, after amending the soils with 75% of the recommended inorganic fertilizers and 2.5 t/ha vermicomposts (Mrinal *et al.*, 1998). Sunflowers (*Hilianthus annuus*) yielded most after soil treatments with 50% of the recommended application rates of inorganic fertilizer and 5t/ha or 10t/ha of vermicomposts (Devi *et al.*, 1998). Peas (*Pisum sativum*) increased yields and production after amending soils with 100% of the recommended application rate of inorganic fertilizers, in combination with vermicomposts produced from farm manures applied at rates of 10t/ha to soils (Ramachandra *et al.*, 1998). Zende *et al.* (1998) reported increased yields of sugarcane after amending soils with vermicomposts at rates of 5t/ha together with 100% of the recommended application rate of inorganic fertilizers. Mulberry (*Muros sp*) growth increased after amending soils with vermicomposts applied at rates of 10t/ha together with 100% of the recommended application rates of inorganic fertilizers to soils (Murakar *et al.*, 1998). Flowering of China aster (*Callistephus chinensis* L.) increased when it was grown in soils amended with 10 t/ha vermicomposts, produced from farm manures, together with 100% of the recommended application rate of inorganic fertilizers (Nenthra *et al.*, 1999). Similarly application of vermicompost resulted in increased fruit yield and quality characteristics of tomato crop under field conditions (Haribhushan *et al.*, 2008).

Conclusion and Future Outlook

Vermicompost constitutes one of the key components of organic farming. It has multifarious advantages over agrochemical in regard to soil and plant health. Use of vermicompost makes soil a living system by enhancing the microbial activities and thereby strengthening the inherent capacity of the soil. Living soil is the only supporting system for sustainable agriculture. Degradation of organic refuse through the use of vermin-worms in the production of vermicompost sustains and maintains the natural cycle or biological process. Thus recycling of plant nutrients on continual basis in the environment in the accent of organic farming. Vermicompost is the product of this natural process which is eco-friendly and safer method of converting wastes into wealth. Aristotle in words stated that if we want to feed plant feed the soil with matter of organic origin since soil is the stomach of plants.

REFERENCES

Acevedo I C, Pire R (2004) Effects of vermicompost as substrate amendment on the growth of papaya (*Carica papaya* L.) *Interciencia* 29(5): 274-279

Aira M, Gómez-Brandón M, Lazcano C, Bååth E, Domínguez J (2010) Plant genotype strongly modifies the structure and growth of maize rhizosphere microbial communities. *Soil Biology and Biochemistry* 42 (12): 2276-2281

Aira M, Monroy F, Domínguez J (2007) Earthworms strongly modify microbial biomass and activity triggering enzymatic activities during vermicomposting independently of the application rates of pig slurry. *Science of the Total Environment* 385: 252-261

Al-Dahmani JH, Abbasi PA, Miller SA, Hoitink HAJ (2003) Suppression of bacterial spot of tomato with foliar sprays of compost extracts under greenhouse and field conditions. *Plant Disease* 87: 913-919

Anonymous (2001) Vermicompost as Insect Repellent; *Biocycle*, Jan. 01: 19

Ansari AA, Rajpersaud J (2012) Physicochemical Changes during Vermicomposting of Water Hyacinth (*Eichhornia crassipes*) and Grass Clippings. *ISRN Soil Science* 2012: 1-6 doi:10.5402/2012/984783

Anwar M, Patra DD, Chand S, Kumar A, Naqvi AA, Khanuja SPS (2005) Effect of organic manures and inorganic fertilizer on growth, herb and oil yield, nutrient accumulation, and oil quality of French basil. *Communications in Soil Science and Plant Analysis* 36 (1-14): 1737-1746

Arancon NQ, Galvis P, Edwards CA, Yardim E (2003) The trophic diversity of nematode communities in soils treated with vermicompost. *Pedobiología* 47: 736-740

Arancon NQ (2004) An Interview with Dr. Norman Arancon; In Casting Call, Vol: 9 (2)

Arancon NQ, Edwards CA, Atiyeh RM, Metzger JD (2004a) Effects of vermicomposts produced from food waste on greenhouse peppers. *Bioresource Technology* 93: 139-144

Arancon NQ, Edwards CA, Bierman P, Welch C, Metzger JD (2004b) Influences of vermicomposts on field strawberries-1: Effects on growth and yields *Bioresource Technology* 93: 145-153

Arancon NQ, Galvis PA, Edwards CA (2005a) Suppression of insect pest populations and damage to plants by vermicompost. *Bioresource Technology* 96: 1137-1142

Arancon NQ, Edwards CA, Bierman P (2005b) Influences of vermicomposts on field strawberries: Part 2. Effects on soil microbiological and chemical properties. *Bioresource Technology* 97: 831-840

Arancon NQ, Edwards CI, Bierman P (2006) Influences of vermicomposts on field strawberries-2: Effects on soil microbiological and chemical properties *Bioresource Technology* 97: 831-840

Arancon NQ, Edwards CA, Yardim EN, Oliver TJ, Byrne RJ, Keeney G (2007) Suppression of two-spotted spider mite (*Tetranychus urticae*), mealy bug (*Pseudococcus* sp.) and aphid (*Myzus persicae*) populations and damage by vermicomposts. *Crop Protection* 26: 29-39

Arancon NQ, Edwards CA, Babenko A, Cannon J, Galvis P, Metzger JD, (2008) Influences of vermicomposts, produced by earthworms and microorganisms from cattle manure, food waste and paper waste, on the germination, growth and flowering of petunias in the greenhouse *Applied Soil Ecology* 39: 91-99

Argüello JA, Ledesma A, Núñez SB, Rodríguez CH, Díaz Goldfarb MDC (2006) Vermicompost effects on bulbing dynamics nonstructural carbohydrate content, yield, and quality of 'Rosado Paraguayo' garlic bulbs. *Hortscience* 41(3), 589-592

Athani SI, Hulamanai NC, Shirol AM (1999) Effect of vermicomposts on the maturity and yield of banana. *South Indian Horticulture* 47: 1-6, 4-7

Atiyeh RM, Dominguez J, Sobler S, Edwards CA (2000a) Changes in biochemical properties of cow manure during processing by earthworms (Eiseñia andrei) and the effects on seedling growth. *Pedobiologia* 44: 709-724

Atiyeh RM, Subler S, Edwards CA, Bachman G, Metzger JD, Shuster W (2000b) Effects of Vermicomposts and Composts on Plant Growth in Horticultural Container Media and Soil. *Pedobiologia* 44: 579-590

Atiyeh RM, Arancon NQ, Edwards CA, Metzger JD (2000c) Influence of earthworm processed pig manure on the growth and yield of greenhouse tomatoes. *Journal of Bioresource Technology* 75: 175-180

Atiyeh RM, Edwards CA, Subler S, Metzger JD (2001) Pig manure vermicompost as a component of a horticultural bedding plant medium: Effects on physicochemical properties and plant growth. *Bioresource Technology* 78: 11-20

Atiyeh RM, Edward CA, Arancon NQ, Metzger JD (2002) The influence of humic acids derived from earthworm-processed organic wastes on plant growth. *Bioresource Technology* 84: 7-14

Bachman GR, Metzger JD (2007) Physical and chemical characteristics of a commercial potting substrate amended with vermicompost produced from two different manure sources. *HortTechnology* 17(3): 336-340

Bevacqua RF, Mellano V (1993) Sewage sludge compost's cumulative effects on crop growth and soil properties. *Compost Science and Utilization. Spring 1993:34-37*

Bhattacharjee G, Chaudhuri PS, Datta M (2001) Response of paddy (Var. TRC-87- 251) crop on amendment of the field with different levels of vermicompost. Asian *Journal of Microbiology, Biotechnology and Environmental Sciences* 3 (3): 191-196

Biradar AP, Sunita ND, Teggelli RG, Devaranavadgi SB (1998) Effect of vermicomposts on the incidence of subabul psyllid. *Insect Environment* 4: 55-56

Brown GG (1995) How do earthworms affect microfloral and faunal community diversity? *Journal of Plant and Soil*, 170: 209-231

Bryan HH, Lance CJ (1991) Compost trials on vegetables and tropical crops. *Biocycle 32:36-37*

Buckerfield JC, Webster KA (1998) Worm-worked waste boosts grape yields: prospects for vermicompost use in vineyards. *The Australian and New Zealand Wine Industry Journal* 13: 73-76

Buckerfield JC, Flavel TC, Lee KE, Webster KA (1999) Vermicompost in Solid and Liquid Forms as a Plant-Growth Promoter *Pedobiologia* 43: 753-759

Bulluck LR, Brosius M, Evanylo GK, Ristaino JB (2002) Organic and synthetic fertility amendments influence soil microbial, physical and chemical properties on organic and conventional farms. *Applied Soil Ecology* 19:147-160

Cabanas-Echevarría M, Torres García A, Díaz-Rodríguez B, Ardisana EFH, Creme-Ramos Y (2005) Influence of three bioproducts of organic origin on the production of two banana clones (*Musa* spp AAB.) obtained by tissue cultures. *Alimentaria* 369: 111-116

Campitelli P, Ceppi S (2008) Chemical, physical and biological compost and vermicompost characterization: A chemometric study. *Chemometrics and Intelligent Laboratory Systems* 90: 64-71

Canellas LP, Olivares FL, Okorokova AL, Facanha RA (2000) Humic Acids Isolated from Earthworm Compost Enhance Root Elongation, Lateral Root Emergence and Plasma Membrane H+-ATPase Activity in Maize Roots. *Indian Journal of Plant Physiology* 130: 1951-1957

Canellas LP, Olivares FL, Okorokova AL, Facanha AR (2002) Humic acids isolated from earthworm compost enhance root elongation, lateral root emergence, and plasma H+ ATPase activity in maize roots. *Plant Physiology* 130: 1951-1957

Cavani N, Mimmo T (2007) Rhizodeposition of *Zea mays* L. as affected by heterosis. *Archives of Agronomy and Soil Science* 53: 593 - 604

Chadha S, Ashlesha Rameshwar, Saini JP, Paul YS (2012) *Vedic Krishi*: Sustainable livelihood option for small and marginal farmers. *Indian Journal of Traditional Knowledge* 11(3): 480-486

Chan PLS, Griffiths DA (1988) The vermicomposting of pre-treated pig manure. *Biological Wastes* 24: 57-69

Chand S, Pande P, Prasad A, Anwar M, Patra DD (2007) Influence of integrated supply of vermicompost and zinc-enriched compost with two graded levels of iron and zinc on the productivity of geranium. *Communications in Soil Science and Plant Analysis* 38: 2581-2599

Chaoui HI, Zibilske LM, Ohno T (2003) Effects of earthworms casts and compost on soil microbial activity and plant nutrient availability. *Soil Biology and Biochemistry* 35 (2): 295-302

Coria-Cayupán YS, De Pinto MIS, Nazareno MA (2009) Variations in bioactive substance contents and crop yields of lettuce (*Lactuca sativa* L.) cultivated in soils with different fertilization treatments. *Journal of Agricultural and Food Chemistry* 57(21) : 10122-10129

D'Errico FP, Di Maio F (1980) Effets des boues residuares compostees sur la culture et le development de la pedofaune associee a l'oellet. *Rev. Zool. Agric.* 79: 23-28

Dell'Agnola G, Nardi S (1987) Hormone-like effect and enhanced nitrate uptake induced by depolycondensed humic fractions obtained from *Allolobophora rosea* and *A. caliginosa* feces. *Biology and Fertility of Soils* 4:115-118

Devi D, Agarwal SK, Dayal D (1998) Response of sunflower (*Helianthus annuus*) to organic manures and fertilizers. *Indian Journal of Agronomy* 43(3): 469-473

Domínguez J, Aira M, Gómez Brandón M (2010) *Vermicomposting: earthworms enhance the work of microbes.* In: H. Insam, I. Franke-Whittle and M. Goberna, (Eds.), Microbes at Work: From Wastes to Resources (pp. 93-114) Springer, Berlin Heidelberg,

Donald DGM, Visser LB (1989) Vermicompost as a possible growth medium for the production of commercial forest nursery stock. *Applied Plant Sciences* 3:110-113

Edwards CA, Neuhauser E F (1988) *Earthworms in Waste and Environmental Management,* SPB Acad. Publ., The Hague, The Netherlands

Edwards CA, Burrows I (1988) *The potential of earthworm composts as plant growth media.* In: C.A. Edwards and E.F. Neuhauser (Eds) Earthworms in Waste and Environmental Management. (pp. 211-219) SPB Academic Publ. Co. The Hague, Netherlands

Edwards CA, Arancon N (2004) Vermicompost Supress Plant Pests and Diseases Attacks. In REDNOVA NEWS: http://www.rednova.com/display/ ?id =55938

Edwards CA, Arancon NQ, Greytak S (2006) Effects of vermicompost teas on plant growth and disease. *BioCycle* 47: 28-31

Edwards CA, *et al.* (2004) Earthworm ecology, 2nd ed. CRC Press, Boca Raton, Florida

Edwards CA, Dominguez J, Neuhauser EF (1998) Growth and reproduction of *Perionyx excavatus* (Perr.) (Megascolecidae) as factors in organic waste management. *Biology and Fertility of Soils* 27:155–161

El Harti A, Saghi M, Molina JAE, Téller G (2001a) Production d'une substance rhizogène à effet similaire à celui de l'acide indole acètique par le ver de terre *Lumbricus terrestris. Canadian Journal of Zoology* 79:1911-1920

El Harti A, Saghi M, Molina JAE, Téller G (2001b) Production des composes indoliques rhizogénes par le ver de terre *Lumbricus terrestris. Canadian Journal of Zoology* 79:1921-1932

Ferreras L, Gomez E, Toresani S, Firpo I, Rotondo R (2006) Effect of organic amendments on some physical, chemical and biological properties in a horticultural soil. *Bioresource Technology* 97: 635-640

Frankenberger Jr. WT, Arshad M (1995) Phytohormones in Soils: Microbial Production and Function. Marcel and Deckker Pub., New York, pp: 503

Gajalakshmi S, Abbasi SA (2004) Neem leaves as a source of fertilizer-cum-pesticide vermicompost. *Bioresource Technology* 92:291-296

Garai TK, Datta JK Mondal NK (2013) Evaluation of integrated nutrient management on*boro* rice in alluvial soil and its impacts upon growth, yield attributes, yield and soil nutrient status. *Archives Agronomy Soil Science in press* DOI 10.1080/03650340.2013.766721

Garcia-Gil JC, Plaza C, Soler-Rovira P, Polo A (2000) Long-term effects of municipal solid waste compost application on soil enzyme activities and microbial biomass. *Soil Biology and Biochemistry* 32 (13): 1907-1913

Gopinath KA, Supradip S, Mina BL, Pande H, Kundu S, Gupta HS (2008) Influence of organic amendments on growth, yield and quality of wheat and on soil properties during transition to organic production. *Nutrient Cycling in Agroecosystems* 82 (1) : 51-60

Grappelli A, Tomati V, Galli E, Vergari B (1985) Earthworm Casting in Plant Propagation. *Horticultural Science* 20: 874-876

Grappelli A, Galli E, Tomati U (1987) Earthworm casting effect on *Agaricus bisporus* fructification. *Agrochimica* 31 (4-5): 457-461

Gunadi B, Edwards CA, Arancon NA (2002) Changes in trophic structure of soil arthropods after the application of vermicomposts. *European Journal of Soil Biology* 38: 161-165

Gutiérrez-Miceli FA, Santiago-Borraz J, Montes Molina JA, Nafate CC, Abdud- Archila M, Oliva Llaven MA, Rincón-Rosales R, Deendoven L (2007) Vermicompost as a soil supplement to improve growth, yield and fruit quality of tomato (*Lycopersicum esculentum*) *Bioresource Technology* 98: 2781-2786

Gutiérrez-Miceli FA, García-Romero RC, Rincón-Rosales R, Abud-Archila M, Oliva Llaven MA, Guillen-Cruz MJ, Dendooven, L (2008) Formulation of a liquid fertilizer for sorghum (*Sorghum bicolor* (L.) Moench) using vermicompost leachate. *Bioresource Technology*

Han J, Sun L, Dong X, Cai Z, Yang H, Wang Y, Song W (2005) Characterization of a novel plant growth-promoting bacteria strain Delftia tsuruhatensis HR4 both as a diazotroph and a potential bio-control agent against various pathogens. *Syst. Applied Microbiology* 28: 66-76

Handreck K A (1986) Vermicomposts as components of potting media. *BioCycle* 27:58-62

Haribhushan A, Athokpam HS, Singh NB, Chongtham N, Singh NG (2008) Effect of vermicompost and boron availability, yield and uptake by tomato (*Solanum lycopersicum*) and soil fertility. *Environment Ecology* 26(1): 187-190

Hidalgo PR, Harkess RL (2002a) Earthworm casting as a substrate amendment for Chrysanthemum Production. *Hortscience* 37 (7), 1035-1039

Hidalgo PR, Harkess RL (2002b) Earthworm casting as a substrate for Poinsettia production. *Hortscience* 37 (2) : 304-308

Hidalgo PR, Matta FB, Harkess RL (2006) Physical and chemical properties of substrates containing earthworm castings and effects on marigold growth. *HortScience* 41: 1474-1476

Hoitink HAJ , Fahy P (1986) Basis for the control of soilborne plant pathogens with composts. *Annual Review of Phytopathology* 24 : 93-114

Hoitink HAJ (1993) Proceedings Review: International Symposium on composting research. *Compost Science and Utilization. Spring 1993: 37*

Hointik HAJ, Bohem MJ (1999) Biocontrol within the context of soil microbial communities: a substrate-dependent phenomenon. *Annual Review of Phytopathology* 37: 427-46

Hutchison ML, Walters LD, Avery SM, Munro F, Moore A (2005) Analyses of livestock production, waste storage, and pathogen levels and prevalence in farm manures. *Applied and Environmental Microbiology* 71: 1231–1236

Kabir Z, O'Halloran IP, Fyles JW, Hamel C (1998) Dynamics of the mycorrhizal symbiosis of corn (*Zea mays* L.): effects of host physiology, tillage practice and fertilization on spatial distribution of extra-radical mycorrhizal hyphae in the field. *Agriculture, Ecosystems and Environment* 68: 151-163

Kale RD, Mallesh BC, Kubra B, Bagyaraj DJ (1992) Influence of vermicompost application on the available macronutrients and selected microbial populations in a paddy field. *Soil Biology and Biochemistry* 24: 1317-1320

Karmegam N, Alagumalai K, Daniel T (1999) Effect of vermicompost on the growth and yield of green gram (*Phaseolus aureus* Roxb.) *Tropical Agriculture* 76: 143-146

Knapp BA, Ros M, Insam H (2010) Do Composts Affect the Soil Microbial Community? In: H. Insam, I. Franke-Whittle and M. Goberna, (Eds.), Microbes at Work: From Wastes to Resources. (pp 93-114) Springer, Berlin Heidelberg

Lazcano C, Arnold J, Tato A, Zaller JG, Domínguez J (2009) Compost and vermicompost as nursery pot components: Effects on tomato plant growth and morphology. *Spanish Journal of Agricultural Research* 7: 944-951

Lazcano C, Sampedro L, Zas R, Domínguez J (2010a) Vermicompost enhances germination of the maritime pine (*Pinus pinaster* Ait.) *New Forest* 39: 387-400

Lazcano C, Sampedro L, Zas R, Domínguez J, (2010b) Assessment of plant growth promotion by vermicompost in different progenies of maritime pine (*Pinus pinaster* Ait.) *Compost Science and Utilization* 18: 111-118

Lazcano C, Domínguez J (2010c) Effects of vermicompost as a potting amendment of two comercially-grown ornamental plant species. *Spanish Journal of Agricultural Research* 8 (4) : 1260-1270

Lazcanoa C, Domínguez J (2011)The use of vermicompost in sustainable agriculture: impact on plant growth and soil fertility. In Editor: Miransari M (Ed) Soil Nutrients. Nova Science Publishers, Inc.pp1-23

Lazcano C, Revilla P, Malvar RA, Domínguez J (2011) Yield and fruit quality of four sweet corn hybrids (*Zea mays*) under conventional and integrated fertilization with vermicompost. *Journal of the Science of Food and Agriculture* 91(7): 1244-1253

Lores M, Gómez-Brandón M, Pérez-Díaz D, Domínguez J (2006) Using FAME profiles for the characterization of animal wastes and vermicomposts. *Soil Biology and Biochemistry* 38: 2993-2996

Marinari, S, Masciandaro G, Ceccanti B, Grego S (2000) Influence of organic and mineral fertilizers on soil biological and physical properties. *Bioresource Technology* 72: 9–17

Masciandaro G, Ceccanti B, Garcia C (1997) Soil agro-ecological management: fertirrigation and vermicompost treatments. *Bioresource Technology* 59: 199-206

Maynard A (1993) Evaluating the suitability of MSW compost as a siol amendment in field-grown tomatoes. *Compost Science and Utilization. Spring 1993: p. 34-26*

Maynard A (1989) Agricultural composts as amendments reduce nitrate leaching. *Frontiers of Plant Science* 24:2-4

Mba C C (1983) Utilization of *Eudrilus eugeniae* for disposal of cassava peel. In: Satchell, J.E. (ed) Earthworm Ecology: From Darwin to Vermiculture. Chapman and Hall, London, pp. 315-321

Morra L, Palumbo AD, Bilotto M, Ovieno P, Picascia S (1998) Soil solarization: organic fertilization and grafting contribute to build an integrated production system in a tomato-zucchini sequence. *Colture-Protette* 27: 63-70

Mrinal S, Rajkhowa DJ, Saikia M (1998) Effect of planting density and vermicomposts on yield of potato raised from seedling tubers. *Journal of the Indian Potato Association* 25(3-4):141-142

Munroe G (2007) Manual of On-farm Vermicomposting and Vermiculture; Pub. of Organic Agriculture Centre of Canada, pp: 39

Murarkar SR, Tayade AS, Bodhade SN, Ulemale RB (1998) Effect of vermicomposts on mulberry leaf yield. *Journal of Soils and Crops* 8(1): 85-87

Nair J, Mathew K, Goen Ho (2007) Earthworms and composting worms-Basics towards composting applications. Paper at 'Water for All Life-A Decentralised Infrastructure for a Sustainable Future'; March 12-14, 2007, Marriott Waterfront Hotel, Baltimore, USA

Nakasone AK, Bettiol W, de Souza RM (1999) The effect of water extracts of organic matter on plant pathogens. *Summa Phytopathologica* 25 : 330-335

Nardi S, ArnoldiG, Dell'Agnola G (1988) Release of hormone-like activities from *Alloborophora rosea* and *Alloborophora caliginosa* feces. *Journal of Soil Science* 68:563-657

Nardi S, Morari F, Berti A, Tosoni M, Giardini L (2004) Soil organic matter properties after 40 years of different use of organic and mineral fertilisers. *European Journal of Agronomy* 21: 357-367

Neilson RL (1951) Earthworms and Soil Fertility; In Proc. Of 13th Conf. Of Grassland Assoc., New Plymouth, US, pp: 158-167

Neilson RL (1965) Presence of Plant Growth Substances in Earthworms, Demonstrated by the Paper Chromatography and Went Pea Test, Nature, (Lond.), 208: 1113-1114

Nenthra NN, Jayaprasad KV, Kale RD (1999) China aster (*Callistephus chinensis (L.) Ness*)

cultivation using vermicomposts as organic amendment. *Crop Research Hisar* 17: 209-215

Noble R, Coventry, E (2005) Suppression of soil-borne plant diseases with composts: A review. *Biocontrol Science and Technology* 15: 3-20

Orlikowski LB (1999) Vermicompost extract in the control of some soil borne pathogens. International Symposium on Crop Protection, 64: 405-410

Parle JN(1963) A microbiological study of earthworm casts. *Journal of General Microbiology* 31: 13-23

Patil MP, Humani NC, Athani SI, Patil MG (1998) Response of new tomato genotype Megha to integrated nutrient management. *Advances in Agricultural Research in India* 9: 39-42

Peyvast G, Olfati JA, Madeni S, Forghani A (2008) Effect of vermicompost on the growth and yield of spinach (*Spinacia oleracea* L.) *Journal of Food Agriculture and Environment* 6: 110-113

Piper S, DuPonte M, Sipes B, Dinges K (2005) Small-Scale Vermicomposting. *Cooperative Extention Service, University of Hawaii, Manao*:1-4

Pramanik PGK, Ghosh PK, , Banik G P (2007) Changes in organic-C, N, P and K and enzyme activities in vermicompost of biodegradable organic wastes under liming and microbial inoculants. *Journal of Bioresource Technology* 98: 2485-2494

Ramachandra R, Reddy MAN, Reddy YTN, Reddy NS, Anjanappa M, Reddy R (1998) Effect of organic and inorganic sources of NPK on growth and yield of pea (*Pisum sativum*) *Legume Research* 21: (1) 57-60

Ramesh P (2000) Effects of vermicomposts and vermicomposting on damage by sucking pests to ground nut (*Arachis hypogea*) *Indian Journal of Agricultural Sciences* 70 (5) : 334

Rao KR (2002) Induced host plant resistance in the management of sucking insect pests of groundnut. *Annals of Plant Protection Science* 10: 45-50

Reddy MV, Ohkura K (2004) Vermicomposting of rice-straw and its effects on sorghum growth. *Tropical Ecology* 45 (2) : 327-331

Ribeiro CF, Mizobusi EH, Silva DG, Pereira JCR, Zambolim L (1998) Control of Meloidogyne javanica on lettuce with organic amendments. *Fitopatologia Brasileira* 23: 42-44

Roberts P, Jones DL, Edwards-Jones G (2007) Yield and vitamin C content of tomatoes grown in vermicomposted wastes. *Journal of the Science of Food and Agriculture* 87:1957-1963

Rodda MRC, Canellas LP, Façanha AR, Zandonadi DB, Guerra JGM, De Almeida DL, De Santos GA (2006) Improving lettuce seedling root growth and ATP hydrolysis with humates from Vermicompost. II- Effect of Vermicompost source. *Revista Brasileira de Ciencia do Solo* 30 : 657-664

Rodriguez H, Fraga R (1999) Phosphate solubilizing bacteria and their role in plant growth promotion. J. of Biotechnological. *Advancement* 17: 319-339

Rodríguez JA, Zavaleta E, Sanchez P, Gonzalez H (2000) The effect of vermicompost on plant nutrition, yield and incidence of root and crown rot of Gerbera (Gerbera jamesonii H Bolus), *Fitopathologia* 35: 66-79

Ros M, Pascual JA, García C, Hernández MT, Insam H (2006) Hydrolase activities, microbial biomass and bacterial community in a soil after long-term amendment with different composts. *Soil Biology and Biochemistry* 38: 3443-3452

Saha S, Mina BL, Gopinath KA, Kundu S Gupta HS (2008) Relative changes in phosphatase activities as influenced by source and application rate of organic composts in field crops. *Bioresource Technology* 99: 1750-1757

SailajaKumari MS, Ushakumari K (2000) Effect of vermicompost enriched with rock phosphate on the yield and uptake of nutrients in cowpea *(Vigna unguiculata* l. walp) *Journal of Tropical Agriculture* 40: 27-30

Scheu S (1987) Microbial Activity and Nutrient Dynamics in Earthworms Casts. *Journal of Biological Fertility Soils* 5: 230-234

Scheu S (1993) There is an Earthworm Mobilizable Nitrogen Pool in Soil. *Pedobiologia;* 37: 243-249

Scott MA (1988) *The use of worm-digested animal waste as a supplement to peat in loamless composts for hardy nursery stock.* In: C.A. Edwards and E.F. Neuhauser, (Eds.). Earthworms in Environmental and Waste Management. (Pp. 231-229). SPB Acad. Publ. The Netherlands

Siddagangaiah BA, Vadiraj MR, Sudharshan MR, Krishnakumar V (1996) Standardization of rooting media for propagation of vanilla (*Vanilla planifolia* Andr.). *Journal of Spices* and *Aromatic Crops* 5: 131-133

Singh R, Sharma RR, Kumar S, Gupta RK, Patil RT (2008) Vermicompost substitution influences growth, physiological disorders, fruit yield and quality of strawberry (*Fragaria x ananassa* Duch.). *Bioresource Technology* 99: 8507-8511

Singh K (2009) Microbial and Nutritional Analysis of Vermicompost, Aerobic and Anaerobic Compost. 40 CP Honours Project for Master in Environmental Engineering; Griffith University, Brisbane, Australia; (Supervisors: Dr. Rajiv K. Sinha & Dr. Sunil Heart)

Singh R, Gupta RK, Patil RT, Sharma RR, Asrey R, Kumar A, Jangra KK (2010) Sequential foliar application of vermicompost leachates improves marketable fruit yield and quality of strawberry (*Fragaria x ananassa* Duch.). *Scientia Horticulturae* 124: 34-39

Singh R D (1993) Harnessing the Earthworms for Sustainable Agriculture. Institute of National Organic Agriculture, Pune, India, pp: 1-16

Singh UP, Maurya S, Singh DP (2003) Antifungal activity and induced resistance in pea by aqueous extract of vermicompost and for control of powdery mildew of pea and balsam. *Zeitschrift fur Pflanzenkrankheiten und Pflanzenschutz* 110: 544-553

Suhane RK (2007) Vermicompost (In Hindi); Pub. Of Rajendra Agriculture University, Pusa, Bihar; pp: 88

Sunil K, Rawat CR, Shiva D, Suchit KR (2005) Dry matter accumulation, nutrient uptake and changes in soil fertility status as influenced by different organic and inorganic sources of nutrients to forage sorghum (*Sorghum bicolor*). *Indian Journal of Agricultural Science* 75 (6): 340-342

Swathi P, Rao KT, Rao PA (1998) Studies on control of root-knot nematode *Meloidogyne incognita* in tobacco miniseries. *Tobacco Research* 1: 26-30

Szczech M, Rondomanski W, Brzeski M W, Smolinska U, Kotowski J F (1993) Suppressive effect of commercial earthworm compost on some root infecting pathogens of cabbage and tomato. *Biological Agriculture and Horticulture* 10: 47-52

Szczech M (1999) Supressiveness of vermicompost against *Fusarium wilt* of tomato. *Journal of Phytopathology* 147: 155-161

Szczech M, Smolinska U (2001) Comparison of suppressiveness of vermicompost produced from animal manures and sewage sludge against *Phytophthora nicotianae* Breda de Haar var. *nicotianae. Journal of Phytopathology* 149: 77-82

Tejada M, Gonzalez JL, Hernandez MT, Garcia C (2008) Agricultural use of leachates obtained from two different vermicomposting processes, *Bioresource Technology* 99: 6228-6232

Teotia SP, Duley FL, McCalla TM (1950) Effect of stubble mulching on number and activity of earthworms. Agricultural Experiment Research Station Bulletin, University of Nebraska College of Agriculture, Lincoln, N.E., pp: 165

Termorshuizen AJ, van Rijn E, van der Gaag DJ, Alabouvette C, Chen Y, Lagerlöf J, Malandrakis AA, Paplomatas EJ, Rämert B, Ryckeboer J, Steinberg C, Zmora Nahum S (2006) Supressiveness of 18 composts against 7 pathosystems: Variability in pathogen response. *Soil Biology and Biochemistry* 38: 2461-2477

Thankamani CK, Sivaraman K, , Kandiannan K (1996) Response of clove (*Syzygium aromaticum* (L.) Merr. & Perry) seedlings and black pepper (*Piper nigrum* L.) cuttings to propagating media under nursery conditions. *Journal of Spices and Aromatic Crops*. 5: 99-104

Thanunathan K, Natarajan S, Senthilkumar R, Arulmurugan K (1997) Effect of different sources of organic amendments on growth and yield of onion in mine spoil. *Madras Agricultural Journal* 84: 382-384

Tilman D, Cassman KG, Matson PA, Naylor R, Polasky S (2002) Agricultural sustainability and intensive production practices. *Nature* 418: 671-677

Tiwari SC, Tiwari BK, Mishra RR (1989) Microbial populations, enzyme activities and nitrogen phosphorus- potassium enrichment in earthworm casts and in surrounding soil of a pineapple plantation. *Journal of Biology and Fertility of Soils*; 8: 178-182

Tomati U, Galli E (1995) Earthworms, Soil Fertility and Plant Productivity. Proceedings of the International Colloquium on Soil Zoology. *Acta Zoologica Fennica* 196:11-14

Tomati U, Galli E, Grapppelli A, Dihena G (1990) Effect of earthworm casts on protein synthesis in radish (*Raphanus sativum*) and lettuce (*Lactuca sativa*) seedlings. *Biology and Fertility of Soils* 9: 288-289

Tomati V, Grappelli A, Galli E, (1983) Fertility Factors in Earthworm Humus. In Proc. of International Symposium on 'Agriculture and Environment: Prospects in Earthworm Farming; Rome, pp: 49-56

Tomati V, Grappelli AE, Galli (1987) The Presence of Growth Regulators in Earthworm-Worked Wastes. In Proceeding of International Symposium on 'Earthworms'; Italy; 31 March-5 April, 1985; pp: 423-436

Tomati V, Grappelli A, Galli E (1988) The Hormone like Effect of Earthworm Casts on

Plant Growth. *Biology and Fertility of Soils* 5: 288-294

Trevisan S, Pizzeghello D, Ruperti B, Francioso O, Sassi A, Palme K, Quaggiotti S, Nardi S (2010) Humic substances induce lateral root formation and expression of the early auxin-responsive IAA19 gene and DR5 synthetic element in Arabidopsis. *Plant Biology* 12: 604-614

Trillas MI, Casanova E, Cotxarrera L, Ordovás J, Borrero C, Avilés M (2006) Composts from agricultural waste and the *Trichoderma asperellum* strain T-34 suppress *Rhizoctonia solani* in cucumber seedlings. *Biological Control* 39: 32-38

Ushakumari K, Prabhakumari P, Padmaja P (1999) Efficiency of vermicomposts on growth and yield of summer crop okra (Abelmoschus esculentus Moench). *Journal of Tropical Agriculture*. 37: 87-88

Vadiraj B, Siddagangaiah A, Potty SN (1998) Response of coriander (Coriandrum sativum L.) cultivars to graded levels of vermicomposts. *Journal of Spices and Aromatic Crops* 7 (2) : 141-143

Vasanthi D, Kumaraswamy K (1999) Efficacy of vermicomposts to improve soil fertility and rice yield. *Journal of the Indian Society of Soil Science* 47(2): 268-272

Venkatesh, Patil PB, Patil CV, Giraddi RS (1998) Effect of in situ vermiculture and vermicomposts on availability and plant concentration of major nutrients in grapes. *Karnataka Journal of Agricultural Sciences*. 11: 117-121

Vinotha SP, Parthasarathi K, Ranganathan LS (2000) Enhanced phosphatase activity in earthworm casts is more of microbial origin. *Current Science* 79: 1158-1159

Wang D, Shi Q, Wang X, Wei M, Hu J, Liu J, Yang F (2010) Influence of cow manure vermicompost on the growth, metabolite contents, and antioxidant activities of Chinese cabbage (*Brassica campestris* ssp. chinensis). *Biology and Fertility of Soils* 46: 689-696

Warman PR, AngLopez MJ (2010) Vermicompost derived from different feedstocks as a plant growth medium. *Bioresource Technology* 101: 4479-4483

Weber J, Karczewska A, Drozd J, Licznar M, Licznar S, Jamroz E, Kocowicz A (2007) Agricultural and ecological aspects of a sandy soil as affected by the application of municipal solid waste composts. *Soil Biology and Biochemistry* 39: 1294-1302

Wilson DP, Carlile WR (1989) Plant growth in potting media containing worm-worked duck waste. *Acta Horticulture* 238: 205-220

Yardim EN, Arancon NA, Edwards CA, Oliver TJ, Byrne RJ (2006) Suppression of tomato hornworm (*Manduca quinquemaculata*) and cucumber beetles (*Acalymma vittatum* and *Diabotrica undecimpunctata*) populations and damage by vermicomposts. *Pedobiologia* 50: 23-29

Zaller JG (2006) Foliar Spraying of Vermicompost Extracts: Effects on Fruit Quality and Indications of Late-Blight Suppression of Field-Grown Tomatoes. *Biological Agriculture and Horticulture* 24: 165-180

Zaller JG (2007) Vermicompost as a substitute for peat in potting media: Effects on germination, biomass allocation, yields and fruit quality of three tomato varieties. *Scientia*

Horticulturae 112: 191-199

Zandonadi DB, Canellas LP, Rocha Façanha A (2006) Indolacetic and humic acids induce lateral root development through a concerted plasmalemma and tonoplast H+ pumps activation. *Planta* 225: 1583-1595

Zende GK, Ruikar SK, Joshi, SN (1998) Effect of application of vermicomposts along with chemical fertilizers on sugar cane yield and juice quality. *Indian Sugar* 48: 357-369

7

Breeding for Specific Adaptation to Increase the Yield Potential of Rice for Hill Regions

Subhash C. Kashyap and *Gulzar S. Sanghera*

Modern intensive cropping in irrigated agro-ecosystem has brought a substantial increase in the production of food grains. However, with increasing population, limited resources, uncertain climatic conditions, food security needs becomes more daunting for agricultural scientists. The technologies are required which can increase the grain yield of the crops and raise the income of farmers without deteriorating the quality of environment. Many factors of green revolution like irrigation, input responsive genotypes, fertilizers and other agrochemicals etc. have almost exhausted in most part of country. To achieve, 109 million tons target by 2025, we need to concentrate our research efforts on long term and short term strategies, where the later ones place more emphasis on effective transfer of the available technology to the farmers. We must understand the importance of technological innovations, thus increasing productivity of different crops in a sustained manner, thus bringing about an 'evergreen' revolution. To retain and make use of traditional wisdom, farmers and agriculturists (crop breeders) must optimally utilize technological advancements in agriculture for site specific varietal improvement programme. The objectives and priorities of genetic improvement of crop varieties should be defined for each region and cropping system, using the outputs from a preliminary characterization of the systems conducted in different sites. To harness sustainable production and productivity of different crops, adaptation to environmental conditions and crop management will be regarded as the most important objectives to ensure future adoption of new varieties. Increase in yield is always a priority for farmers. Moreover, grain characteristics must satisfy local requirements for household consumption and for sale of harvest surpluses. Finally, resistance to major insect pests and diseases is often a priority for site specific varietal improvement, especially where small farmers cannot access, or have difficulties in accessing, credit to buy pesticides. When seekin a specific adaptation the direct incorporation of new germplasm, whether landrace or exotic into an existing improved population could induce a setback in genetic progress in productivity, plant height, earliness or grain quality, thus moving away from the objectives However, the possibility of incorporating well-adapted local varieties into these population was considered in terms of contributing, for example, to adaptation to specific environmenta conditions, stable resistance to diseases, or a specific grain quality for household consumption An integrated pre-breeding procedure leading to the production of novel geneti combinations (designer genotypes) and participatory breeding jointly with farming familie for the development of location-specific varieties would help to combine genetic efficienc and diversity in a mutually reinforcing manner. This will help to avoid the danger inheren in spreading single genotypes over large areas and also sustainable agriculture needs fo sustenance of site specific varieties. Therefore, in present chapter attempt has been made to compile various conventional and non-conventional breeding approaches that are require for development of location specific crop varieties along with some case studies.

Introduction

Rice (*Oryza sativa* L.) is a staple food for 17 countries in Asia and the pacific, eight countrie in Africa, seven countries in Latin America and Caribbean and one in Near East. Rice provide two-thirds of calories intake of more than three billion people in Asia and one-third calori intake of nearly 1.5 billion people in Africa and Latin America (FAO, 1995), when all developin countries are considered together rice provides 27% dietary energy supply and 20% dietar protein. Globally 150 m ha are planted to rice with an average production of 400 m t/year. I India, it occupies about 24% of total gross cropped area thereby contributing 43% of tota food grain production and 46% of total cereal production. India produces about 97 mt of ric over an area of about 43 m ha with a productivity level of 2.25 t/ha which is much below th world average of 3.9 t/ha (Anonymous, 2010). China which is next to India in rice area (28. m ha) has an average productivity of 6.2 t/ha. In the State of Jammu & Kashmir rice, bein the principal food crop, is cultivated across diverse agro-ecological situations extendin from sub-tropical region (>1000 m asl) of Jammu to temperate cold high altitude areas (180C 2500 m amsl) of Kashmir (Ahmad *et al*, 1999). During 2009, 0.56 million tonnes of rice wer grown from 0.26 m ha with an average yield of about 0.22 t/ha (Anonymous, 2009).

In Kashmir the production environment is typically temperate, with cold winters an mild summers, characterized by short growing season, long days, cool nights and occurrenc of wide differences between maximum and minimum temperatures (Sanghera *et al.*, 2011 Compared with average yields of temperate regions of world, the rice productivity i disappointingly very low (0.26 t/ha) in Kashmir (Sanghera and Wani, 2008). The rice area i Kashmir can be broadly categorized into two distinct zones with entirely different agro-climati conditions. The higher belts, comprising areas between 1800-2500 m amsl, constitute 10-12% c total rice area. The growing season is restricted to 125-135 days. The lower belt or valley basi enjoys relatively favourable growing conditions of 140-145 days and lies between 1500-1800 r amsl. The varietal requirement of two zones is different with *Japonica* rices being mor predominantly grown in upper belts and *Indica* in the valley basin (Sanghera and Wani, 2008

The objectives and priorities of genetic improvement of crop varieties should be defined for each region and cropping system, using the outputs from a preliminary characterization of the systems conducted in different sites. To harness sustainable production and productivity of different crops, adaptation to environmental conditions and crop management will be regarded as the most important objectives to ensure future adoption of new varieties. ncrease in yield is always a priority for farmers. Moreover, grain characteristics must satisfy ocal requirements for household consumption and for sale of harvest surpluses. mprovements in rice farming will help not only farmers, farm workers, and their families out also the millions of poor people who are totally dependent on rice. An increase in rice production is strongly needed, and it should come primarily from higher yields from existing ice fields in the irrigated and favorable rainfed lowland ecosystems, as opportunities to ncrease rice planting area are limited. It is even more difficult to achieve higher rice yields n less favorable agricultural environments. The good news is that a wealth of valuable nowledge on rice farming exists from the research efforts. However, if technologies leveloped by rice research are not effectively delivered to farmers, their livelihoods are nlikely to improve. The challenge is to move this research-based rice farming knowledge rom the research centers to the doorsteps of rice farmers. When seeking a specific adaptation he direct incorporation of new germplasm, whether landrace or exotic, into an existing mproved population could induce a setback in genetic progress in productivity, plant height, arliness or grain quality, thus moving away from the objectives. However, the possibility f incorporating well-adapted local varieties into these populations was considered in terms f contributing, for example, the adaptation to specific environmental conditions, stable esistance to diseases, or a specific grain quality for household consumption. An integrated re-breeding procedure leading to the production of novel genetic combinations (designer enotypes) and participatory breeding jointly with farming families for the development of ocation-specific varieties would help to combine genetic efficiency and diversity in a mutually einforcing manner. This will help to avoid the danger inherent in spreading single genotypes ver large areas and also sustainable agriculture needs for sustenance of site specific varieties. herefore, in present chapter attempt has been made to compile various conventional and on-conventional breeding approaches that are required for development of location specific rop varieties with special reference to rice grown under temperate conditions.

Vhy Site Specific Varieties: Constraints in Hill Rice Production

The rice production constraints in Kashmir as noted above are essentially those ncountered elsewhere in temperate regions of the world. Apart from low temperature, ater shortages as a result of aberrant weather conditions, susceptibility to biotic stresses articularly blast and socio economic, technical, management, institutional, technology transfer nd adoption/ linkage problems limit the rice production (Sanghera and Wani, 2008). The IYVs bred for situations with assured nutrient and water supply could not largely replace ne traditional land races having tolerance to local adverse growing conditions in the region. ven promising HYVs for favourable conditions could not be grown as seed of the desired arieties are in short supply and are not available on time. High humidity prevailing in the gion during most part of the year causes quality deterioration of the seed (Bandey and anghera, 2002). Diverse and variable rice-growing ecologies prevail even in a small geographical area of a village Panchayat or a Block level. Upland, favourable shallow rainfed lowland, unfavourable deep-water and flood prone area, swampy land, and hilly area co-exist together in a block requiring completely different problem-solving approaches (Sanghera and Zarger, 2002). Heavy, erratic and torrential rain causes recurrent floods in many parts of the region. Inherent poor response of the local land races to high dose of fertilizer, its poor recovery due to various type of losses restrain the farmers from investing more on fertilizer. The major problems which hinder the rice production in Kashmir are essentially those of temperate regions associated with condition of low temperature (cold), high altitude, short growing season and insect pest and diseases (Sanghera and Zarger, 2002). The low temperature is the greatest concern of rice growers in the region. Rice crop is exposed to injury by cold from early seedling stage to the ripening stage. The problem is more severe in higher belts. The effects of cold are manifested as poor and delayed germination, stunted growth, seedling discoloration, tip burning and reduced tillering during vegetative phase. At the reproductive stage stress effects are considered more severe which result in delayed heading, poor panicle exertion, prolonged flowering period due to irregular heading spikelet degeneration , impaired seed fertility and abnormal grain formation (Sanghera *et al.*, 2011, Kaneda and Beachell, 1974). The critical low temperature for inducing sterility ranges between 15-17^{0}C in highly cold tolerant varieties and 17-19^{0}C for cold sensitive varieties at the meiotic stage of crop growth (Sanghera *et al.*, 2011). Although farmers have tried to deploy varieties possesses tolerance to low temperature and recommended cultural practices such as planting date and water depth during panicle development rice crop still suffers from high grain sterility due to variable climate. The dependence on yield for achieving the level of production needed to meet the household food requirements makes it imperative to develop varieties having high yield potential with tolerance to low temperate and other stresses.

Rice crops are found in various growth stages favouring perpetuation of different insects and pathogens. Among biotic stresses, diseases and weeds are of greater significance, whereas insect pests have a minor influence on rice production in Kashmir. Blast (*Magnaporthe grisea*) brown spot *(Helminthosporium oryzae* Breda diffean) sheath blight (*Rhizactoria solani* Kuch) are important rice diseases in the Valley. However, blast is the most devastating and of wide spread occurrence when favourable climatic conditions prevail for the pathogen and, if left unchecked, could inflict heavy yield losses.This frequently causes severe incidences of diseases like blast, sheath blight and sheath rot (Bhat *et al.*, 2001). Continuous cultivation of just one variety (monoculture) with high seed rate and nitrogen application coupled with high relative humidity (>90%) and moderate temperature (22-26^{0}C) promote disease development. Management practices aiming at controlling collateral hosts (*Echinochola causgalli, cypress* sp. etc.) and a combination of low population density and nitrogen fertilization are advocated to check the disease. Seed treatment and economic spray schedules devised to minimize the impact of disease are not without economic, health and environmental implications. Further, insecticides and fungicides to control these are not popular and are in short supply in the region.

Therefore, use of resistant varieties offers an easy, economic and most eco-friendly approach for managing the rice blast (Barman *et al.*, 2004). Nevertheless, because of the high variability of the pathogen genetic resistance generally breaks down shortly after new varieties are released. This envisages the germplasm enhancement through a systematic and well planned resistance breeding programme to cope with the problem. Using international blast differentials (Dular, Raminad Str.3, Caloro, Kanto-52, Zenith, NP 125, Usen and Sha-tiao-tsao) two races of blast ID-I and IC 17 have been identified in the valley (Anwar *et al.*, 2011). Identification of resistant / tolerant donors against blast and their utilization in varietal improvement continues to be the integral part of rice breading programme. It is thus imperative to find ways and means to lift the present yield level, optimize the use of various inputs such as water and fertilizer, in order to make the rice production efficient, cost effective, and suitable for resource poor farmers, sustainable and environment friendly.

Development strategies for sustainable production

Improving crop management will have major impacts on production but there are still several countries that do not have adequate varieties or access to improved germplasm. In addition, there are countries that have access to elite germplasm but lack the means to evaluate genetic material. Furthermore, few countries have established technology transfer programmes. Most of the small regions and province are deficient in all of the above. These have limited human and financial resources for rice research and development and it would be cost-ineffective for them to work independently. The solution is collective action in which states join forces and share the costs of rice research and development. It is apparent that immediate production gains can be obtained by bridging the yield gap in irrigated rice. The yield gap is significant in all countries and is mainly a function of inadequate crop management. Bridging the yield gap requires an effective technology transfer programme. Variety improvement is also important, especially in countries that have weak national programmes and limited access to improved germplasm. However, the possibility of incorporating well-adapted local varieties into these populations was considered in terms of contributing, for example, to adaptation to specific environmental conditions, stable resistance to diseases, or a specific grain quality for household consumption. These choices were based on the results of the diagnosis and the preliminary *in situ* PVS trials of diverse lines and varieties. The incorporation of local varieties as contributors of local adaptation genes into the exotic populations may be carried out in two ways, according to the type of population and the time needed to develop new varieties. It should be emphasized that, when seeking a specific adaptation or insect pest resistance, the direct incorporation of new germplasm, whether landrace or exotic, into an existing improved population could induce a setback in genetic progress in productivity, plant height, earliness or grain quality, thus moving away from the objectives as defined by (Chaves, 1997). Hence, new materials should first be incorporated through intermediate populations and then, after several crosses between these materials and the improved population, introduced into the principal population (Gallais, 1990). Morais *et al.*, (2000) presented the alternative of crossing each new parent with the population and evaluating the individual crosses, before mixing the seeds of all individual combinations. The yield potential of modern rice varieties has reached the plateau. There are also signs of yield decline in rice as found at various experimental farms and some research locations. Earlier, rice yields reaching 8-9 t/ha at Farm, but now it has not exceeded 6-7 t/ha in dry season and 5-6 t/ha in the wet season (Casman and Pingali, 1993). Similarly findings have been reported from other rice growing areas of the world therefore the challenge before rice scientists is how to sustain long term rice production. Fortunately, a yield gap between potential rice yields and actual on-farm yields still exists and improved site specific varieties will make an important contribution to closing this gap.

Elements of a breeding strategy

Genetic improvement is a basic component of national policies for raising agricultural production in many countries. Although breeding programmes run at international centres of the CGIAR (Consultative Group on International Agricultural Research) system have made an important contribution to major food crops, national breeding programmes (public or private) maintain a fundamental role for crop improvement in their target region (usually applicable to the whole country), particularly insofar as the exploitation of specific adaptation and yield stability characteristics are concerned, owing to their better knowledge of, and easier access to, local germplasm and cropping environments. These programmes aim to make correct decisions on a number of issues which comprise a breeding strategy (Simmonds, 1991). Decisions may concern, in particular:

- adaptation strategies, yield stability and other (e.g. crop quality) targets;
- genetic resources forming the genetic base (indigenous or exotic, from traditional or improved varieties);
- techniques for the recombination and introgression of useful genetic variation;
- variety type (e.g. single-cross hybrid, double-cross hybrid, improved population or synthetic variety, with regard to outbred species); and
- breeding plan and selection procedures (selection environments, indirect selection criteria, presence and extent of participatory breeding, experimental designs etc.).

The definition of a strategy with respect to G x E interactions may require decisions on most of these elements, namely: adaptation strategy and stability targets, genetic resources, variety type, breeding plan and selection procedures. Initial decisions may change with time as a consequence of new opportunities offered by scientific progress, experimental evidence, available funding, food security policies, changes in national seed systems, international cooperation etc., but they should remain consistent with the breeding objective. For example, the inconsistency between targeting also unfavourable areas and adopting genetic resources and selection procedures producing material specifically adapted to favourable environments has contributed to the partial failure of a number of breeding programmes carried out in the Green Revolution context (Simmonds, 1991; Ceccarelli, 1994).

Breaking current yield ceiling

According to several estimates there is a need to produce 30% more rice by 2030 to satisfy the growing demand without negatively affecting the sustainability of rice production (Brown, 1996). It is projected that the population will increase 93% during this period in the developing world, where most countries depend on rice as a staple food. At present, the rice production in Jammu and Kashmir is just 0.59 million tonnes per year as against the total requirement of 1.24 million tonnes. This is not sufficient to feed 0.85 million people

(Anonymous, 2006). This population growth will increase rice demand, which will have to be met with less land, water, labour and chemicals. A huge task is now proposed to rice breeders: to develop varieties with higher yield potential so that more rice can be produced from suitable lands. On the other hand, rice cultivars will have to tolerate several abiotic stresses like low or high temperature, submergence, drought and salinity. Further, studies were performed aiming to explore the natural diversity regarding stress tolerance (Ali *et al.*, 2006). This may be another strategy allowing the adaptation of rice farming to environments formerly not suitable to its use. The fundamental concern of rice breeders is to produce varieties with superior yields. Grain yield is the result of several components that are determined at various stages in the growth of rice (Yoshida, 1981), while changes to any components can effect yield. Therefore, breeder is concerned with special agronomic, morphological and physiological characteristics that make a plant suitable for a particular production method or growing environment.

Rice production was increased remarkably in the 1960s with the introduction of the semi-dwarfing gene, into traditional *indica* and *japonica* cultivars resulting in the "green revolution" in rice and probably the most well known triumph of agricultural research (Khush, 1999a). This recessive gene, *sd1* or *OsGA20ox2*, was detected in the Taiwanese native semi-dwarf variety, *Dee-geo-woo-gen*. The modification of plant architecture through the introduction of semi-dwarfism resulted in lodging resistance, and increased harvest index, and two to three times higher yield potential than that of varieties available prior to green revolution. The development and release of the variety IR8 derived from Dee-geo-woo-gen had twice the yield potential of the traditional, tall varieties (Khush, 2005). This and other new high-yielding varieties also presented multiple resistances to diseases and insects and consequently higher yield stability (Khush, 1999b).

In Kashmir major increases in rice production have occurred during the last 25 years, due to adoption of high yielding semi-dwarf varieties and improved management practices (Sanghera and Zarger, 2002). However, the rate of increase of rice production is lower (1.5% per year) than the rate of increase of population (1.8% per year). If this trend is not reversed, severe food shortage will occur in 1st quarter of century as it has been estimated that demand of rice will exceed production by next 25-30 years (Anonymous, 2006). To further increase yield potential of rice under temperate regions, prospects of new strategies like hybrid rice, ideotype breeding and some other approaches that are still not well explored are described below.

New plant type for higher genetic gain

To increase yield potential of rice beyond what was achieved by the high-yielding semi-dwarfs, breeders conceptualized a new plant type in the late 1980s. Modern semi- dwarf varieties derived from IR8 produce a large number of unproductive tillers and excess leaf area, causing mutual shading and reducing sink size and canopy photosynthesis (Dingkuhn *et al.*, 1991; Virk *et al.*, 2004). Plant physiologists have suggested that physical environment in the tropics is not a limiting factor for increasing rice yields. Maximum yield potential was estimated to be 9.5 t ha^{-1} during wet season and 15.9 t ha^{-1} during dry season (Yoshida, 1981).

It is well known that yield is a function of total dry matter or biomass and the harvest index. Therefore, to further increase the yield potential of rice, we have to either increase the total biomass production or harvest index or both. The IRRI Scientists conceptualized a plant type to increase the biomass to about 23 t ha^{-1} and harvest index to 0.55. Such a plant should produce a grain yield of about 12.5 t ha^{-1} or an increase of 25 % over the yield of existing high yielding varieties (Khush, 1995). The harvest index can be increased by increasing the production of energy stores in the grain or by increasing the sink size. The sink size can be increased by:

- Large number of spikelets per panicle or ear,
- Increased spikelet filling (High spikelet fertility),
- Slow leaf senescence,
- Maintenance of healthy root system, and
- Increased lodging resistance.

Biomass can be increased by both genetic manipulations as well as by better management practices. Varietal characteristics for increasing the biomass include:

- Establishment of desirable leaf canopy structure.
- Rapid leaf area development.
- Rapid nutrient uptake, and
- Increased lodging resistance.

Breeding efforts to develop a "New Plant-Type" (NPT) started in 1990, with the objective of developing an improved germplasm with low tillering capacity; few unproductive tillers; 200–250 grains per panicle; plant height of 90–100 cm; thick and sturdy stems; dark green, thick and erect leaves; vigorous root system; 100–130-d growth duration; and increased harvest index from 0.3 to 0.5 (Khush, 1995). This work started with tropical *japonica* germplasm identified as donors of these traits from Indonesia, Malaysia, Thailand, Myanmar, Laos, Vietnam, and the Philippines (Virk *et al.*, 2004). These NPT donors had the desired ideotype; however most were susceptible to tropical diseases and insects, characteristics that they inherited to their progeny. An additional difficulty was that farmers and consumers in tropical rice-growing countries prefer varieties with long and slender grains and intermediate to high amylose content (*indica* subspecies) (Virk *et al.*, 2004). Consequently, new modern high-yielding *indica* varieties/ elite lines were included in the hybridization program, and this 2nd generation of NPT lines had improved yield and higher resistance to diseases and insects. A major breakthrough in the NPT project has been the inclusion of *japonica* germplasm in a breeding program for tropical regions. Crosses between the two subspecies had so far very limited success, but there are notable exceptions, such as Tongil rice in Korea and Mahsuri in the rainfed lowland tropics (Virk *et al.*, 2004). Typically, traits from *indica* germplasm are being incorporated into tropical *japonica* breeding programs due to narrow genetic bases in *japonica* cultivars and the higher yield potential of *indica*. From the ideotype breeding, it became evident that the introgression of *japonica* genetic background into *indica* gene pool can lead to improvements in secondary traits such as lodging resistance and nutritional quality. Similarly to what has been done using tropical *japonica,* should we also start using temperate *japonica* to "mingle" for NPT breeding? May be this strategy could be exploited to improve rice breeding in several temperate areas world wide (Negrao *et al.*, 2008).

The main aim to work on new plant type (super rice) was to improve physiological aspect of modern rice for increasing yield potential. The potential yield of this would reach 15 t/ha, compared with the present potential yield of 10 t/ha. It is expected that new plant type (NPT) lines with all the desirable traits will help to feed 300 million more rice consumers (Khush, 1999b), when planted widely in tropics and subtropics regions. However, these lines could not be adopted for growing in the temperate areas as they lack cold tolerance, one of the most important adaptability traits of temperate rices. Thus, NPT lines could be useful parents for increasing the yield potential of temperate rice. In this regarcollaboration has been established with the rice improvement programme of international Rice research Institute (IRRI), Philippines. Preliminary studies on the adoption and stability of different NPT lines from international Network for genetic Evaluation of Rice (INGER) nurseries viz International Rice Cold Tolerant Nursery (IRCTN), International Irrigated Rice Observational nursery (IIRON) and International Rice Temperate Observational Nursery (IRTON) have been reflected under Kashmir conditions (Sanghera *et al.*, 2011) which can be used to breed high yielding NPT varieties adapted to local conditions of Kashmir valley.

Hybrid breeding towards increased yield potential

Hybrid rice offers yet another opportunity to boost the yield of rice. Hybrid rice has a yield advantage of 15-20 % over the conventional high yielding varieties (Virmani *et al.*, 1993). Commercial exploitation of hybrid rice has been demonstrated in China, which enabled it to reduce its rice area from about 34.4 M ha in 1978 to about 31.98 M ha in 1988 and at the same time increased its rice production from 136.9 Mt to 169. 1 Mt during the same period (Ma and Yuan, 2003; Tran and Nguyen, 1998). China is the leading producer of hybrid rice in the world (Swaminathan, 2006).

The success story of hybrid rice in China aroused spurts of interests at the IRRI, Philippines and in many national rice research programmes to intensify research on hybrid rice. The trials conducted at IRRI, Philippines and in several national programmes, viz. India, Vietnam and Malaysia, the rice hybrids out yielded the best check variety (Virmani, 1996). Many heterotic rice hybrids have been released for commercial cultivation in countries other than China (Takita, 2003; Mishra, 2003).

Rice is being an important cereal crop and staple food of Kashmir. Its demand is increasing day by day with the increase in human population. It is grown in both temperate and sub temperate zones of valley within an altitude of 1650m to 2200m amsl (Ahmad *et al.*, 1999). Though a few high yielding rice varieties have been released in the past, rice productivity in the valley has reached a plateau and chances of further yield enhancement are scanty due to low genetic variability in hill cultures (Sanghera and Zarger, 2002). However, hybrid rice offers an opportunity to boost the yield of rice under valley conditions as hybrid rice has a yield advantage of 15-20% over the conventional high yielding varieties (Virmani *et al.*, 1996). A good number of hybrids have been released in India (Mishra, 2003) but the initial evaluation conducted on these hybrids and their parental lines introduced from Directorate of Rice Research, Hyderabad has shown that hybrids developed from tropical and sub tropical areas as such were not suitable for cultivation under temperate condition of Kashmir (Sanghera *et al.*, 2003). The positive standard heterosis in grain yield reported in rice hybrids is attributed to the increased dry matter production due to increased leaf area index, higher crop growth rate and harvest index, high spikelet number and increased 1000 grain weight (Peng ***et al.***, 2003). Rice hybrids showing positive heterosis for adverse temperature, soil conditions and water regime will be of immense importance in developing rice hybrids for stress environments. The heterosis in the hybrids between modern high yielding *indica* varieties and the new plant type *japonicas* is even high approaching 25% (Virmani *et al.*, 1996). Such hybrids may yield as much as 15 t /ha. Thus, the combined approaches of varietal modification and hybrid vigour exploitation should help us to increase the yield potential of irrigated rice by 50 % (Virmani and Kumar, 2004). Recently, Sanghera *et al* (2010a) had developed SKAU 7A and SKAU 11A temperate CMS linesthat can be utilized for development of hybrids for temperate conditions. Further, some putative restorers and maintainers to these cyto-sterile lines have been identified (Sanghera *et al.*, 2010b; Sanghera and Hussain, 2012, 2013) so the following research strategies can be explored to further increase rice production in temperate areas like Kashmir.

Combination of ideotype and heterosis

Since Johan and Jeroen (2003) proposed the ideotype concept, several models have been proposed for super high-yielding rice: the low-tillering and large panicle model by Khush (1996), the bushy type and rapid-growing model by Huang (2003), the ideal plant type and huge rice model by Yang et al (1996), and the heavy panicle model by Zhou *et al.* (1997). The new plant type (NPT) being developed by IRRI might raise current yield by 20–25%. These models, yet to be proven in practice, provide the leading concepts for super high-yielding programs since they are based on certain theories and practical experiences. Based on the characteristics of in hybrid rice breeding, Yuan (1997) proposed a morphological model of super high-yielding rice in terms of (1) plant height, (2) the uppermost three leaves, (3) plant type, (4) panicle weight and number, (5) leaf area index (LAI) and ratio of leaf area to grains, and (6) harvest index of above 0.55.

Exploring inter-subspecific heterosis

As the heterosis of intersubspecific hybrids is much stronger than that of intervarietal hybrids, its use is one of the most feasible approaches for realizing super high yield. To exploit intersubspecific super high-yielding hybrid rice, the development of various lines with wide compatibility (WC), especially lines with a broad spectrum of compatibility, is important (Wan and Ikehashi, 1996). The differentiation of the rice subspecies *indica* and *japonica* is ancient, and there is a sterility barrier associated with the hybrids between these two subspecies. However, the discovery of wide compatibility gene has great potential to overcome *indica-japonica* hybrid sterility (Ikehashi and Araki, 1984). In addition, the fine location of the hybrid sterility locus *S-5* responsible for hybrid sterility in crosses between *japonica* and *indica* subspecies will be very useful for marker-assisted selection (MAS) of wide compatibility varieties (Zhang *et al.*, 2007). Development of *indica/japonica* hybrid varieties have shown immense yield potential compared with intra-subspecific rice hybrids due to the genome diversity (Gao *et al.*, 2005). It appears that this manmade "genepool blend" is enhancing the genetic variability deployed throughout the decades, in the majority of rice lands. In the *japonica* subspecies the advances in hybrid rice is problematic because there are few effective restorer lines, and the yield advantage is not as large as in *indica* inbred rice which may be due to the narrow genetic base of *japonica* gene pool and unstable sterility of CMS lines (Tao *et al.*, 2002). Therefore, in order to promote *japonica* hybrid rice development, it is essential to identify suitable *japonica* restorer lines. The genetic analysis of fertility restoration genes (*Rf*) in *indica* group is difficult because of variations in number and mapping positions of the various *Rf* genes involved. Probably *indica* varieties may contain additional unidentified genes that influence pollen fertility (Tada, 2007, Najeeb *et al.*, 2011). By incorporating WC genes into restorer lines and male sterile lines of indica, japonica, or intermediate type, each with different growth duration, various super high yielding hybrids will be developed for different ecological environments.

Utilization of favourable genes from wild Species

Wide hybridization is an important research strategy used for broadening the rice gene pool. It is often difficult to improve rice cultivars further by using the *O. sativa* genetic resources. Rice yields have been stagnant for 15–20 years in the rice producing regions where farmers have successfully adopted green-revolution technologies (Tilman *et al.*, 2002). The genetic diversity present in the wild relatives was crucial for the green revolution. An *sd1* gene variant that does not induce short-stature was preserved in the wild ancestor as hidden variation. Farmers selected this allele together with genes interacting with it, in order to obtain a higher yield, an identical variant has been maintained in several landraces (Nagano *et al.*, 2005). Additionally, several wild species having the AA genome (*O. sativa* L. f. *spontanea, O. perennis* and *O. glumaepatula*) have been an important source of CMS, a major resource to breed commercial rice hybrids (Brar and Khush, 1997). Nevertheless, several constraints such as low crossability, increased sterility, and limited recombination between chromosomes of wild and cultivated rice have inhibited the transfer of useful genes (Brar and Khush, 2002). Xiao *et al.*, (1996, 1998) analysed BC_2 test cross families from the interspecific cross (*O. sativa x O. rufipogon*) and found that *O. rufipogon* alleles at marker loci RM5 on chromosome 1 and RG256 on chromosome 2 were associated with an 18 and 7 percent increase in grain yield per plant. A total of 68 significant QTL were identified and of these, 35 (51%) had beneficial alleles derived from the phenotypically inferior *O. rufipogon* parent. Moncada *et al.*, (2001) used advanced backcross QTL analysis on *O. sativa x O. rufipogon* derivatives and found that certain regions of rice genome harbour genes which are useful in a range of environments. Molecular markers can be used to identify QTL from wild species responsible for transgressive segregation. Tagging of specific genes/QTLs introgressed from wild species allows the use of MAS (Rahman *et al.*, 2007). In the past years, several studies have reported the identification and tagging of valuable genes and QTLs. These include genes for yield enhancement (Xiao *et al.*, 1996), resistance to bacterial blight including *Xa21* (Khush *et al.*, 1990), *Xa23* and *Xa27(t)* (Song *et al.*, 1995; Zhang *et al.*, 1998), resistance to BPH (Ishii *et al.*, 1994) and resistance to blast through the gene *Pi-9t* (Amantebordeos *et al.*, 1992).

Hybrids between cultivated rice and AA genome wild species can be produced through normal procedures. Hybrids between rice and distantly related wild species, on the other hand, are usually difficult to produce; low crossability and abortion of hybrid embryos are common features in such crosses. Hybrids have been produced through embryo rescue between elite breeding lines or varieties and several accessions of wild species representing BBCC, CC, CCDD, EE, FF, GG, HHJJ and HHKK genomes. A number of useful genes for resistance to brown planthopper (BPH), white backed plant hopper (WBPH), bacterial blight (BB), blast and tungro disease have been transferred from wild species to rice (Jena *et al.*, 1991; Brar and Khush, 2002).

With the available tools offered by genomics we should now explore intensively the reservoir of wild relatives and incorporate useful genes into commercial *japonica* varieties grown in temperate regions. These alien genes are often very different from the ones available in the cultivated rice and therefore can expand the gene pool of rice which is of particular interest in the *japonica* subspecies.

Genotype x Environment Interactions in Breeding Stable varieties

The present superior-yielding varieties exhibit variable performance because of a high proportion of G x E interaction. There is a need to identify and release stably yielding varieties even on a specific-area basis, instead of relatively less stable varieties on a wide-area basis. There are strong genotypic differences among varieties for this interaction as well as methods for selecting varieties that are more stable across environments. Prior to releasing varieties, it is possible to select varieties with a stable performance even in unfavorable environments or management regimes.

Defining Adaptation Strategies

Setting adaptation strategies for breeding programmes and defining recommendation domains for cultivars are distinct objectives. As such, they may require partly different analytical approaches and provide different results with regard to the definition of subregions. The same data set may be analysed with both objectives in mind. However, the adaptation strategy objective focuses on the responses of a set of genotypes to obtain indications and generate predictions relative to future breeding material that may be produced from the genetic base of which the tested genotypes are assumed to be a representative sample. Assessing the value of a specific adaptation strategy, implying a distinct selection programme for each subregion rather than a unique selection programme for the whole target region, is of obvious interest to globally-oriented breeding programmes of large seed companies or international research centres, where the target region may include more than one country and very different environments. In this case, each subregion may include several countries. Specific adaptation, however, may also prove a valuable target for national breeding programmes, for which the yield gain derived from exploitation of GL interaction effects within the country can also help face the increasing competition exerted on local seed markets by international seed companies. For public institutions, the breeding of diversified, specifically adapted germplasm can be a major element of a research policy enforcing sustainable agriculture (Bramel-Cox *et al.*, 1991; Ceccarelli, 1994) by:

- maximizing the potential of different areas by fitting cultivars to an environment instead of altering the environment (possibly with costly or environment-unfriendly inputs, such as pesticides, fertilizers and irrigation) to fit widely adapted cultivars; and

- safe guarding crop biodiversity by increasing the number of varieties under cultivation, with positive implications for the stability of production at national level.

Furthermore, specific breeding may facilitate the technological adaptation of varieties by fixing characteristics of specific interest to sub-regions (for small-grain cereals, short straw for intensive cereal farming and long straw for extensive cereal-livestock systems; for cereal or food legume crops, different grain quality characteristics etc.).

Stable Performing Varieties

Agricultural techniques, following an assessment of their economic performance in regional yield trials, can be recommended: either widely over the target region or specifically for one subregion (Perrin, 1976; Shaner *et al.*, 1982). For specific recommendation, the recommendation domain of a given technique may be defined on the basis of geography alone, or also on the basis of farming practices (e.g. irrigated or rainfed cropping) or socio-economic constraints. In all cases, the information obtained from previous testing is exploited for predicting yield responses in coming years and, most frequently, in new locations. Superior yielding varieties are available (Chaudhary, 1996), which can take farmers' yield to 8.0 tonnes/ ha if grown properly. But their performance is variable due to higher proportion of Genotype X Environment (G X E) interaction. G X E interaction is a variety dependent trait (Kang, 1990; Gauch, 1992; Chaudhary, 1996). While the genetic reasons of stability in the performance may be difficult to understand, resistance to biotic and abiotic stresses, and insensitivity to crop management practices are the major reasons. There is a need to identify and release stable yielding varieties even on a specific area basis, as against relatively less stable but on a wide area basis. There are strong genotypic differences among varieties for this interaction, providing opportunities for selecting varieties which are more stable across environments and methods are available to estimate these (Kang, 1990; Gauch, 1992). Thus, two varieties with similar yield may have different degrees of stability. During the final selection process, before release, it is possible to select varieties which are more stable and thus giving stable performance even in poorer environments or management regimes.

Farmer's Participatory Varietal Selection

Peculiarity of Kashmir province lies in its peculiar rice-growing ecology and ethnic even in a small geographical area. The taste and requirements are specific as they prefer medium bold rice varieties with high amylose content. Therefore, it is wise to involve farmers in selecting the appropriate varieties from a few selected good performing ones so as to suit their local need. The first 5-10 varieties from co-ordinated trials can be multiplied and a mini-kit consisting of 2 kg seed of each variety along with local checks for various duration groups can be distributed at block levels and farmers' group can be invited and associated to select the best ones. This will allow the selection of location, specific best varieties and also their spread with minimal effort. In addition, it will provide feedback to the researchers about the problems and requirements of farmer. The present day's technologies are not percolating to the farmers, as their problems are not properly understood by the researchers. The farmer's participatory varietal selection approach opens the frontier to the researcher in properly accessing the difficulties of the farmers and their requirements. Thus the research gaps whenever exist can easily be filled up.

Cultivar Recommendations

For definition of an adaptation strategy and yield stability targets, the contribution of any of the tested genotypes (considered as a sample of the relevant genetic base) is equally important. Emphasis is therefore placed on the estimation of genotypic and genotype-environmental components of variance, and location similarity is assessed on the basis of adaptation patterns for all genotypes. Conversely, the most important information for variety recommendation concerns the response of, and comparison between, high-yielding genotypes. There are four possible conclusions, implying a general recommendation for the target region or a specific recommendation for distinct sub-regions and, in cases, the inclusion or exclusion of yield stability in the assessment of genotype merit.

Conclusion

A quantum jump in rice yield was possible using semi dwarf cultivars that responded to increased use of fertilizers, pesticides, herbicides, and a most of other chemicals, along with water. In India, rice production grew at 1.11% from 1994 to 2001 (Venkataramani, 2002). The biological pathways for raising the ceiling to yield included both an increase in total biomass and higher harvest index. It is essential that the capacity of the plant to produce higher biomass per day is enhanced, because the scope for yield improvement through the harvest index pathway has been practically exhausted. But to raise the yield ceiling by breaking the yield barrier, new approaches need to be implemented vigorously. These could be feasible by using the concepts of hybrid rice and the new plant type (Yang *et al.*, 2007). In addition to the morphological and physiological attributes, tolerance or resistance to a wide range of biotic and abiotic stresses will also be necessary. Several donors from tropical germplasm are available in cultivars with good genetic backgrounds (Negrao *et al.*, 2008). The greatest source of rice improvement research is the availability of a wide range of germplasm in the International Rice Gene Bank at IRRI as well as in the *ex situ* gene banks of several countries like India, China, and Japan. With the advent of molecular breeding technologies, it is also becoming possible to transfer genes from wild *Oryza* species. Stabilization of already achieved yield levels by developing host plant resistance through the integration of conventional and molecular breeding approaches; raising yield frontiers further in irrigated areas through exploitation of hybrid vigour and restructuring of plant types and special efforts to improve the production potential of ecologically handicapped rice areas should continue to receive major research thrust to achieve the envisaged enhanced rice production.

REFERENCES

Ahmad N, Sanghera GS, Zarger MA and Rather MA (1999) Status of rice production in World, India and Kashmir. Paper presented in training on "Rice Production Technology" at Directorate of Extension Education, SKUAST (K) - Shalimar. pp 1-15

Ali AJ, Xu JL, Ismail AM (2006) Hidden diversity for abiotic and biotic stress tolerances in the primary gene pool of rice revealed by a large backcross breeding program. *Field Crops Res* **97**: 66–76

Amantebordeos A, Sitch LA and Nelson R (1992) Transfer of bacterial-blight and blast resistance from the tetraploid wild rice *Oryza minuta* to Cultivated Rice, *Oryza sativa*. *Theor Appl Gen* 84: 345–54

Anonymous (2010) Annual Progress Report, Crop Imrovement, Directorate of Rice Research, Rajendranagar, Hyderbad, AP

Anonymous (2009) Statistical Digest. Directorate of Economics and Statistics, Government of Jammu and Kashmir

Anonymous (2006) A bridged account of area, production and productivity since independence. Government of Jammu and Kashmir, Directorate of Agriculture, Kashmir. pp 5-7

Anwar A, Teli MA, Bhat GN, Parray GA, Wani SA (2011) Characterization of physiological races of *Pyricularia grisea* in temperate agro-ecosystem of Kashmir, India. *Indian Phytopathology* 64 : 52-56

Ashikari M, Matsuoka M (2002) Application of rice genomics to plant biology and breeding. Bot Bull Acad Sin 43: 1-11

Bandey AH and Sanghera GS (2002) Seed quality assurance during production, processing and storage. Presented in a short course on "Recent Advances in Rice Production Technology in the Hills" held on 1-10 August, 2002, at Division of Plant Breeding and Genetics, SKUAST (K)- Shalimar. pp 32-41.

Barman SR, Gowda M, Venu RC, Chattoo BB (2004) Identification of a major blast resistance gene in rice cultivar "Tetep". *Plant Breeding* 123: 300-302

Bhat GN, Anwar A, Sanghera GS and Rather MA (2001) Occurrence of fungal diseases and their severity on rice under Kashmir conditions. National Seminar on "Agriculture and Environment" at SKUAST, Shalimar, March 28-29, 2001. pp 57.

Bramel-Cox P1, Barker T, Zavala-Garcia F and Eastin JD (1991) Selection and testing environments for improved performance under reduced-input conditions. Pp. 29-56, In: Plant Breeding and Sustainable Agriculture, Considerations for Objectives and Methods (D.A. Sleeper, T.C. Barker and P.J. Bramel-Cox, eds.). *CSSA Special Publication No* 18

Brar DS and Khush GS (1997) Alien introgression in rice. *Plant Mol Biol* **35**: 35–47

Brar DS and Khush GS (2002) Transferring genes from wild relatives into rice In: Kang MS (ed) quantitative genetics, genomics and plant breeding CAB International Wallingford (UK), 197–217

Brown LR (1996) Tough Choices: facing the challenge of food scarcity. New York, USA

Casman K and Pingali P (1993) Extrapolating trends from long-term experiments to farmers fields: the case of irrigated rice systems in Asia, Paper presented at the Working Conference on Measuring Sustainability using Long Term Experiments, Rothamsted Experiment Station, April.

Ceccarelli S (1994) Specific adaptation and breeding for marginal conditions. *Euphytica* 77: 205-219.

Chaudhary RC (1996) Internationalization of elite germplasm for farmers: Collaborative mechanisms to enhance evaluation of rice genetic resources. In: New Approaches for Improved use of Plant Genetic Resources; Fukuyi, Japan; pp. 26.

Chaves LJ (1997) Criterios para escoger progenitores para un programa de selección recurrente. In E.P. Guimarães, ed. Selección recurrente en arroz, pp. 13-24. Cali, Colombia, CIAT.

Dingkuhn M, Penning de Vries FW Tand De Datta SK (1991) Concepts for a new plant type for direct seeded flooded tropical rice In: Institute IRR (ed) Direct-seeded flooded rice in the tropics Los Ban˜os (Philippines), pp 17–38

FAO (1995) World Rice Information, issue No.1. FAO, Rome, Italy.

Gao LZ, Zhang CH, Chang LP (2005) Microsatellite diversity within *Oryza sativa* with emphasis on *indica- japonica* divergence. *Genet Res* 85: 1–14.

Glaszmann JC (1987) Isozymes and classification of Asian rice varieties. *Theor Appl Gen* **74:** 21–30

Gallais A (1990) Théorie de la sélection en amélioration des plantes. Collection des sciences agronomiques. Paris, Masson. 588 pp.

Huang Y (2003) Construction and advancement of rice ecological breeding system. *World Sci Tech Res* 4: 1-8 (In Chinese.)

Ikehashi H and Araki H (1984) Varietal screening of compatibility types revealed in F1 fertility of distant crosses in rice. *Jap J Breed* 34: 304–13.

Ishii T, Brar DS, Multani DS and Khush GS (1994) Molecular tagging of genes for brown planthopper resistance and earliness introgressed from *Oryza australiensis* into cultivated rice, *O. sativa. Genome* 37: 217-21

Jena KK, Multani DS and Khush GS (1991) Monosomic alien addition lines of *Oryza australiensis* and alien gene transfer. *Rice Gen* 2: 728-29

Johan DP, Jeroen RV (2003) Breeding by design. *Trends Plant Sci* 8: 330-334

Kaneda C; Beachell HM (1974) Response of indica/ japonica rice hybrids to low temperatures. *SABRAO J* 6: 17–32.

Kang MS (1990) Genotype-by-environment interaction in plant breeding. Louisiana State University, Agricultural Centre, Baton Rouge, Louisiana, USA.

Khush GS (1995) Breaking the yield frontier of rice. *Geo Journal* 35 (3): 324-32

Khush GS (1999a) New plant type of rice for increasing the genetic yield potential. In: Rice Breeding and Genetics, Research priorities and challenges, Nanda, J.S. (ed.) pp. 99-108

Khush GS (1999b) Green revolution: preparing for the 21st century. *Genome* **42**: 646–55

Khush GS (2005) What it will take to feed 5.0 billion rice consumers in 2030. *Plant Mol Biol* **59**: 1–6

Khush GS, Bacalangco E, Ogawa T (1990) A new gene for resistance to bacterial blight from *O. longistaminata. Rice Genet Newsl* **7**: 121–123

Khush GS (1996) Prospects of and approaches to increasing the genetic yield potential of rice. In: Evenson RE et al, editors. Rice research in Asia: progress and priorities. Wallingford (UK): CAB International and IRRI. p 59-71

Ma GH and Yuan LP (2003) Hybrid rice achievements and development in China. In: Virmani S S, Mao C X, Hardy B, eds. Hybrid rice for food security, poverty alleviation, and environmental protection. Proceedings of the 4th International Symposium on Hybrid Rice, Hanoi, Vietnam, 14-17 May 2002. Los Baños (Philippines): International Rice Research Institute. pp 247-256.

Mishra B (2003) Rice research in India- Major achievements and future thrusts. Presented in training on "Advances in hybrid rice technology" Directorate of Rice Research, Hyderabad, 10-30, Sept. 2003, pp1-15

Moncada P, Martinez CP, Borrero J, Chatel M, Gauch H Jr., Guimaraes E, Tohme J, McCouch SR (2001) Quantitative trait loci for yield and yield components in an *Oryza sativa x Oryza rufipogon* BC2F2 population evaluated in an upland environment. *Theor Appl Genet* 102: 41-52

Morais OP, Castro E da M, Sant-Ana EP and Neto FP de M (2000) Evaluación y selección de los progenitores: Población CG2 de arroz de terras altas. *In* E.P. Guimarães, ed. *Avances en el mejoramiento poblacional en arroz*, pp. 210-220. San Antônio de Goiás, Brazil, Embrapa Arroz e Feijão.

Nagano H, Onishi K and Ogasawara M (2005) Genealogy of the "Green Revolution" gene in rice. *Genes Genet Syst* 80: 351–56

Najeeb S, Zargar MA, Rather AG, Hassan B, Sheikh FA, Ahanger MA, Razvi SM, Dar ZA and Sanghera GS (2011) Hybrid sterility and role of wide compatibility variety in different genetic backgrounds of rice (*Oryza sativa* L.) under temperate conditions of Kashmir. *Applied Biological Research* 13(1): 32-37

Negrao S, Oliveira MM, Jena KK and Mackill D (2008) Integration of genomic tools to assist breeding in the *japonica* subspecies of rice. *Mol Breed* 23: 143-154

Peng S, Yang J, Laza RC, Sanico AL, Visperas RM, Son TT (2003) Physiological bases of heterosis and crop management strategies for hybrid rice in the tropics. In: Virmani SS, Mao CX, Hardy B, editors. Hybrid rice for food security, poverty alleviation, and environmental protection. Proceedings of the 4th International Symposium on Hybrid Rice, Hanoi, Vietnam, 14-17 May 2002. Los Baños (Philippines): International Rice Research Institute. pp 153-172

Perrin RK (1976) From agronomic data to farmer recommendations: An Economic Training manual. CIMMYT, Mexico City.

Rahman ML, Chu SH and Choi MS (2007) Identification of QTLs for some agronomic traits in rice using an introgression line from *Oryza minuta. Mol Cells* 24: 16–26

Sanghera GS Ahmad N, Zarger MA and Rather MA (2003) Studies on the performance of some CMS lines under temperate conditions of Kashmir. *SKUAST J Res* 5: 121-24

Sanghera GS and Wani SH (2008) Innovative approaches to enhance genetic potential of rice for higher productivity under temperate conditions of Kashmir. *The Journal Plant*

Science and Research 24: 99-113

Sanghera GS and Zarger MA (2002) Breeding strategies for Improvement of Yield, Maturity, Cold and Drought in Rice. Presented in a Seminar on "Strategies to Combat Biotic and Abiotic Stresses in Rice" at Directorate of Research, SKUAST (K) - Shalimar, January 9, 2002

Sanghera GS, Hussaini AM, Anwer A and Kashyap SC (2011) Evaluation of some IRCTN rice genotypes for cold tolerance and leaf blast disease under temperate Kashmir conditions. *Journal of Hill Agriculture* 2(1): 28-32

Sanghera GS, Husaini AM, Rather AG, Parray GA and Shikari AB (2010a) SKAU 7A and SKAU 11A: New cold tolerant CMS lines from Kashmir, India. *International Rice Research Notes* 35: 1-3

Sanghera GS, Husaini AM, Parray GA, Rather AG, Shikari AB and Wani SA (2010b) Generation of cold tolerant CMS lines of rice and identification of maintainers/ restorers for hybrid rice development in Kashmir. *The Indian Journal of Crop Sciences* 5(1-2): 143-146

Sanghera GS and Hussain (2012) Heterosis in relation to combining ability *per se* Performance in temperate rice (Oryza sativa L.). *Pakistan J Agric Res* 25: 186-195

Sanghera GS and Hussain (2013) Gene action and combining ability studies using CMS system for developments of hybrid rice under temperate conditions. 1: 27-44 *American Journal of Agricultural Science and Technology*

Shaner WW, Phillip PF and Schmehl WR (1982) Farming Systems Research and Develognent: Guidelines for Developing Countries. Westview Press. Boulder, Colorado 1982.

Simmonds NW (1991) Selection for local adaptation in a plant breeding programme. *Theor Appl Genet 82:* 363-367

Song WY, Wang GL, Chen LL (1995) A receptor kinase like protein encoded by the rice disease resistance gene, *Xa21*. *Sci* 270: 1804–1806

Swaminathan MS (2006) An evergreen revolution. *Crop Sci* 46: 2293-2303

Tada Y (2007) Effects of *Rf-1*, *Rf-3* and *Rf-6(t)* genes on fertility restoration in rice (*Oryza sativa* L.) with WA- and BT-type cytoplasmic male sterility. *Breed Sci* 57: 223–229

Takita T (2003) Hybrid rice research and development in Japan. In: Virmani S S, Mao C X, Hardy B, editors. Hybrid rice for food security, poverty alleviation, and environmental protection. Proceedings of the 4th International Symposium on Hybrid Rice, Hanoi, Vietnam, 14-17 May 2002. Los Baños (Philippines): International Rice Research Institute. pp 337-340.

Tao D, Xu P, Li J (2002) Inheritance of *japonica* upland rice restoration lines and their restoration gene mapping. *Rice Gen News* 19: 23–26

Tilman D Cassman KG, Matson PA (2002) Agricultural sustainability and intensive production practices. *Nature* 418: 671–677

Tran DT, Nguyen VN (1998) Global hybrid rice: Issues and challenges. Paper presented 19th session Int. Rice Comm., FAO, Cario, Egypt, Sept. 7-9

Venkataramani G (2002) Policies need to be farmer-friendly. pp. 5–7. In: Survey of Indian agriculture. The Hindu, Tamil Nadu, India.

Virk PS, Khush GS, Peng S (2004) Breeding to enhance yield potential of rice at IRRI: the ideotype approach. *IRRN* 29: 5–9

Virmani SS (1996) Hybrid rice. *Adv. Agron.* 57: 377-462

Virmani SS, Kumar I (2004) Development and use of hybrid rice technology to increase rice productivity in the tropics. *IRRN* 29: 10-20

Virmani SS, Prasad MN, Kumar I (1993) Breaking yield barrier of rice through exploitation of heterosis. In: Muralidharan, K and Siddiq, E A (Eds.). *New frontier in rice research*, pp. 76-85

Wan J, Ikehashi H (1996) Two new loci for hybrid sterility in cultivated rice. *Theor Appl Genet* 92: 183-190

Xiao JH, Grandillo S, Ahn SND and McCouch SR (1996) Genes from wild rice improve yield. *Nature* 384: 223–224

Xiao J, Li J, Grandillo S, Ahn SN, Yuan L, Tanksley SD, McCouch SR (1998) Identification of trait-improving quantitative trait loci alleles from a wild rice relative. *Genetics* 150: 899-909

Yang S, Zhang B, Chen W, Xu Z, Wang J (1996) Theories and methods of rice breeding for maximum yield. *Acta Agron Sin* 22(3): 295-304

Yang W, Peng S, Laza RC, Visperas RM, Dionisio-Sese ML (2007) Grain yield and yield attributes of new plant type and hybrid rice. *Crop Sci* 47: 1393–1400

Yoshida S (1981) Fundamentals of rice crop science. IRRI, P.O.Box 933, Manila Philippines.

Yuan L (1997) Hybrid rice breeding for super high yield. *Hybrid Rice* 12(6): 1-6

Zhang Q, Lin SC, Zhao CL (1998) Identification and tagging a new gene for resistance to bacterial blight (*Xanthomonas oryzae pv oryzae*) from *O. rufipogon*. *Rice Gen News* 15: 138-42

Zhang XJ, Chen YZ, Wei YP (2007) Fine location of the *S-5* locus responsible for wide compatibility in rice using SSR markers. *Cereal Res Com* 35: 1–10

Zhou K, Wang X, Li S, Li P, Li H, Huang G, Liu T, Shen M (1997) The study on heavy panicle type of inter-subspecific hybrid rice (*Oryza sativa* L.). *Sci Agric Sin* 30(5): 91-93

8

Carbon Metabolism in Transgenic Plants

Pushp Sharma

Genetically modified crops have been developed for crop improvement. Carbon metabolism in the photosynthetically active cells of higher plants is of prime importance for allocation of carbohydrates formed in the tissues that are dependent on the supply of carbon and energy. Many control mechanisms influence carbohydrate formation and partitioning. Manipulation of chloroplast, the site of photosynthesis; biosynthetic pathways involved in carbon fixation, Rubisco, supercharging photosynthesis are briefly discussed. These results lead to the reassessment of ideas about the regulation of carbon metabolism and have consequences for design of bioengineering strategies to increase crop productivity.

Introduction

Transgenic plants are crops which have been genetically modified with genes from another organism to make the plants more agriculturally productive. Transgenic plants are only those with genes from other species whereas genetically modified plants can have both new genes and a re-arrangement of genes already found in plant. Traditional breeding methods are one form of genetic modification. Genetic engineering has given an opportunity to manipulate genes for getting desired plants or animals. Recent advances in plant molecular biology, tissue culture, recombinant DNA technology and also with the advent of modern genetic engineering new avenue have opened in transformation of higher plants, thus overcoming various problems and have given transgenic plants that contain desired characteristics. Transgenic plants have a single or multiple genes transferred from a different species. In the natural processes DNA from another species can be integrated into a plant's genome. The term "transgenic plants" refers to plants which are created in laboratory using recombinant DNA technology. The aim of creating transgenic plants is to design plants which have specific characteristics by using artificial insertion technique of genes from other species.

For improvement in crop yield genetically modified crops have been developed because carbon metabolism in photosynthetically active cells of higher plants is of central importance for the allocation of carbohydrates formed to the tissues that are dependent on the supply of carbon and energy. Many controlled mechanisms influence carbohydrate formation and partitioning into different pools: Photosynthetic process occur exclusively within the chloroplast and can be divided into light dependent and light independent reactions. The light dependent reactions convert electromagnetic energy into biologically convertible energy forms like ATP and NADPH. These are used to reduce CO_2 in the reductive pentose phosphate cycle (Calvin cycle) reactions which are chemically light independent but can be light regulated.

Chloroplast manipulation

The manipulation of organelle DNA has become a realistic possibility now. The first report of chloroplast transformation was due to co-cultivation of *Nicotiana tabacum* and *Agrobacterium*. Some of the transformants obtained with pGV 3850: pNOSCA17 showed a material inheritance of the introduced chlorophenicol acetyl transferase (CAT) activity. This activity could be localized in the chloroplast; southern blots revealed that the integrated DNA fragment was not stably maintained. This was one of the major problems and many laboratories started to develop more sophisticated vectors. The first step was to fuse chloroplast promoters to the selection marker to ensure a high level of expression in the chloroplast. Next step was the equipment of vectors with plastid DNA sequences that allow homologous recombination without disturbing functional regions. With a similar construct, PGV 2260 containing NPT II gene fused to the psbA promoter flanked by a 3rbcL gene fragment, *Agrobacterium* co-cultivation experiments with tobacco protoplasts were performed. However, the unexpected nuclear activity of this construct, no possible putative chloroplast transformation was detected.

Current methods for DNA delivery into chloroplast include particle bombardment and polyethylene glycol. The particle bombardment is very expensive and polyethylene glycol method requires preparation of protoplasts and fermtosyringe method that allows microinjection of foreign DNA and other substances into prokaryotic cells and eukaryotic organelles. The essence of the method is the heat induced expansion of a liquid metal called galistan (alloy of gallium, indium and tin) within a glass syringe to expel sample through a capillary tube with a diameter of about 0.1μm. Khan and Maliga (1999) described the use of new fluorescent antibiotic resistance enzyme conferring resistance to spectinomycin/ streptomycin (FLARE-S) to detect chloroplast transformation FLARE-S was created by fusing the antibiotic resistance enzyme to the GFP from jellyfish *Aequorea victoria*.

Commercial useful chloroplast transgenic plants, expressing high levels of herbicide resistance (Daniell *et al.*, 1998) or insect resistance (Kota *et al.*, 1999) have been engineered. Gud *et al.* (2000) reported exploitation of the chloroplast transformation approach for molecular farming by the genetic engineering of a value added trait that is not advantageous or required by transgenic plants. Bioelastic protein-based polymers (PBP) have several medical and non-medical applications. They have compared the expression levels of PBP genes (synthetic) integrated into the nuclear genome or the large single copy (LSC) or inverted repeat (IR) region of the chloroplast genome in transgenic tobacco plants. Higher levels of polymer transcripts were observed only in the IR transgenic plants.

Manipulation of photosynthetic carbon fixation

Optimizing photosynthetic carbon assimilation with the environment plays a crucial role in determining fitness and survivals. Studies of biodiversity have shown that the relationship between photosynthetic capacity of leaves and investment of biomass in leaf area is conserved across species (Reich *et al.*, 1997). Thus, plants have evolved mechanisms to ensure co-ordination of leaf development with metabolism. Photosynthesis in mature source leaves determines the availability of carbohydrates essential for plant growth; it is likely that products of primary metabolism may also play a role in mediating development.

The first enzyme in the Calvin cycle to be studied using antisense technology was Rubisco. Antisense plants with reductions in photosynthesis, accumulated less carbohydrate, resulting in changes in the root -to- shoot ratios and specific leaf area (SLA) defines leaf area per unit of leaf Carbon mass (m^2kg^{-1}) (Stiff and Schulze, 1994).

A linear relationship between photosynthetic rate, starch levels and SLA over a wide range of reductions in Rubisco activity was observed by Fichtner *et al.* (1993). The decrease in carbohydrate availability seen in the Rubisco antisense plants also caused a delay in the normal timing of developmental events. In the juvenile phase of development, leaves produced are smaller and simpler than those in adult plant and the Rubisco antisense plants continued to produce this type of leaves until node 11, in contrast with the wild type plants which had adult leaves by node 8 (Tsai *et al.*, 1997). This suggests that plants had an extended period in the juvenile vegetative phase. Increased longevity suggests that senescence was delayed in the antisense Rubisco plants (Tsai *et al.*, 1997, Miller *et al.*, 2000). This indicates the reductions in photosynthetic carbon assimilation, resulting in reduced availability of carbohydrate (source strength) not only had an impact on plant yield but also played a role in modulating the developmental programme of the shoot. Confounding factor that the antisense plants used only 20% of the wild type Rubisco activity and as a consequence, large changes in nitrate and the nitrogen status of leaves, as Rubisco constitutes approximately 25% of leaf protein. Thus, the changes in development observed in the antisense Rubisco plants were not only due to an altered carbon status but the changes in N status may also be having impact (Fichtener *et al.*, 1993).

In contrast to Rubisco, the enzyme Sedoheplulase 1, 7-bisphosphatase (SBPase) accounts for less than 1% of leaf protein, therefore, antisense down-regulation of this enzyme would not lead to large perturbation in nitrogen balance and hence direct effects of photosynthesis on leaf development. SBPase functions in the regenerative phase of Calvin cycle where it catalyses the dephosphorylation of sedohelptulase 1-7-bisphosphates. Transgenic tobacco plants with small reductions in SBPase activity have decreased rates of photosynthetic carbon fixation (Harrison *et al.*, 2001, Olcer *et al.*, 2001). In order to determine the limiting steps of the Calvin cycle and factors that influences the carbon allocation, a considerable number of studies have been undertaken on the regulation of carbohydrate metabolism in the photosynthetic CO_2 fixation in plant leaves (Table 1). Transgenic plants with reduced levels of glyceraldehyde-3-phosphate dehydrogenase (GAPDH), phosphoribulokinase (PRK),

Rubisco and aldolase showed little effects on photosynthesis (Price *et al.*, 1995; Paul *et al.*, 1995); Hudson *et al.*, 1992; Haake *et al.*, 1998). These data indicated that a number of enzymes involved in the Calvin cycle are present at protein levels well in excess of that required to sustain a continued rate of CO_2 fixation.

According to Tamoi *et al.* (2005) data from metabolite analysis of transformant suggests that an increase in the chloroplastic FBPase or SBPase level correlates with an increase of the RuBP level through the regeneration of RuBP, and thus affects the photosynthetic capacity and the growth in transgenic plants, while a slight increase in the FBPase activity seems to contribute to starch synthesis rather than to RuBP regeneration in chloroplasts (Fig.1).

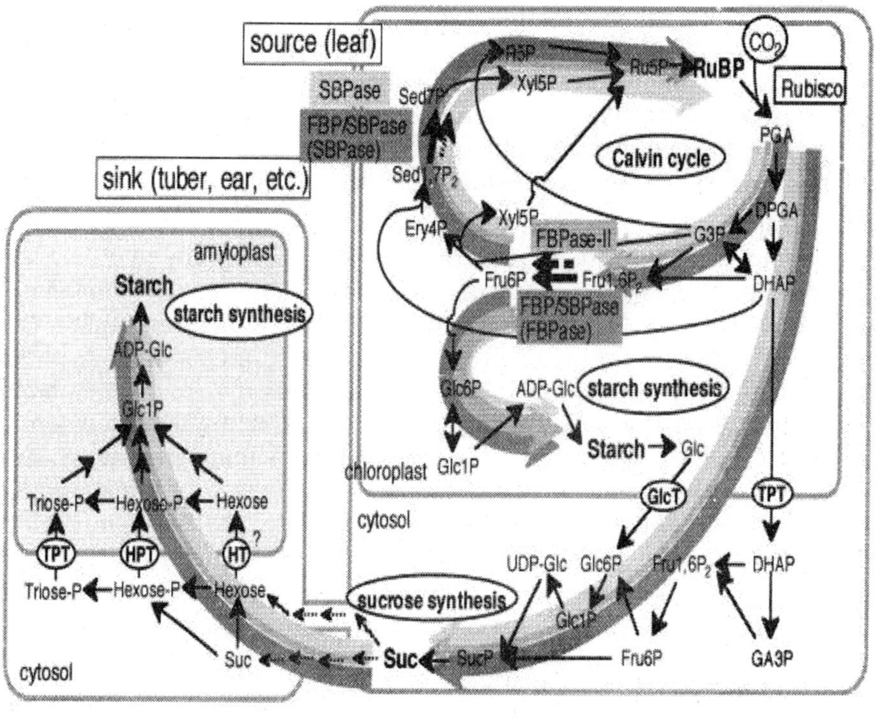

Fig. 1: Carbon metabolism of source and sink organs in the wild-type and transgenic tobacco plants. The arrows indicated carbon enhanced by the introductions of FBPase-II, SBPase, and FBP/SBPase in the chloroplast (Source: Tamoi *et al.* 2005)

Analysis of the SBPase antisense plants revealed that small changes in photosynthesis resulted in a reduction in starch accumulation in the source leaves and changes in leaf and shoots development (Lawson *et al.*, 2006). As with the Rubisco antisense plants, a change in SLA was evident in SBPase antisense plants, but no delay in the development of the juvenile phase of vegetative growth was seen. Antisense plants with small reductions in SBPase activity and concomitantly, small decrease in the end of day levels of starch were shorter and produced smaller, thicker leaves (reduced specific leaf area) when compared with wild type plants. In contrast, plants with large reductions in SBPase activity had low levels of both starch and sucrose and in these plants, the specific leaf area was increased and the height of the plants was either similar to, or taller, with thinner stems when compared with wild type plants. Similar changes in shoot morphology have been observed in some species in response to light levels in the growth environment (Evans, 1996). In growth conditions, where light is limiting, and the carbohydrate status of the leaf is reduced, plants produce thinner, larger leaves and longer, thinner stems to maximum light capture. Recently, evidence has been provided indicating a role for photoassimilated sucrose in the regulation of leaf growth patterns in response to shade (Kozuka *et al.*, 2005; Sharma *et al.*, 2012). It is possible that the reductions in source capacity in the SBPase antisense plants alter the metabolic signals that mimic the acclamatory responses to different light environments, but as yet no candidate molecules mediating the response have been identified. At the whole plant level increase SLA would attenuate the effects of decreased photosynthesis. This is likely to be beneficial to the plant as resources are invested in leaf area to increase light capture, rather than photosynthetic rates was observed in Calvin cycle antisense plants (Banks *et al.*, 1999, Rainies, 2003). The conservation of response in the diverse set of transgenic plants suggests a common signal coming either directly or indirectly from photosynthesis that can influence shoot development.

The literature cited here, together with the findings reported so far, suggest that the increase in respective enzyme levels involved in the Calvin cycle and sucrose biosynthetic pathway correlates with photosynthetic capacity and carbon partitioning of source and sink organs in higher plants. Regulating the balance of intermediate contents of carbon metabolisms including the Calvin cycle in higher plant is well answered with the generation of transgenic plants with significantly increased levels of FBPase and SBPase in the Calvin cycle by a chloroplast transformation technique. The transplastomic tobacco plants with 20–30-fold higher FBPase and SBPase activities in the chloroplasts showed enhanced photosynthetic capacity and growth compared with the wild-type. Tamoi *et al.* (2005) have analyzed the metabolic profiling of transgenic tobacco plants having significantly enhanced activities of enzymes involved in the Calvin cycle to discuss the regulation of carbon flux among various metabolic pathways, including the Calvin cycle. The transgenic plants with improved photosynthetic capacity and growth seem to be controlled by various enzymes and metabolites involved in carbon metabolism. Accordingly, in order to identify the kinds of genes that participate in increased Rubisco activation, photosynthesis and growth of the transgenic plants, they have generated FBP/SBPase-introduced transgenic *Arabidopsis*, in which the complete genome sequence has been analyzed and the microarray technology has-been established. It was found that FBP/SBPase introduced *Arabidopsis* plants showed enhanced photosynthesis and growth, in agreement with those of transgenic tobacco plants. This group is also analyzing the up regulated and down regulated genes in the transgenic plants by comprehensive analysis of transcriptional levels by DNA microarray analysis. In the near future, we may be faced with a food shortage as a result of the explosive increase in world population and environmental deterioration. The findings reported here suggest that it may be possible to use a cyanobacterial gene to manipulate photosynthetic carbon metabolism and improve the crop yield of various plant species. Plant productivity is also limited by various environmental stresses. Much of the injury to plants imposed by environmental stresses is associated with oxidative damage at the cellular level, including lipid hydroperoxidation leading to membrane damage and DNA damage (Shigeoka *et al.*, 2002). With the generation of transgenic plants expressing the antioxidative enzymes, such as catalase, ascorbate peroxidase or glutathione peroxidase (Miyagawa *et al.*, 2000; Yabuta *et al.*, 2002; Yoshimura *et al.*, 2004) could mitigate the stress. These transgenic plants have an increased tolerance to photo-oxidative stress imposed by various types of environmental stress. Accordingly, it seems likely that attempting multigene transfer for the simultaneous increase in several components involved in the Calvin cycle and the active oxygen species-scavenging systems seems to be necessary to obtain a substantial increase in plant productivity.

Quest for better Rubisco

By interacting with atmospheric CO_2, Rubisco initiates the chain of biochemical reactions that creates the carbohydrates, proteins and fats to sustain plants. But, the enzyme has another distinction. According to T.J. Andrews, Australian National University, Canberra "Rubisco is nearly the world's worst, most incompetent enzyme – it is almost certainly the most inefficient enzyme in primary metabolism that there is". First, it is the slowest known enzyme and its turnover in higher plant is as little as 2 or 3 reactions per second. Second, Rubisco triggers an additional reaction which interferes with the first. Rubisco catalyzes a second, opposing reaction i.e. photorespiration, the enzyme combines with oxygen rather than CO_2 to create a compound that is subsequently converted partly into CO_2. In other words, Rubisco catalyzes one reaction that incorporates carbon into plants and another that ultimately strips them of carbon.

Rubisco's ineffectiveness has been a spur to scientists since it became fully apparent in 1970s. According to George Lorimer, the quest for a better Rubisco is a Holy Grail in Plant Biology. Everyone always goes in with the hope of changing the face of agriculture. In spite of more than two decades efforts, the hopes have not yet paid off. But recent advances in molecular biology and the unexpected discovery of more efficient Rubisco in red algae have given now impetus to the long struggle to modify the enzyme. Scientists across the world are trying to improve the Rubisco in food crops by either replacing the existing enzyme with red algae from or bolting on, what could be thought of as molecular superchargers.

Hopes reawakened in 1992, when some diatoms and red-algae were discovered to have more specific Rubisco than the higher plants. In 1997, red algae were found to have Rubisco about three times more efficient. The scientists are attempting to insert Rubisco genes from red algal chloroplastic DNA developed by Rutger University biochemist Pal Maliga. The Scientists hope to see the result in about ten years from now. To alter Rubisco, 16 part molecule is to be encoded by many genes in both the cell nucleus and the chloroplast where photosynthesis take place. Rubisco is also dependent on supporting enzymes, some of which probably need to be revamped if former is changed.

Supercharging photosynthesis

Most of the researchers are trying to modify Rubisco; a few are trying another, even more ambitious strategy. Just as small engines can go faster, if they are equipped with supercharger which force-feeds them with fuel, some plants have their own photosynthetic supercharger, known as C_4 cycle. In C_4 plants, the bundle sheath cells where photosynthesis take place are surrounded by specialized mesophyll cells, which temporarily fix Co_2 and join it into the bundle cells at such high concentration that oxygenation reaction is effectively blocked. The C_4 cycle requires so much energy that C_4 plants can not grow in dim light but in well illuminated conditions. C_4 crops photosynthesize more efficiently and about 5% of all terrestrial higher plant species use the cycle. Matsuoka and his team are attempting to reproduce the C_4 cycle in rice. They are identifying non-functional equivalents of C_4-type gene in rice and selectively replacing them with their active counterparts from maize. Phosphoenolpyruvate carboxylase (PEPc) which catalyze the beginning of C_4 cycle. In rice, the mesophyll like cells are not photosynthetically active and these needs to be activated. The transgenic rice with only C_4-enzyme NADP malic tends to die quickly. The active PEPc in rice slows the destructive reaction upto the tune of one third. Ectopic expression of phosphoenol pyruvate carboxykinase (PCK) in rice chloroplast was able particularly to change the C-flow in mesophyll cells into C_4 like photosynthetic pathway. Such strategy appears to provide a possible method for enhancing photosynthetic capacity of C_3 (Suzuki *et al.*, 2000). The relationship between carbon assimilation and high level expression of the maize PEP_c in PEP_c transgenic rice, lead to increase in stomatal conductance.

Rubisco in transgenic plants

Tobacco plants with reduced Rubisco content were analyzed with respect to their photosynthetic activity, growth, carbon allocation under different environmental conditions. Under ambient light regime (310 µmol photos $m^{-2}s^{-1}$) Rubisco content can be decreased to 50-60% of wild type levels before any effects on the photosynthetic activity of tobacco leaves could be observed, thus, indicating that Rubisco is present in excess. It is now well documented that Rubisco is present in excess. Experimental evidences indicate that photosynthetic activity was not affected, nor was starch synthesis over the same range of inhibition. When Rubisco was reduced in a way that photosynthetic rates declined, lower biomass production and lowered starch synthesis in leaves were observed. Rubisco activity is positively regulated through carboxylation by Rubisco activase. Furthermore, it is possible that activase promotes the release of inhibitor carboxyarabinitol 1-phosphate from the enzyme in tobacco plants. However, lower Rubisco activation was compensated by higher Rubisco levels in the transgenic plants.

Triose phosphate translocator activity

It is debated for a long time whether photosynthates are exported as triose-phosphate (TP) or as hexoses from the chloroplast into the cytosol. Further, it is an open question whether this export is rate limiting for photosynthesis. Biochemical studies with Spinach chloroplasts clearly indicates that triose phosphate export is the major route for carbon channeled into sucrose for further transport into sink organs, as other activities could not account for the sufficient translocation of metabolites across the chloroplast membrane during photosynthesis. Photoassimilates enter the cytosol as triose phosphates and two major possibilities exist for the subsequent utilization. Triose phosphates can be used either in respiratcry reactions or for sucrose biosynthesis. Several attempts have been made to analyze the influence and function of key enzymes involved in these pathways by creating transgenic plants. The objective behind these experiments was to understand the mechanism controlling partitioning and allocation of carbohydrates in higher plants. To increase crop productivity, it is desirable to shift the direction of triose phosphate flow from glycolysis to sucrose synthesis, as the amount of sucrose present in the source leaf is rate-limiting factor with respect to the capacity of a source leaf.

Manipulation of Biosynthetic Pathways

Glycolytic pathway

After transport into the cytosol triose-phosphate may enter the glycolytic pathway. Glycolytic pathway ends in the cytosol at the level of pyruvate which subsequently enters the mitochondria for further metabolism e.g. respiration, amino acid or fatty acid synthesis. Pyruvate-kinase (PK) catalyses the formation of Pyruvate and ATP from phosphoenolpyruvate (PEP) and ADP.

Multiple isozymes of pyruvate kinase (PK) occur in plastids (PKp) and in the cytosol (PKc) and there are complex regulatory mechanisms controlling their expression. In potato, PKc was fused to a Rubisco transit peptide to target the protein to plastids in transgenic tobacco plants. Surprisingly, expression of both resident and transgenic PKc was eliminated in some plants due to co-suppression. Plant lacking any PKc grew and reproduced normally. Lack of PK would be fatal in other organisms. It is assumed that plants have some means of bypassing the reaction catalysed by PKc perhaps via the action of PEP phosphatase or by the combined action of PEP carboxylase, malate dehydrogenase and malic enzymes. Better understanding of glycolytic pathway can be achieved via the modification of the levels of the various PK forms in transgenic plants and subsequent biochemical and physiological analysis.

Sucrose biosynthesis

In higher plants, the interconversion of fructose 1,6-bisphosphate and fructose-6-phosphate is the major controlling steps in sucrose biosynthesis for the formation of sucrose-6-phosphate from UDPG and fructose-6-phosphate and the dephosphorylation of sucrose 6 phosphate yielding the final product, sucrose. Three enzymes are involved in catalyzing the first reaction: the forward reaction is catalysed by fructose 1,6 bisphosphatase(FBPase), the backward reaction towards glycolysis is controlled by phsophofructokinase (PFK) and the third enzyme pyrophosphate : fructose-6-phosphate-1-phosphotransferase (PEP) is able to catalyse the reaction in both directions. Each of these enzymes is altosterically regulated by intermediates of carbohydrate metabolism; furthermore FBPase and PEP are highly regulated by regulatory molecule fructose 2, 6, bisphosphate (F2, 6P2).

Two approaches have been employed to manipulate the interconversion of F1, 6P and F6P using transgenic plants. The precise functional significance of PEP is still unclear and controversial. The transformed potato plants with antisense constructs to the genes encoding the α and β submits of PEP. In several transgenic plants PEP expression was decreased upto 99 per cent. These plants did not show any phenotypical changes, reduced growth, or a decrease in tuber yield. The reduced PEP activity leads to a small reduction in starch and to small increase in sucrose in potato tubers. These results were unexpected because the essentially complete reduction of an enzyme that is highly regulated and expressed in most tissues was expected to influence plant carbohydrate metabolism significantly UDP-glucose (UDPG) is derived from glucose-1-phosphate by the action of UDP-glucose pyrophosphorylase (UGPase) and represents a branch point in sugar metabolism: UDPG is needed for sucrose biosynthesis and as a precursor of cellulose. Transgenic plant with only 4 per cent of wild-type UGPase activity showed no phenotypic changes, and sugar metabolism was not substantially affected (Zrenner *et al.*, 1995). Neither the suppression of PEP nor suppression of UGPase leads to major phenotypic effects has two meanings. Either plants express certain enzymes in such an excess than even the low remaining activity is sufficient for normal development and growth or these enzymes even when they are expressed in a more or less constitutive way are not needed under certain, as yet unknown, environmental or stress situations.

The reversible reaction catalysed by PEP may be shifted towards sucrose synthesis by removal of the phosphate group donor inorganic pyrophosphate (PP). Sonnewald (1992) introduced *E.coli* inorganic pyrophosphatase (PPase) into the cytosol of transgenic tobacco and potato plants. The expression of PPase lead to dramatic increase in the soluble sugar in source leaves. However, an accumulation of starch was observed in older leaves, the ratio between soluble sugars and starch increased upto three times. This result could mean that expression of PPase leads to expected alteration of carbon partitioning. The irreversible F1, 6P and F6P is shifted towards F6P in the direction of sucrose synthesis. The formation of UDPG catalysed by enzyme UGPase becomes irreversible due to removal of PP_i and based on an increase in free P_i in the cytosol; the export of triose phosphate from the chloroplast is triggered. Growth and development of these plants are charged drastically viz. stunted growth and loss of chlorophyll. Potato plants showed an increase in tuber number, arguing for a reduction of sink capacity. Since PPase was expressed in all tissues, several explanations are possible for the charges. Expression in sink tissues may prevent the utilization of UDPG, which is cleaved into sucrose by sucrose synthase. Further, PPase activity in companion cells may block phloem loading. It is well documentated that sucrose phosphate synthase (SPS) regulates the synthesis of sucrose in the cytosol, and thus, play a major role in the availability of photoassimilates for export from source leaves (Stitt and Quick, 1989). The maximal SPS activity is regulated in a light dependent manner via phosphorylation or dephosphorylation and further through G6P and P_i (Huber and Huber, 1992).

Lefebvre *et al.* (2005) reported the activity of sedoheplutase 1,7 bisphosphatase, increased by over expression of *Arabidopsis*. CDNA in tobacco (*Nicotiana tabacum*) plants with increased SBPase activity, photosynthetic rates increased, higher levels of sucrose and starch accumulation during the photoperiod, and an increase in leaf area and biomass of upto 30% were also evident.

Different species show a variation with respect to light activation of SPS. Maize SPS activity, for example, is strongly down-regulated during dark but enzyme does not respond to G6P or P_i concentration in the light (Huber *et al.*, 1989). The expression of maize SPS in transgenic tomato plants caused a reduction of starch and increased sucrose in mature leaves (Worrell *et al.*, 1991). This clearly indicates that SPS is involved in the regulation of carbon partitioning in leaves.

Trehalose metabolism

A striking impact on specific leaf area and its relationship to photosynthesis and carbohydrate metabolism has been observed in plants expression *E. coli* genes encoding enzymes in the trehalose pathway. The trehalose pathway, once believed to be of curiosity interest in plants, in now known to be indispensable and probably universal playing a control role in regulatory carbon use for growth (Eastmond *et al.*, 2002, Schlupmann *et al.*, 2003). It is thought that a key role is played by the signal metabolite trehalose-6-phosphate (T-6-P, Lunn *et al.*, 2006). Remarkably, elevated T-6-P in transgenic tobacco leads to a decrease in SLA and upto a 40% increase in photosynthetic carbon assimilation per unit leaf area, mainly due to increased Rubisco activity (Pellny *et al.*, 2004). However, because SLA and leaf area per plant is lower, photosynthesis at the whole plant level does not differ from the wild type. In plants expressing transgenes encoding trehalose phosphate phosphatase or trehalose phosphate hydrolase to reduce T-6-P content.

SLA was increased and in this case, photosynthetic capacity per unit leaf area was lower but, at the whole plant level, greater leaf area compensated for the decrease in photosynthetic capacity. The changes in leaf development observed in plants with altered T-6-P levels may be a consequence of the altered allocation of carbon between starch and sucrose. The recent finding that high levels of T-6-P stimulate the post-translational activation of ADP-glucose pyrophosphorylase, thereby increasing starch biosynthesis, provides a link between cytosolic metabolism and the chloroplast (Kolbe *et al.*, 2005; Lunn *et al.*, 2006). This work supports the central role of T-6-P in regulatory carbon metabolism in plants and the possibility that T-6-P effects on leaf development can be explained through this route. T-6-P may link cytosolic metabolism with that of the chloroplast, but the full mechanistic basis for this has to be established.

Sucrose-loading

In photosynthetically active tissues a surplus of fixed carbon produced is transported to sink e.g. seeds, tubers and sink leaves. Solutes are transported within the phloem from source to sink. However, we will concentrate on the transport of sucrose from mesophyll cells into the sieve element/companion cell complex (SECC-complex). Phloem loading is exclusively symplastic or if apoplastic transport is involved has been the subject of many discussions (Lucas and Madore, 1988). The distribution of plasmodesmata has provided evidence for symplastic transport, at least between mesophyll cells. In some plants of family Laminaceae, high symplastic connectivity between mesophyll cells and SECC-Complex (Type 1) may allow phloem loading via plasmodesmata, whereas in species like potato, sugar beet and tobacco (Type 2) with comparatively low symplastic connectively (Van Bel *et al.*, 1993), an apoplastic step is proposed.

The apoplastic loading predicts an energy dependent transport step of sucrose across of plasma membrane SECC Complex, generating a sucrose gradient with almost tenfold difference in concentration. Transport is driven by a proton gradient generated by plasmalemna H^+-ATPase. Sucrose/proton symport activity has been detected in plasma-membrane vesicles and protoplasts of several plant species (Bush, 1993). There are two hypotheses for such a transport activity. Firstly, the transport protein is integral part of general sucrose retrieval system. Secondly, the protein is responsible for the active transport of sucrose into the SECC complex. Transgenic generated recently gave a deep insight into the *in vivo* processes related to the questions concerning phloem loading.

Yeast invertase has been expressed in various organs and different subcellular compartments of transgenics (Sonnewald *et al.*, 1991) for example. the apoplast of potato leaves (Heineke *et al.*, 1992). Since, the phloem loading in the leaves involves an apoplastic step, sucrose is cleaned before long distance transport. As a result export is blocked and high levels of soluble sugars and starch accumulates, leading to the repression of photosynthetic genes. At the same time, the sinks are starved of carbon and the plants produce fewer tubers.

Unloading and transport of sucrose in potato tubers is thought to occur symplastically. Surprisingly, expression of yeast invertase in the apoplast of transgenic tubers leading a reduction in tuber number per plant and an increase in tuber size (Sonnewald *et al.*, 1997). Total tuber yield per plant on a fresh weight basis increases whereas production of tuber biomass remains constant. Expression of the yeast enzyme in the cytosol of transgenic potato tubers is detrimental to both size and yield. These studies reveal that the activity of a sucrose cleaving enzyme is at wrong time and place, which has pronounced effects drastically altering carbon metabolism and partitioning as well as gene expression and development.

Reduced expression of sucrose transporters

New insights into mechanisms *in vivo* due to reduced activity caused by the antisense technology. A pre-requisite for this approach is the isolation of genes to be suppressed. Sucrose transporter cDNA clones from spinach and potato were isolated by complementation of the yeast mutant (Riesmeir *et al.*, 1993). The transporter proteins from spinach (SoSUT) and potato (StSUT) showed characteristics consistent with those reported for the sucrose/ proton symport in isolated plasma membrane vesicles. RNA *in situ* hybridization reveals that the transporter gene is especially expressed in the minor veins of mature leaves.

The physiological role of isolated cCDNAs, in potato plants was transferred with an antisense construct; which lead to drastic phenotypic effects. The growth of the plants was strongly retarded; mature leaves curled bleached and accumulated anthocyanins at the rims. However, the sink leaves showed no phenotypic changes. Export of soluble sugars through the petiole provides direct evidence that the flow of sucrose is reduced in the antisense plants. The reduced efflux strongly affects the supply of carbohydrates to sink organs clearly visible as a reduced root system. No tubers are formed in the case of plants showing the highest reduction of StSut mRNA. The plants were analysed more intensively with respect of physiological changes. The soluble carbohydrate content in the mature leaves increased upto twenty fold and the starch amount upto ten fold. The maximum rate of photosynthesis decreased to 50% in strongly affected plants as compared to wild type. Yoon *et al.* (2004) reported sucrose phosphate synthase (SPS) activity in leaves of transgenic *A. thaliana* transformed with the cynobacterial sucrose phosphate synthase gene, and the wild type plants was not significantly different. Level of sucrose and starch in leaves slightly decreased whereas glucose and fructose contents increased upto two fold as compared to wild type during light period. They speculated the decrease in sucrose level was due to high acid invertase in transgenics.

Starch Synthesis

Regulation is quite complex. Probably only a part of controlling mechanisms has been identified. Starch is produced in the chloroplasts, where part of the carbon fixed by calcium cycle is channelized along with pathway starting of fructose-6-phosphate (F-6-P). F-6-P is converted to ADP-glucose by the enzyme phosphoglucoisomerase, phosphoglucomutase, and ADP-glucose pyrophosphorylase. ADP-glucose serves as a direct precursor for starch production; which involves a complex array of enzymes including different starch synthase (soluble and bound forms); branching enzyme and some unidentified proteins. During the day, when temporary storage of starch occurs, only a part of the recently fixed carbon is immediately available for regeneration of RUBP in the Calvin cycle or conversion to sucrose in the cytosol. At night, degradation products are converted to sucrose, allowing transport of carbon metabolites to heterotrophic tissues. A precise control of leaf carbohydrate metabolism is needed to organize these processes efficiently and to ensure efficient growth of the plant.

ADP-glucose pyrophosphorylase

ADP glucose pyrophosphorylase, a major enzyme controlling starch synthesis has been demonstrated in many plant species e.g. lower plants, monocots and dicots. The enzyme is also present in bacteria, where ADP-glucose is the substrate for glycogen synthesis. Two long known features of plant AGPase are the substrate for glycogen synthesis, allosteric activation by 3 PGA (3-phosphoglycerate) and the inhibition by inorganic phosphate (P_i). In contrast, this enzyme is regulated by fructose-6-bisphosphate (activator) and AMP (inhibitor) (Planton and Preiss, 1987).

AGPase is located in chloroplasts and amyloplasts (starch storing organelles). *In vitro* activity of AGPase plays an important role *in vivo* rate of starch accumulation. Kinetic models highlight that 3 PGA and P_i influence AGPase activity *in vitro*, thus, controlling starch synthesis *in vivo*. Further, the mutants deficient for AGPase have reduced levels of starch. Finally, the transgenic plants in which AGPase activity was modulated showed altered levels of starch. Plant AGPase has been described as heterotetrameric protein composed of two different but similar polypeptides. The size of the two smaller subunits ranges from 50-55 KD_a, the size of the larger subunit from 51 to 60 KD_a depending on the plant. At least 10 different AGPase polypeptides from angiosperms e.g. wheat, maize, rice and potato have been characterized at the DNA level (Anderson *et al.*, 1989). Based on biochemical and DNA sequence data, the larger subunit polypeptides appear to be more variable than the small subunit polypeptides existing in the potato tuber (Okita *et al.*, 1990).

Altered activity of AGPase

Down regulation in starch levels, AGPase activity was decreased by the antisense RNA technique in transgenics. The cDNA encoding the small subunits of the potato enzyme (Muuller-Rober *et al.*, 1990) was inserted in its reverse orientation behind the cauliflower mosaic-virus (CaMV) 35S RNA promoter and were transformed into potato. The leaf AGPase levels of some of the transgenic plants were decreased to 5-30 per cent of the wild type activity. A good correlation was observed between the amount of residual mRNA and protein expression and the degree of inhibition in AGPase.

Regulation of leaf development

Role of starch/sucrose ratio and T-6-P : The similarities in leaf phenotypes (altered SLA), together with changes in starch-to-sucrose ratios, found in the plants with reduced Calvin cycle activity and with altered T-6-P levels, indicates a common mechanism responsible for these changes. These results suggest that a balance between the utilization and storage of fixed carbon plays a role in the modulation of leaf development. The importance of sucrose-to-starch ratio's for growth was demonstrated by analysis of tomato plants over expressing sucrose phosphate synthase (SPS). This manipulation had the effect on directing the newly fixed carbon into sucrose, at the expense of starch accumulation. A small change in this balance in favour of sucrose accelerated floral development and increased fruit yield in tomato, but, if an excess of newly fixed carbon was used to synthesize sucrose, then no increase in growth was evident (Sharkey *et al.*, 2004). The mechanism underlying the increase in plant yield in the over expressing SPS still are not clear but it was not due to stimulation in photosynthesis. The major effect was to partition a higher proportion of newly fixed carbon directly into sucrose, resulting in an increase in the sucrose-to-starch ratio.

Changes in the starch-to-sucrose ratio were evident in the SBPase antisense plants and in source leaves, starch levels were decreased in response to smaller reductions in SBPase activity. Interestingly, in these plants with small reductions in SBPase activity, sucrose accumulated in the young leaves, despite having lower levels of starch in the source leaves. To explain this result, it was suggested that the plant was able to sense the source limitation and to compensate the growth of the shoot was reduced, imposing a temporary sink limitation in the young leaves (Olcer *et al.*, 2001; Lawson *et al.*, 2006). A more recent study where daily starch accumulation was limited by altering the length of the photoperiod provides support for this hypothesis. Reductions in the levels of starch accumulation during the photoperiod resulted in a depletion of sugar levels at the end of the night but surprisingly, when sugar became available at the start of the next photoperiod, it was not used and starch accumulated (Gibon *et al.*, 2004).

It is likely that this accumulation of carbohydrate occurs in response to reduced growth caused by a shortage of sugars for growth in the dark phase. The importance of carbon allocation over the diurnal cycle was also shown using a range of transgenic plants with the altered metabolism (Kehr *et al.*, 1998). These studies highlight the potential importance of the availability of carbohydrate for utilization in the dark, in the regulation of growth and development. The mechanism by which plants sense changes in the availability of carbohydrate and how this feeds through to effect on growth are not known. A number of recent studies have provided evidence showing that majority of starch degradation at night occurs through the hydrolytic pathway, resulting in the formation of maltose and glucose in the cytosol (Fig. 2). This opens up the possibility that starch to sucrose conversion and the resulting altered starch sucrose ratios can be sensed by tracking flux through the cytosolic hexokinase (Sharkey *et al.*, 2004). This mechanism is likely to form part of a complex regulatory network involving T-6-P that maintains a balance between carbon storage and utilization. How such changes in carbon balance are then linked to changes in development is not known, but hexokinase may play a central role in this process by integrating the effects of light, hormones and nutrient status (Rolland and Sheen, 2005).

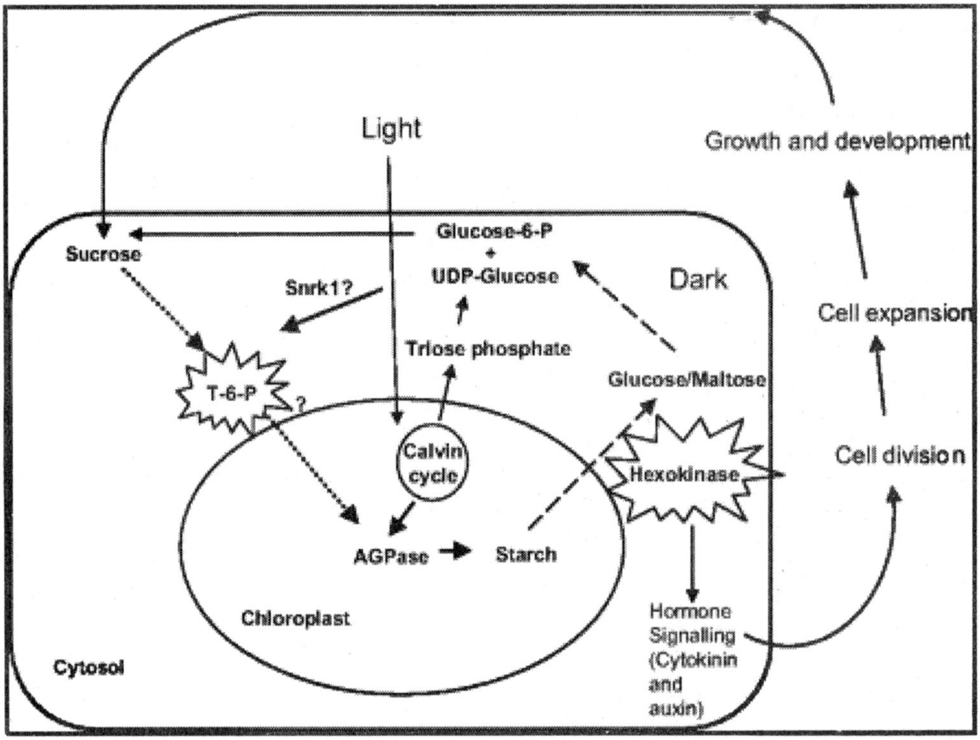

Fig 2: The central role of metabolism in mediating the regulation of growth and development. During the day light absorbed is used to form carbon in the Calvin cycle producing intermediates for both starch and sucrose biosynthesis (solid arrows). Starch produced during the photoperiod is broken down at night into glucose and maltose via the hydrolytic pathway to provide sucrose for growth at night(broken lines). Levels of sucrose in the cytosol are communicated to the chloroplast by T-6-P by an, as yet unknown process and increasedT-6-P activates AGPase increasing starch accumulation (dashed line). The mechanism by which T-6-P modulates AGPase activity is also not known but may be mediated by thioredoxin. It has been proposed that the glucose sensor hexokinase ,integrate light hormone and nutrient signaling to control plant development. Under conditions where starch levels are reduced, sucrose supply is restricted and growth is slowed to compensate. Conversely, when growth is restricted sucrose accumulates and starch levels increase.

Changes in non-photosynthetic metabolism: In addition to the impact of changes in photosynthetic carbon fixation on development, evidence is accumulating to suggest that development in plants is also sensitive to changes in the flux of carbon in the chloroplast, between the chloroplast and the cytosol and also in the cytosol. A number of mutants have been produced in which the flux of carbon has been altered and this has led to dramatic and unexpected changes in leaf shoot morphology.

The enzyme ketohexokinase (KHK) is not found in plants but in *E. coli*, it catalyses the conversion of fructose to fructose-1-phosphate. Ectopic expression of KHK in transgenic potato plants creates an alternative pathway of fructose metabolism and leads to severe abnormalities in leaf area and shape. The main metabolic effect of this manipulation is to increase the levels of triose phosphates and glyceraldehyde, KHK expression also resulted in a reduction in photosynthetic carbon fixation by as much as 50% and flux of carbon to both starch and sucrose was also decreased. An increase in respiratory activity was evident and the activity of glucose-6-phosphate dehydrogenase, the first enzyme in the oxidative pentose phosphate pathway, was increased by 50% suggesting an increased demand for NAD(P)H in these plants (Geigenberger *et al.*, 2004). A severe leaf phenotype with characteristics similar to that observed in the KHK over expressing potato plants, was observed in antisense tobacco plants with reduced levels of a small chloroplast protein CP12. The CP12 antisense phenotype was not expected as the proposed role of CP12 is in the regulation of two enzymes in the calvin cycle phosphoribulokinase (PRKase) andNADP-glyceraldehyde-2-phosphate dehydrogenase (GAPDH). Neither the PRKase (Paul *et al.*, 1995) nor the GAPDH (Price *et al.*, 1995) antisense plants exhibited such a phenotype; it seems likely that CP12 plays a wider role in chloroplast metabolism over an above regulation the activity of PRK and GAPDH. Analysis of the CP12 antisense plants revealed an increase in the activity of glucose-6-phosphate dehydrogenase and a decrease in plastid malate dehydrogenase activity in the light. These data indicate that with the KHK overexpressing potato, the NADPH status of the CP12 plants was altered suggesting a change in redox state may be due to underlying cause for the abnormal leaf phenotypes. An alternative explanation may be that changes in carbon fluxes are occurring, altering the availability of hormones, essential for the control of leaf development.

Large changes in carbon fluxes with a reallocation of photosynthesis from sucrose biosynthesis into organic acids also occurred in plants with reduction in the levels of the plastidic-2-oxoglutarate/malate translocator (D_iT1). These plants had altered leaf morphology, irregular leaf margins and reduced apical dominance (Schneidereit *et al.*, 2006).

C4 plants and genetic engineering

Genetic engineering of C_4 plants for improved agronomic performance, has led to the development of transformation procedures for major crops such as maize and sugarcane and the model C_4 dicot *Flavarria bidentis*. Modern high yielding varieties of sugarcane have been produced by conventional breeding but there is a limit of further improvements in yields by this approach. The accumulation of sucrose in sugarcane stems is determined by the relative activities of the vacuolar acid invertase and sucrose phosphate synthase (Zhu *et al.*, 1997). Therefore, a decline in vacular acid invertase activity by anti-sense or the over expression of SPS in the stem would seem to be a promising approach to increase sucrose accumulation. Over expression of SPS activity in leaves might lead to higher rates of sucrose synthesis in the leaves and export to the stems (Grof *et al.*, 1998).

A positive correlation between leaf SPS activity and growth rate in maize plants. Sarquis *et al.* (1998) reported an increase in SPS activity as maize plants were selected for higher yields. However, attempts to over-express SPS in species often other than tomato have met with limited success due to inactivation of the over-expressed enzyme probably by protein phosphorylation. In the transgenic tomato plants the maize SPS was not recognized by the endogenous tomato SPS protein kinase, so the over-expressed enzyme remained active. One approach to overcome the problem of enzyme inactivation would be to express a prokaryotic SPS gene in plants, such as from the cyano bacterium.

Engineering the C4 cycle

The engineering of C_4 photosynthesis into C_3 crops has been a goal of plant biologists. Within 3 years of the discovery of the C4 pathway in 1966, efforts had begun to cross C_3 and C_4 species in the hope of identifying important genes that control the expression of C_4 photosynthesis (Brown and Bouton, 1993). With the advent of molecular biology and efficient transformation technologies, it became feasible insert C_4 genes into C_3 plants and in the 1990s concerted efforts were underway to create a single–celled C_4 rice plant (Burnell, 2011, Miyao *et al.*, 2011). C_4 plants are able to have elevated photosynthetic capacities at warmer temperatures compared with ecologically similar C_3 plants and realize instantaneous water use efficiency (WUE) and nitrogen use efficiency(NUE)that are typically 1.3-4 times greater than in C_3 plants(Ghannoum *et al.*, 2011). There is no doubt that introducing a multigenic trait such as C_4 photosynthesis into rice is an enormous challenge. However,one compelling reason to accept this challenge is the strong humanitarian case for significantly increasing RUE and some biological reasons . First ,C_4 photosynthesis has evolved multiple times indicating that C_3 species may in some ways preconditioned for the C_4 pathway (Sage, 2004). Secondly ,according to Kajala *et al.* (2011)no new genes are associated with the C_4 pathway, all enzymes of C_4 cycle are found in C_3species although they often accumulate to low levels and are thought to fulfill housekeeping functions. In fact, C_3 plants possess characterisistic of the C_4 pathway in cells close to the veins of stems and petioles (Brown *et al.*, 2010). Third, although there is only one study to date the current estimate is that approximately 3% of the mature leaf tanscriptome differs between closely related C_3 and C_4 species (Brautigam *et al.*, 2010). Fourth, it is possible that a good proportion of these changes to the leaf transcriptome are due either to secondary effects associated with the alterations to primary metabolism in the C_4 leaf or other processes associated with phylogenetic distance between those species. This can be tested to some extent by assessing the impact of partitioning a significant portion of the C_4 cycle into M and BS cells of the rice. Lastly, recent advances in sequencing technologies proteomics and the availability of multiple C_4 genomics (Li *et al.*, 2010) has opened the C4 pathway to experimental dissection using a system biology approach. Comparative transcriptomic and proteomics have provided insights into the partitioning of C_4 gene expression and protein accumulation through analysis of M and BS cells of maize (Friso *et al.*, 2009) along the maize leaf developmental gradient (Li *et al.*, 2010) and between both distantly related (Brautigam *et al.*, 2008) and closely related C_3 and C_4 species (Brautigam *et al.*, 2010). In future ,the systems biology approach will be extended to multiple lineages of closely related C_3 and C_4 species to reveal the core elements of C_4 biology . All the genes encoding components of the C_4 pathway that are introduced into the rice need to be expressed correctly in either a M or BS –specific fashion . The M-specific expression pattern of maize promoters in rice have been characterized for ZmPEPC and ZmPPDK (Nomura *et al.*, 2000) and these promoters have been successfully used to drive strong M-specific expression in rice (Suzuki *et al.*, 2006). Although the intact genes for PEPC and PPDK have also been reported to generate cell specificity . In addition, for other genes ,it is clear that cell specificity inC4 leaves can be mediated by elements that are not present in the promoter (Hibberd and

Covshoff, 2010). Therefore ,in order to maximize the chances of obtaining correct cell specificity as well as other types of regulation (e.g. light) for each gene the full length maize gene has been cloned (including the native promoter ,5# and 3#UTRs ,exons and introns) and placed into rice. It is possible that some of these genes will not generate coding sequence (CDS) has also been fused to either the M-specific ZmPEPC promoter (Matsuoka *et al.*, 1993) or the BS promoter OsPcKI(Nomura *et al.*, 2005) with the nopaline synthase(nos) terminators at the 3# end. This approach should help to ensure that the correct expression pattern will be obtained in each gene. The prototype of transgenic C_4 rice will require detailed characterization. The biochemistry will be assayed for correct enzyme function and C_4 cycle activity will be confirmed carbon fixation studies. Physiological processes such as gas exchange and efficiency of photochemistry will be analized for improvement in compensation point and quantum yield expected for C_4 photosynthesis. If these results provide cause for optimism it is likely that emerging technologies such as site specific recombination or artificial chromosome engineering will have to be used to reduce the number of transgenic events associated with these plants and these second generation approaches used to cross traits into elite lines. Confined field trails would then be conducted to characterize the effects of C_4 engine on rice yield and the parameters contributing to it. Subsequently, C_4 photosynthesis would be transferred to elite rice varieties adapted to a wide range of environmental and agronomic conditions by conventional breeding

Stress tolerance is another area for engineering sucrose metabolism in C_4 plants. Transgenic tobacco plants, engineered to accumulate compatible solutes (trehalose, glycine betaine, sugar alcohol) showed some increase in water stress tolerance over wild-type plants (Hare *et al.*, 1998). Thus, the diversion of photosynthate from sucrose synthesis towards the synthesis of compatible solutes might also improve water-stress tolerance in other species. However, improved water stress tolerance might be more difficult to achieve in C_4- plants, as their water use efficiency is higher than C_3 plants. Cold tolerance is a multifactorial trait that is unlikely to be affected by changes to sucrose metabolism (Sarhan and Danyluk, 1998). Therefore, the diversion of photosynthate from sucrose to more valuable products such as fructan or polyhydroxybutyrate could be worthwhile goal.

Engineering C4 leaf anatomy

Bundle sheath cells of C_4 plants possess thickened and suberized cell walls which act as a diffusion barrier to help decrease the loss of CO_2 from the bundle sheath cells. Consequence of this anatomical adaptation is that the bundle sheath cell walls of C_4 grasses are resistant to degradation in the digestive trait of grazing animals. In addition, since most of the starch in the leaves of C_4 plants is within the bundle sheath cells, it is also inaccessible to digestion. An improvement in the digestibility of C_4 pasture grasses might be achieved either by bringing about a higher accumulation of soluble carbohydrates e.g. sucrose or by transferring starch synthesis from the bundle sheath cells (BS) to the mesophyll (M) cells. These results lead to a reassessment of ideas about the regulation of carbon metabolism and thus redesign bioengineering strategies to increase crop productivity in plants.

According to Kajata *et al.* (2011) none of the genes controlling C_4 leaf anatomy have been identified although the changes in leaf anatomy are well characterized and integral to two celled C_4 photosynthesis. Similarly, uncharacterized are the genetic determinants for the majority of C_4 cell biology and the ultra structure, including the genes coding controlling suberization of BS cells, increased plasmodesmatal connectivity between M and BS cells, and the production of dimorphic chloroplasts and their intracellular positioning. In order to engineer these structural C_4 changes into rice, the genetic determinants need to be identified. The patterns of leaf vascular systems are highly varied in both C_3 and C_4 species (Mc Known and Dengler, 2007). The leaves of two celled C_4 systems exhibit increased vein density (Uneno *et al.*, 2006) which decreases the diffusion distance of the photosynthate into the vasculature (Hatch, 1987). It has been suggested that the veins of C_4 species play a role in determining tissue differentiation and gene expression. Comparisons of the vein densities of leaves from C_3,C_3-C_4 intermediates and C_4 *Flaveria* species suggest that the increased vein density appeared prior to the evolution of C_4 biochemistry (Mc Known and Dengler, 2007). A theoretical optimum for vein density has been suggested (Noblin *et al.*, 2005) but there is probably adaptive value for the variation in vein density in C_3 leaves in response to environmental conditions (Sack and Frole, 2006). In order to understand the variation in the vein density, two genetic screen approaches have been chosen: screen for the emergence of C_4 characteristics in the activation-tagged rice population and a screen for the loss of C_4 characteristics in C_4 sorghum mutant populations.

To study the emergence of C_4 traits in C_3 rice, gain of function genetics has been employed. Activation tagging utilizes enhancer elements randomly inserted into the genome which creates lines with increased expression or knock out of specific genes. Activation–tagged rice population for Indica and Japonica rice (Jeong *et al.*, 2006) are being screened for decreased vein spacing and C_4-like leaf anatomical characteristics. 60,000 plants have been screened and vein density was affected in 0.1%of the activation–tagged plants. Preliminary results are promosing as some of the mutants lines have increased vein densities and importantly reduced numbers of M cells between the veins. These rice lines with increased vein density can be directly used as acceptors lines for the engineered C_4 biochemistry. Complementarily, classical loss of function mutant population of sorghum was generated with ethyl methane sulphonate (EMS) and gamma ray radiation treatments, and over 70,000 M_2 mutants have been screened. Mutants with decreased vein density and an increasing number of M cells between veins have been successfully identified. These potential C_4 vein density genes from both rice and sorghum mutants can be identified and used for C_4 engineering.

The rice and sorghum mutant populations were also screened for the emergence and loss of physiological C_4 characteristics. The carbon concentrating mechanism of C_4 plants allow them to carry out net carbon assimilation at low CO_2 concentrations, indicated by a low carbon compensation point. Hence sorghum mutants screened for growth at low CO_2 (20ppm) where mutants that exhibit stunted growth were rescued in a high CO_2 environment(10,000ppm). Preliminary results show the coincidence of the increased carbon compensation point and decreased vein density in the sorghum mutants. The efficiency of C_4 plants also coincided with the alterations in the leaf chlorophyll and a/b ratio (Black and Mayne, 1970) and the activation-tagged rice mutants screened for changes in the chlorophyll content. Interestingly, breeding and evolution has led to a reduced carbon compensation point of rice compared with the typical C_3 values at high temperatures (Sage and Sage, 2009).

This is believed to be due to a non-C_4 photo respiratory compensation mechanism arising from the lobed shape of M cells and it may have as yet unknown implications for C_4 rice engineering (Sage and Sage, 2009). Regardless of the lack of understanding of the genetic determinants of the leaf anatomy the mutant screen approaches are providing results for C_4 like rice leaves. The direct engineering approach involves expressing genes encoding the classical NADP-ME subtype pathway in appropriate cells in the rice leaf. To replicate the C4 cycle RuBisco and GDC expression are being down regulated in M cells. Transgenic plants are currently being produced and screened for C_4 enzyme activity and localization . To build up the C_4 pathway classical breeding will of great help. Further steps include the addition of C_4 transporters as well as uncharacterized genes that are being discovered.

How to build a C4 plant: a module based approach

The approach of Leegood (2002) is build upon by proposing a module based hierarchical framework that guide C_4 engineering and the dissection of C_4 evolution. This hierarchy is comprised of four organizational levels as depicted in Fig. 3.

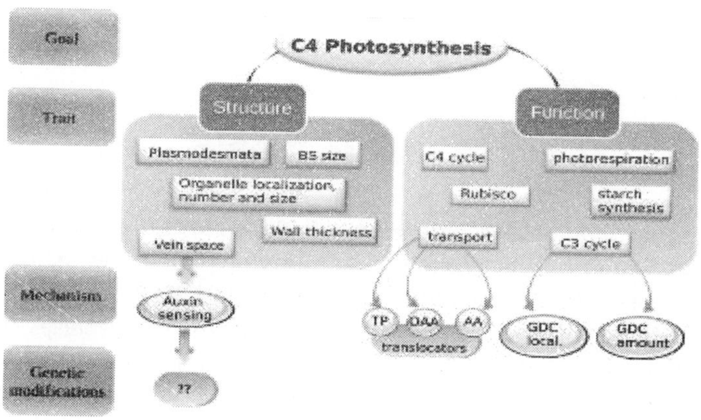

Fig 3: A hierarchical framework for converting a C_3 plant to a C_4 plant via \either bioengineering or natural evolution. The top level identifies the engineering goal (or in an evolutionary context, the end result). The second level identifies some of the key traits that will need to be modified to introduce a C_4 system into a C_3 plant . Each of the listed traits can be further represented as a module which can be expanded to show the mechanisms underlying each trait. For example, the transport function module can be expanded to describe changes in triose phosphate (TP), oxaloacetate (OAA) or amino acid (AA) transporters. The fourth level will describe the genetic changes underpinning the mechanistic changes, as they become known. (Source: Leegood 2002)

Each trait within the structure and function categories is represented as a module that can be expanded first to highlight the mechanistic changes that explain the trait, and then further expanded to delineate the associated genetic changes if known (Fig. 4). If the mechanism and genetic changes are unknown, they can be identified as such to emphasize the need to target these area for future research. As indicated in Fig. 4 these changes were brought about by changes to the amino acid sequences, for example by the substitution of a serine for an alanine near position 774 in the PEPC protein and alterations to the promoter region, notably the formation of a mesophyll expression module in the promoter of ppcA that confers mesophyll specific gene expression in the C_4 leaf. Although not shown in Fig. 4. Similar layers of understanding are being developed for the other major C_4 cycle enzymes (see Ludwig (2011) for CA, Maier et al (2011) for decarboxylating enzymes and Chastain (2011) for PPDK).

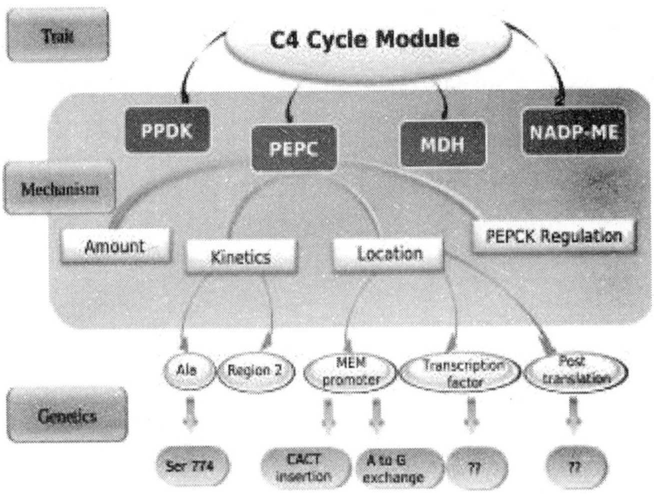

Fig 4: An expanded description of the C_4 cycle module shown in figure 4. Mechanisms shown can be expanded into sub module to identify changes to key enzymes systems in the C_4 metabolic cycle for example pyruvate orthophosphate dikinase (PPDK), PEP carboxylase (PEP_C), malate dehydrogenase (MDH), and the NADP-malic enzyme (NADP-ME) as would occur in an NADP-ME species such as maize. In this example, PEP_C sub module is expanded to show some of the genetic level changes that alter its kinetics and location of expression as described by Peterhansel (2011) and Gowik and Westhoff (2011)

When a detailed engineering schematic, is developed, it would be possible to quickly identify areas of progress versus knowledge gaps that require investigation. It may also be possible to identify knowledge gaps that can be ignored or circumvented should transcription factors and other regulatory elements be shown to control the expression of multiple genes (Westhoff and Gowik, 2010).

REFERENCES

Banks FM, Driscoll SP, Parry MAJ, Lawler DW, Knight JS, Gray JC and Paul MJ (1999) Decrease in phosphoribulokinase activity by antisense RNA in transgenic tobacco. Relationship between photosynthesis, growth and allocation at different nitrogen levels. *Plant Physiol* 119: 1125-1136

Black CC and Mayne BC (1970) P_{700} activity and chlorophyll content of plants with different photosynthetic carbon double fixation cycles. *Plant Physiology* 45: 738-741

Brautigam A, Kajala K, Wullenweber J et al. (2010) An mRNA blueprint for C4 photosynthesis derived from comparative transcriptomics of closely related C_3 and C_4 species. *Plant Physiology* 155: 142-156

Brautigam A and Hofmann-Benning RH (2008) Comparative proteomic of chloroplast envelope from C_3 to C_4 plants reveals specific adaptations of the plastid envelope to C_4 photosynthesis and candidate proteins required for maintaining C_4 metabolite fluxes. *Plant Physiology* 148: 568-579

Brautigam A and Gowik U (2010) what can next generation sequencing do to you? Next generation sequencing as a valuable tool in plant research. *Plant Biology*12: 831-841

Brown NJ, Palmer BG, Stanley S et al. (2010) C_4 acid decarboxylases required for C_4 photosynthesis are active in the mid vein of the C_3 species Arabidopsis thaliana and are important in sugar and amino acid metabolism. *The Plant Journal* 61: 122-133

Brown RH and Bouton J (1993) Physiology and genetics of interspecific hybrids between photosynthetic types. *Annual Review of Plant Physiology and Plant Molecular Biology* 44: 435-456

Burnell JN (2011) Hurdles to engineering greater photosynthetic rates in crop plants C4 rice. In Raghavendra AS, Sage RF eds *C_4 photosynthesis and related CO_2 concentrating mechanisms* Dordrecht, The Netherlands Springer, 316-378

Bussis D, Heineke D, Sonnewald U, Willmitzer L, Raschke K and Heldt HW (1997) Solute accumulation and decreased photosynthesis in leaves of potato plants expressing yeast derived invertase either in the apoplast, vacuole or cytosol. *Planta* 202: 126–136

Chastain CJ, Failing CJ, Manandhar L, Zimmerman M, Lakner M and Nguyen TTHT (2011)Functional evolution of C4 pyruvate orthophosphate dikinase. *Journal of Experimental Botany* 62: 3083-3091

Dai N, Schaffer A, Petreikov M,Shahak Y, Giller Y, Ratner K, Levine A and Granot D (1999) Over expression of *Arabidopsis* hexokinase in tomato plants inhibits growth, reduces photosynthesis, and induces rapid senescence. *Plant Cell* 11: 125-129

Daniell H, Datta R,Varma S, Gray S and Lee SB (1998) Containment of herbicide resistance through genetic engineering of the chloroplast genome. *Nat Biotechnol* 16: 345-348

Demo J, Tingyun K, Xia Li, Qiaoyin GE, Xueging H, Naibin HAO and Kezhi BAI (2003) Physiological characteristics of the primitive CO_2 concentrating mechanism in PEPc transgenic rice. *Science in China* 46 (4): 438-447

Eastmond PJ, Van Dijken AJH and Spielman M (2002) Trehalose-6-phosphate synthase 1, which catalyses the first step in trehalase synthesis is essential for *Arabidopsis* embryo maturation. *The Plant Journal* 29: 225-235

Evans JR and Baker N(1996) Effects of light nutrition on photosynthesis. Photosynthesis and the environment (Kluwer Academic Publishers, The Netherlands) 5: 281-304

Fichtner K, Quick WP, Schulze ED, Mooney HA, Rodermel SR, Bogorad L and Stitt M (1993) Decreased ribulase-1-5 bisphosphate carboxylase oxygenase in transgenic tobacco transformed with antisense rbc S.V., Relationship between photosynthetic rate, storage strategy, biomass allocation and vegetative plant growth at three different nitrogen supplies. *Planta* 199: 1-9

Friso G, Majeran W, Huang MS, Sun Q and Van Wijk KJ (2010) Reconstruction of metabolic pathways,protein expression and homeostasis machineries across maize bundle sheath and mesophyll chloroplasts,large scale quantitative proteomics using the first maize genome assembly. *Plant Physiology* 152: 1219-1250

Geigenberger P, Regierer B, Lytovchenko A, Leisse A, Schauer A, Springer F, Kossman J and Fernie A R (2004) Heterologous expression of a ketohexokinase in potato plants leads to inhibited rates of photosynthesis, severe growth retardation and abnormal leaf development. *Planta* 218: 569-578

Ghannoum O, Evans JR and Von Caemmerer S (2011) Nitrogen and water use efficiency in C4 plants. In Raghavendra AS, Sage RF eds *C_4 photosynthesis and related CO_2 concentrating mechanisms* Dordrecht, The Netherlands Springer, 129-146

Gibon Y, BlasingO E, Palacoisp Roja N, Pankovic D, Hendriks J H M, Fisahn J, Hohne M, Gunther M and Stitt M (2004) Adjustment of diurnal starch turnover to short days : depletion of sugar at night leads to temporary inhibition of carbohydrate utilization accumulation of sugars and post-translational activities of ADP - glucose pyrophosphorylase in the following light period. *The Plant Journal* 30: 847-862

Gowik U and Westhoff P (2011) C_4 phosphoenolpyruvate carboxylase. In Raghavendra AS, Sage RF eds. *C_4 photosynthesis and related CO_2 concentrating mechanisms* Dordrecht,The Netherlands Springer, 257-275

Grof C P L, Knight D P, McNeil S D, Lunn J E and Campbell J A (1998) A modified assay method shown leaf sucrose phosphate synthase activity is correlated with leaf sucrose content across a range of sugarcane varieties. *Aus J Plant Physiol* 25: 499-502

Gud C, Lee S B and Daniell H (2000) Stable expression of biodegradable protein based polymer in tobacco chloroplast. *Plant Cell Rep* 19: 257-262

Haake V, Zrenner R,Sonnewald U and Stitt M (1998) A moderate decrease of plastid aldolase activity inhibits photosynthesis, alters the levels of sugars and starch, and inhibits growth of potato plants. *Plant J* 14: 147–157

Hankes S, Sonnewald U, Badur R, Flachmann R and Stitt M (2001) A small decrease of plastid transketolase activity in antisense tobacco transform ants has dramatic effects on photosynthesis and phenylpropanoid metabolism. *Plant Cell* 13: 535–551

Hare P D, Cress W A and Van Staden J (1998) Dissecting the roles of osmolyte accumulation during stress. *Plant Cell and Environ* 21: 535-553.

Harrison E P, Olcer H, Lloyd J C, Long S P and Raines C A (2001) Small decreases in SBPase cause a linear decline in the apparent RUBP regeneration rate, but do not affect Rubisco carboxylation capacity. *Journal of Exp Botany* 52: 1779-1784.

Hatch MD (1987) C_4 photosynthesis: a unique blend of modified biochemistry,anatomy and ultra structure. *Biochmica et Biophysca Acta* 895:81-105

Heineke D *et al.* (1992) Apoplastic expression of yeast-derived invertase in potato. *Plant Physiol* 100: 301-308

Hibberd JM and Covshoff S (2010) The regulation of gene expression required for C_4 photosynthesis. *Annual review of Plant Biology* 61: 181-207

Huber SC, Kerr PS and Torres WK (1985) Regulation of sucrose synthesis and movement. In: Haeth RL, Preiss J (eds) Regulation of Carbon Partitioning in Photosynthetic Tissue, Williams and Wilkins, Baltimore, M.D., pp 199–214

Hudson GS,Evans JR, Von Caemmerer S, Arvidsson YBC and Andrews TJ (1992) Reduction of ribulose-1,5-bisphosphate carboxylase/oxygenase content by antisense RNA reduces photosynthesis in transgenic tobacco plants. *Plant Physiol* 98: 294–302

Jeong DH, An S, Park S *et al.*, (2009) Generation of a fanking sequence tag database for activation –tagging lines in japonica rice. *The Plant Journal* 45: 123-132

Kajala K, Covshoff S, Karki S *et al.*, (2011) Strategies for engineering a two –celled C_4 photosynthetic pathway into rice. *Journal of Experimental Botany* 62: 3001-3010

Kehr J, Hustiak F, Walz C, Willmitzer L and Fishan J (1998) Transgenic plants changed in carbon allocation pattern display a shift in diurnal growth pattern. *The Plant Journal* 16: 497-503

Kolbe A, Tiessen A, Schluepmann H, Paul M, Ulrich S and Geigenberger P (2005) Trehalase-6-phosphate regulates starch synthesis via post-translational redox activation of ADP-glucose pyrophosphorylase. *Proc of Natl. Aca. of Sci,* USA 102: 11118-11123.

Koßmann J, Sonnewald U and Willmitzer L (1994) Reduction of the chloroplastic fructose-1,6-bisphosphatase in transgenic potato plants impares photosynthesis and plant growth. *Plant J* 6: 637–650

Kota M, Daniell H, Verma, S, Garezynski F, Gould F and Moar WJ (1999) Over expression of the *Bacillus thuringiensis* Cry 2A protein in chloroplast confers resistance to plants against susceptible and Bt-resistant insects. *Proc. Natal. Acad. Sci. USA* 96: 1840-1845

Kozuka T, Horiguchi G, Kim GT, Ohgishi M, Sakai T, Tsukaya H (2005)The different growth responses of Arabidopsis *thaliana* leaf blade and petiole during shade avoidance are by photoreceptors and sugar. *Plant Cell Physiol* 46: 213-223

Lunn JE, Feil R, Hendriks JHM, Gibon Y, Morcuende R, Osuna D, Scheible WR, Carillo P, Hajirezaei MR and Stitt M (2006) Sugar induced increases in trehalase-6-phosphate are correlated with redox activation of ADP-glucase pyrophosphorylase and higher rates of starch synthesis in *Arabidopsis thaliana. Biochemical Journal*

Lawson T, Bryant B, Lefebvre S, Lloyd JC, Raines CA (2006) Decreased SBPase activity alters growth and development in transgenic tobacco plants. *Plant Cell and Envir* 29: 48-58

Lefebve S, Lawson T, Fryer M, Zakhlewuk OV, Lloyd JC and Raines CA (2005) Increased sedoheptulose 1-7 bisphosphate activity in transgenic tobacco plants stimulates photosynthesis and growth from an early stage in development. *Plant Physiology* 138: 451-460

Leegood RC (2002) C4 photosynthesis, principles of CO_2 concentration and prospects for its introduction into C_3 plants. *Journal of Experimental Botany* 53: 581-590

Ludwig M (2011) The molecular evolution of carbonic anhydrase in *Flaveria. Journal of Experimental Botany* 62: 3071-3081

Lunn J E, Price G D and Furbank R T (1999) cloning and expression of a prokaryotic sucrose - phosphate synthase gene from the *cyanobacterium synochocystis* sp. PCC 6803. *Plant Mol Biol* 40: 297-305

Madore M and Webb J A (1981) Leaf free space analysis and vein loading in *Cucurbita pepo. Can J Bot* 59: 2250-2557

Maai E, Shimada S, Yamada M, Sugiyama T, Miyake H and Taniguchi M(2011)The avoidance and aggregate movements of mesophyll chloroplasts in C_4 monocots in response to blue light and abscisic acid. *Journal of Experimental Botany* 62:3213-3221.

Matsuoka M, Tada Y, Fujimura T and Kanomurakami Y (1993) Tissue specific light regulated expression directed by the promoter of a C_4 gene maize pyruvate ortho phosphate dikinase in a C_3plant rice. *Proceedings of the National Academy of Science USA* 90: 9586-9590

McKown AD and Dengler NG (2007) Key innovations in the evolution of the Kranz anatomy and C4 vein pattern in *Flavea* (Asteraceae American). *Journal of Botany* 94: 382-399

McKown P and Dengler NG (2009) Shifts in leaf vein density through accelerated vein formation in C4 Flaveria (Asteraceae). *Annals of Botany* 104: 1085-1096.

Miller A, Schlaguhaufer C, Spalding M and Rodermel S (2000) Carbohydrate regulation of leaf development: prolongation of leaf senescence in Rubisco antisense mutants of tobacco. *Photosynthesis Research* 63: 1-8

Miyagawa Y, Tamoi M and Shigeoka S (2000) Evaluation of the defense system in chloroplasts to photo oxidative stress caused by paraquat using transgenic tobacco plants expressing catalase from *Escherichia coli. Plant Cell Physiol* 41: 311–320

Miyagawa Y, Tamoi M and Shigeoka S (2001) Over expression of a cyanobacterial fructose-1,6-/sedoheptulose-1,7-bisphosphatase in tobacco enhances photosynthesis and growth. *Nature Biotechnol* 19: 965–969

Miyao M, Masumoto C, Miyazawa SI and Fukayama H (2011) Lessons from engineering a single–cell C_4 photosynthetic pathway into rice. *Journal of Experimental Botany* 62: 2999-3011

Nablin X, Mahadevan L, Coomaraswamy IA, Weitz DA, Holbrook NM and Zwieniecki MA (2008) optimal vein density in artificial and real leaves. *Proceedings of the National Academy of Science USA*105: 9140-9144

Nomura M, Sentoku N, Tajima S and Matsuoka M (2000) Expression patterns of cytoplasmic pyruvate orthophosphate dikinase of rice (C_3) and maize (C_4) in a C_3 plant rice. *Australian Journal of Plant Physiology* 27: 343-347

Nomura M. Higuchi T. Ishida Y, Ohta S, Komari T, Imaizumi N, Miyao-Tokutomi M, Matsuoka M and Tajima S (2005) Differential expression pattern of C_4 bundle sheath expression genes in rice a C_3 plant. *Plant and Cell Physiology* 46:754-761

Olcer H, Lloyd, J C and Raines C A (2001) Photosynthetic capacity is differentially affected by reductions in sedoheptulase-1-7 bisphosphatase activity during leaf development in transgenic tobacco plants. *Plant Physiol* 125: 982-989

Paul MJ, Knight JS, Habash D, Parry MAJ, Lawlor DW, Barnes SA, Laynes A and Gray JC (1995) Reduction in phosphoribulokinase activity by antisense RNA in transgenic tobacco; effect on CO_2 assimilation and growth at low irradiance. *The Plant Journal* 7: 535-542

Peterhansel C (2011) Best practice procedures for the establishment of C_4 cycle in transgenic C_3plants *Journal of Experimental Botany* 62:3011-3019.

Pellny TK, Ghannown C, Conroy JP, Schluepmann H, Smeekens S, Andraloje J, Krause KP, Goddiju O and Paul MJ (2004) Genetic modification of photosynthesis with *E. coli* genes for trehalase synthesis. *Plant Biotech Journal* 2: 71-82

Price GD, Evans JR, Von-Caemmerer S, Yu JW and Badger MR (1995) Specific reduction of chloroplast glyceraldehyde-3-phosphate dehydrogenase activity by antisense RNA reduces Co_2 assimilation via a reduction in ribulose bisphosphate regeneration in transgenic plants. *Planta* 195: 369-378.

Price GD, Evans JR, Von Caemmerer S, Yu J-W and Badger MR (1995) Specific reduction of chloroplast glyceraldehyde-3-phosphate dehydrogenase activity by antisense RNA reduces CO_2 assimilation via a reduction in ribulose bisphosphate regeneration in transgenic tobacco plants. *Planta* 195: 369– 378

Raines CA (2003) The Calvin cycle revisited: a transgenic analysis. *Photosynthesis Research* 75: 1-10

Reich PB, Waters MB and Ellsworth DS (1997) From tropics to tundra: global convergence in plant functioning. Proceedings of National Academy of Science, USA 94: 13730-13734

Rolland F and Sheen J (2005) Sugar sensing and signaling networks in plants. *Biochemical Society Transactions* 33: 269-271

Sack L and Frole K (2006) Leaf structure diversity is related to hydraulic capacity in tropical rain forest trees. *Ecology* 87: 483-491

Sage RF (2004) The evolution of C_4 photosynthesis. *New Phytologist* 161: 341-370.

Sage TL and Sage RF (2009) The functional anatomy of rice leaves implications for refixation of photo respiratory CO_2 and efforts to engineer C_4 photosynthesis into rice. *Plant and Cell Physiology* 50: 756-772

Sarhan and Danyluk J (1998) Engineering cold tolerant crops–throwing the master switch. *Trends in Plant Sci.* 3: 289-290

Sarquis JJ, Gonzalez H, deJimenez ES and Dunlap JR (1999) Physiological traits associated with mass selection for improved yield in a maize population. *Field Crops Res* 56: 239-246

Schluepmann H, Pellny T, Van Dijeen A, Aghdasi M, Wobbes B, Paul M, Smeekens S (2003) Trehalase-6-phosphate is indispensable for carbohydrate utilization and growth in *Arabidopsis thaliana. Proc. of the Natl. Acad Sci* USA, 100: 6849-6854

Schneidereit J, Hausles R, Fiene G, Kaiser W M and Weber A P M (2006) Antisense repression reveals a crucial role of the plastidic-2-oxo-glutarate/malate translocator D; T1 at the interface between carbon and nitrogen metabolism. *The Plant Journal* 45: 201-224.

Sharkey D, Laporte M, Lu Y and Weber WPM (2004) Engineering plants for elevated Co_2: a relationship between starch degradation and sugar sensing. *Plant Biology* 6: 1-9

Sharma P, Sardana V and Banga SS (2012) Shading effects on photosynthesis, dry matter accumulation and yield of Indian mustard (*Brassica juncea L.*) genotypes. In Sandhu et al. (eds) Proc. International Conference on Sustainable Agriculture for Food and Livelihood Security, Nov. 27-29, 2012 Ludhiana, India: Crop Improvement Vol. 39(Spl.Issue), pp293-294

Shigeoka S, Ishikawa T, Tamoi M, Miyagawa Y, Takeda T,Yabuta and Yoshimura K (2002) Regulation and function of ascorbate peroxidase isoenzymes. *J Exp Bot* 53: 1305–1319

Signora L,Galtier N, Skot L, Lucas H and Foyer C (1988) Over expression of sucrose phosphate synthase in *Ababidopsis thaliana* results in increased foliar sucrose/starch ratios and favors decreased foliar carbohydrate accumulation in plants after prolonged growth with CO2 enrichment. *J Exp Bot* 49: 669–678

Sonnewald U *et al.*, (1997) Increased potato tuber size resulting from apoplastic expression of a yeast invertase. *Nat. Biotechnol* 15: 794-797

Stitt M and Schulze ED (1994) Does Rubisco control the rate of photosynthesis and plant growth? An exercise in molecular ecophysiology. *Plant Cell and Envi.* 17: 465-487

Suzuki S, Murai N, Burnell JN and Arari M (2000) Changes in photosynthetic carbon flow in transgenic rice plants that express C_4 type phosphoenol pyruvate carboxy kinase from urochloa panicoides. *Pl. Physiol* 124 (1): 163-172

Suzuki S, Murai N, Kasaoka K, Hiyoshi T, Imaseki H, Burnell JN and Arai M (2006) Carbon metabolism in transgenic rice plants that express phosphoenol pyruvate carboxylase and /or phosphoenol carboxylase. *Plant Science* 170:1010-1019.

ThorbjornsenT, Asp T, Jorgensen K and Nielsen TH (2002) Starch biosynthesis from triose-phosphate in transgenic potato tubers expressing plastidic fructose-1,6-bisphosphatase. *Planta* 214: 616–624

Tsai C H, Miller A, Spalding M and Rodermel S (1997) Source strength regulates an early phase transition of tobacco shoot morphogenesis. *Plant Physiol* 115: 907-914

Ueno O, Kawano Y, Wakayama M and Takeda T (2006) Leaf vascular systems in C3and C4 grasses a two dimensional analysis. *Annals of Botany* 97: 611-621.

Westhoff P and Gowik G (2010) Evolution of C_4 photosynthesis –looking for the master switch. *Plant Physiology* 154: 598-601

Woodrow IE and Berry JA (1988) Enzymic regulation of photosynthetic CO_2 fixation in C_3 plants. *Annu Re. Plant Physiol Plant Mol Biol* 39: 533–594

YabutaY, Motoki T, Yoshimura K, Takeda T, Ishikawa T and Shigeoka S (2002) Thylakoid membrane-bound ascorbate peroxidase is a limiting factor of antioxidative systems under photo-oxidative stress. *Plant J* 32: 915–925

Yoon S K, Lee JW, Bhoo S H, Jean J S, Lee Y H and Hahn T R (2004) Analysis of carbon metabolism in transgenic *Arabidopsis thalania* transferred with the cyanobacterial sucrose phosphate synthase gene. *J of Plant Biology* 47 (1): 422-47

YoshimuraK, Miyao K, Gaber A Takeda T,Kanaboshi H, Miyasaka H and Shigeoka S(2004) Enhancement of stress tolerance in transgenic tobacco plants overexpressing Chlamydomonas glutathione peroxidase in chloroplasts or cytosol. *Plant J* 3: 21–33

Zhu V J, Kornar E and Moore P H (1997) Sucrose accumulation in the sugarcane stem is regulated by the differences between the activities of soluble acid invertse and sucrose phosphate synthase. *Plant Physiol* 115: 609-616

Zrenner R *et al.* (1995) Evidence for the crucial role of sucrose synthase3 for sink strength using transgenic potato plants. *Plant J* 7: 97-107

Zrenner R, Krause KP, Apel P and Sonnewald U (1996) Reduction of the cytosolic fructose-1,6-bisphosphatase in transgenic potato plants limits photosynthetic sucrose biosynthesis with no impact on plant growth and tuber yield. *Plant J* 9: 671–681

9

Living Organisms as a Source of 'Nanoscale Engineering', and Its Applications

Himakshi Bhati-Kushwaha and *CP Malik*

Recently, construction of nanomaterials with a characteristic dimension ranging from 1-100 nanometers (nm) are at the leading edge of nanoscience and nanotechnology. With the passage of time nanomaterials, and specifically metal nanoparticles, have received particular interest in diverse field ranging from material science to biotechnology. With the development of different techniques for the biosynthesis of nanomaterials, the important role of biological systems and microorganisms are well demonstrated in nanoscience and nanotechnology. These organisms show a unique potential in eco-friendly production and accumulation of nanoparticles with different shapes and sizes. Therefore, researchers in the field of nanoparticle synthesis are focusing their attention to biological systems. In order to obtain different applied chemical compositions, controlled monodispersity, desired morphologies (e.g., amorphous, spherical, needles, crystalline, triangular, and hexagonal), and interested particle size, they have investigated the biological mechanism and enzymatic process of nanoparticle production. In this review, most of these organisms used in nanoparticle synthesis are shown.

Introduction

Nanotechnology, the modern science is also labelled as molecular nanotechnology or nanotechnology and has its history dating back to 9th century. First scientific description of the properties of nanoparticles was given in "Experimental relations of Gold (and other metals) to light". Richard Feymon (1959) opined that "There's plenty of space at bottom". Thereafter Peter Paul (1960) and his research students were pioneers in investigating first polyacrylamide beads for the oral administration. Soon after, nanoparticles were synthesized for Drug Delivery and Vaccines. In a way nano-revolution began somewhere in 1950's and took off during 1960's. Sugibayashi (1977) bound 5-flourouracil to the albumin nanoparticles and injected intravenously in a mice tail vein. This enhanced the longevity when nanoparticles were injected intraperitonially into Charlich Ascites Carcinoma bearing mice by Kreutter (2007). In the following a brief history is recounted.

1974: Term nanotechnology was used in paper for the first time by Taniguchi

1977: Term Molecular Nanotechnology was proposed by Dexler

1981: The first paper on nanoparticles was published by K. Eric Dexler

1981-1982: Invention of the Scanning Tunnelling Microscopy by Gerd Benning and Heinrich Rocher

1985: Discovery of the "bulk ball" a lovely round molecule consisting of 60 carbon atoms by Sumio Iijima

1986: Invention of the Atomic Force Microscopy by Gerd Benning and Heinrich Rocher

1990: The two strategies to synthesize nanparticles "Top- down" and "Bottom -up" by Don Eigles and Echead Schewieger.

1990: First Nanotechnology Journal was published

1991: Discovery of 'carbon nanotubes' by Sumio Iijima

1996: Discovery of new and improved form of carbon and nano sized product which was named as "super- carbon" by Prof. Richarg Smalley

1996: Potentials of nanotechnology in medical field by Prof. Mauro Ferrari

1997: First Company founded First design of Nanorobots

1999: First Nanomedicine Book published First Safety Guidelines, Congressional hearing on proposed NNI (National Nanotechnology Initiative)

2000: NNI- First state research initiative: $ 100 million in California by President Clinton

2001: First report on nanotech industry USA announces first centre for military application

2002: First nanotech industry conference Regional nanotech efforts multiply

2003: Congressional hearing on societal implications Call for balancing NNI research portfolio Dexler/Smalley debate is published in Chemical and Engineering News

2004: First policy conference on advanced nanotech

2005: Roco announces in a meeting that nanomachine/nanosystem project count has reached 300.

The potential of nanotechnology is amazing in divergent fields' e.g. health care and information and communication technology. Some of its products are already in the market like stain resistant fabrics, scratch resistant paints, transparent sunscreen lotion, antibacterial dressings, etc. (Bhattacharya, 2008).

Nanotechnology: Current status in Medicine and Agriculture

Keeping in mind the enormous potential and its applications, in USA National Nanotechnology Initiative (NNI) began in 2000 with a view to building, characterize and understand nanodevices. Accordingly a programme encompassing $464 million was budgeted during 2001 and raised to $1 billion during 2005 (Roco, 2003b). Concomitantly several other programmes also funded nanotechnological development. The utmost emphasis was laid on the impact of nanotechnology on medicine and accordingly six segments were constituted, which are; cardiovascular, anti-inflammatory, anti-infective, central nervous system, anti-cancerous and other applications.

In brief, USA appears to be the leading country where investments beyond 3.7 billion USD have been made. Considering the importance of this discipline Japan and Europe are other countries which have made gross investments. In recent years China is up-surging in this discipline and is coming up with the products constituting giant market for nanofoods. In the corporate world it is estimated that the number of nano-companies might touch 1000 in the next 10 years and market for nanotechnology was anticipated to touch several trillion USD.

Nanoparticles serve as 'magic bullets', containing herbicides, chemicals, or genes, which target particular plant parts to release their content. Nanocapsules can enable effective penetration of herbicides through cuticles and tissues, allowing slow and constant release of the active substances. For details a reference may be made to a recent review by Kushwaha and Malik (2012).

Illustratively nanotechnology is likely to emerge as an important field of modern research encompassing various disciplines (Fig 1 and 2).

Nanoparticles

Nanoparticles can be utilized at places where high surface areas are critical for success possessing high surface to volume ratio and enhanced catalytic property. They also possess bactericidal effects. Their size ranges from 10- 100 nm which is comparable to the size of a human protein and therefore, they can be effectively used in biology and biochemistry. Some of the properties like surface plasmon resonance, super-paramagnetic and quantum confinement occurs only when the particle size is very small ranging from 1-10 nm (Jain *et al.*, 2007).

The major goals in designing nanoparticles as a delivery system is to control the particle size, surface properties and release of pharmacologically active agents in order to achieve the site-specific action at optimal rate and dose regimen.

Previously, liposomes have been used as potential carriers with unique advantages including protecting drugs from degradation, targeting to site of action and reduction toxicity or side effects but their applications are limited due to inherent problems e.g. as low encapsulation efficiency, rapid leakage of water-soluble drug in the presence of blood components and poor storage stability. Contrarily, polymeric nanoparticles offer some specific advantages over liposomes by increasing the stability of drugs/proteins and having useful controlled release properties (Vila *et al.*, 2002 and Mu and Fang, 2003).

Following are certain advantages of using nanoparticles as delivery system:

- Particle size and surface characteristics of nanoparticles can be easily manipulated to achieve both passive and active drug administration.
- The control and sustain release of the drug and reduction in side effects.
- Controlled release and particle degradation characteristics can be readily modulated by the choice of matrix constituents.
- Drug loading is relatively high and drugs can be incorporated into the systems without any chemical reaction; this is an important factor for preserving the drug activity.
- Site-specific targeting can be achieved.
- The system can be used for various routes of administration including oral, nasal, parenteral, intra-ocular etc.

In spite of these advantages, nanoparticles do have certain limitations for e.g. their small size and large surface area can lead to particle-particle aggregation, making physical handling difficult in liquid and dry forms. In addition, small particles size and large surface area readily result in limited drug loading and burst release. These practical problems have to be overcome before nanoparticles can be used clinically or made commercially available.

Classification of Nanoparticles

Depending on the origin, nanoparticles are classified as follows:

Natural nanoparticles: These types of nanoparticles existed from the beginning of the earth and still persist in the environment (for e.g. volcanic dust, lunar dust, mineral composites etc.)

Incidental Nanoparticles: These types of nanoparticles arise from man-made industrial process (diesel exhaust, coal combustion, welding fumes etc.) and are also termed as waste or anthropogenic particles.

Engineered Nanoparticles: These types of nanoparticles can be engineered easily to synthesize the nanoparticles of desired shape and size as well as functionality. These are further grouped into following categories.

a) Carbon based materials-usually includes fullerene, single walled carbon nanotubes and multiwalled carbon naotubes.
b) Metal based nanoparticles-usually includes quantum dots, nanogold, nanoxide, nano-aluminium and nanoscale metal oxide like $TiO-_2$, ZnO and Al_2O_3.
c) Dendrimers: These are the nano-sized polymers built from branched units, capable to perform specific chemical functions.
d) Composites: These are the nanoparticles which combine with the other nanoparticle or with larger bulk type material (Lin and Xing, 2007). These also posses different morphologies such as sphere, tubes, rods and pins (Yu Nam and Leod, 2008).

Out of all these type of nanoparticles, engineered nanoparticles gained much attention as they have a positive contribution in the field of economy, including consumer products, pharmaceuticals, cosmetics, agriculture, energy, transportation. They also have a wide range of applications in industrial sectors (Novach and Buchell, 2007; Roco, 2003 b). Because of their widespread use in consumer products it is believed that nanoparticles will find into

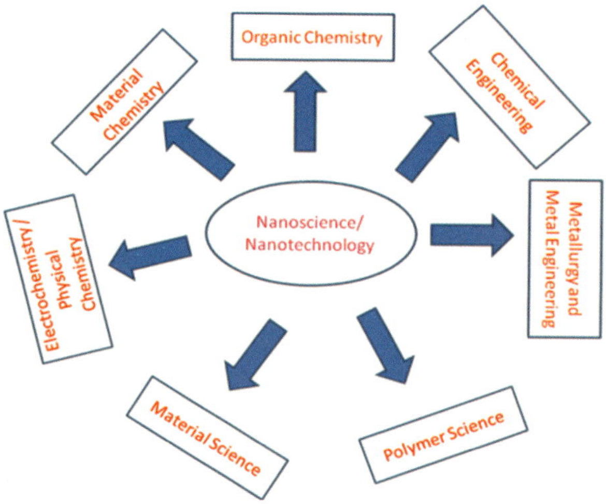

Fig. 1 : Nanoscience/nanotechnology as a multidisciplinary field

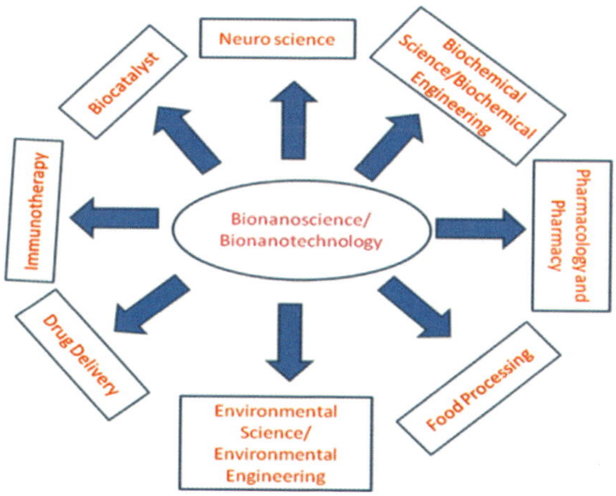

Fig. 2 : Bionanotechnology as Multidisciplinary Field

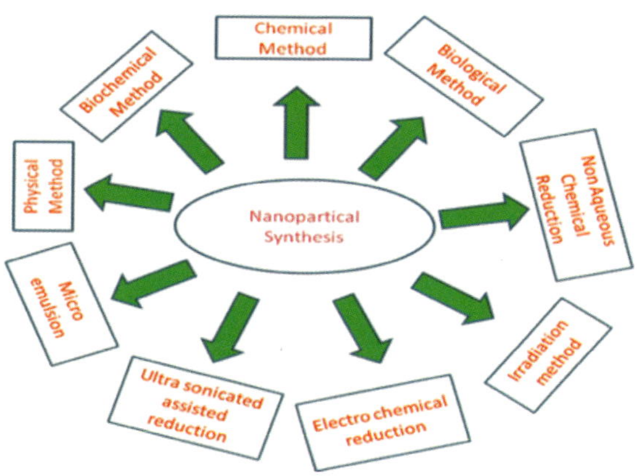

Fig. 3 : Illustrating the synthetic method of nanoparticle synthesis

aquatic, terrestrial and atmosphere environment (Monica and Creminini, 2009).

On the basis of nature, nanoparticles can be broadly grouped into two types: namely organic and inorganic nanoparticles. Organic nanoparticles may include carbon nanoparticles (fullerenes) while some of the inorganic nanoparticles may include magnetic nanoparticles, noble metal nanoparticles (like gold and silver) and semiconductor nanoparticles (like titanium dioxide and zinc oxide).

Inorganic nanoparticles

There is a growing interest in inorganic nanoparticles as they provide superior material properties with functional versatility. Due to their size and advantages over available chemical imaging drugs agents and drugs, inorganic nanoparticles have been examined as potential tools for medical imaging as well as for treating diseases. Inorganic nanomaterials have been widely used for cellular delivery due to their versatile features like wide availability, rich functionality, good biocompatibility, capability of targeted drug delivery and controlled release of drugs (Xu *et al.*, 2006). Gold and silver nanoparticles which are synthesized by various techniques have received a special attention because they have the potential to act as a catalyst, sensors, drug delivery system. On the other hand silver nanoparticles have excellent biocompatibility and have low toxicity.

Organic nanoparticles

Organic nanoparticles are successfully reported using cholesterol and rhovanil or rhodiarome where nanoparticles of size range 3-6 nm and 4.5-6.2 nm are synthesized respectively (Destree and Nagy, 2006). The preparation of stable, monodisperse, quantum-sized silver nanoparticles (NPs) in water soluble form through a simple one-pot procedure using glycerol for reducing Ag^{+} in the presence of o-phenylenediamine (o-PDA) is also reported where nanoparticle of size range 2-3 nm and 5-6 nm is synthesized.

Biosynthesis

This is an exciting and recent addition to the existing field of nanoparticles synthesis involving wide range of their synthesis and consequently these particles have entered another phase of commercial exploration. A perusal of literature reveals that gold and silver nanoparticles are being intensively used for applications in optoelectronic devices, ultrasensitive chemical and biological sensors and as catalysts. In general the techniques concerning their synthesis are divisible on intra cellular and extracellular basis (Ahmad *et al.*, 2005).

Various physical and chemical processes are used to synthesize nanoparticles. Some of the physical methods are solvo-thermal, sol-gel, laser ablation, inert gas condensation techniques (Prathna *et al.*, 2011), thermal decomposition of silver compounds, radiation and microwave assisted method (Parashar *et al.*, 2009). Chemical method involves reverse micells, sonochemical (Parashar *et al.*, 2009) and biochemical method, microemulsion , non aqueous chemical reduction and electro-chemical reduction methods (Kannan *et al.*, 2011) (Figure 3). The toxic chemicals used during the chemical synthesis are surface absorbed and cause adverse effects in medical applications (Jain *et al.*, 2009). Moreover, these methods are expensive and non-ecofriendly (Kannan *et al.*, 2011).

(a) By Chemicals

Currently, chemical reduction method is most commonly used for the synthesis of metal nanoparticles (Lin *et al.*, 2010). Depending on the condition of reaction mixture, metal ions may favour either the process of nucleation or aggregation to form small metal clusters. This synthetic method usually employs chemicals for e.g. hydrazine, sodium borohydride and hydrogen as reducing agents (Leff *et al.*, 1995; Pileni, 1997).

Synthetic or natural polymers such as natural rubber (Abu Bakar *et al.*, 2007), chitosan (Adlim *et al.*, 2004), cellulose (Kotelnikova *et al.*, 2003) and copolymer micelles (Egorova and Revina, 2000) have been used as stabilizing agent. As these chemicals are hydrophobic, organic solvents such as ethanol, dimethyl formamide, ethylene glycol, toluene and chloroform are usually used (Vijayaraghavan and Nalini, 2010).

Moreover, these methods are energy intensive processes and require large amount of capital investment. What is more, nanoparticles produced by these methods agglomerate with time which ultimately compromises with the size factor upon storage. Table 1 below shows examples of nanoparticles synthesized chemically.

Table 1 : Showing the references of chemical synthesis of nanoparticles

S.No.	Products	Size (nm)	Shape	References
1	Ag with histidine and tryptophan	Nil	Nil	Kazachenko *et al.*, 2000
2	Au, Ag, Pb, Pd with sugars	Nil	Nil	Panigrahi *et al.*, 2004
3	Ag with ammonium water	5	Nil	Song *et al.*, 2006
4	Ag with glucose	Nil	Nil	Andreescu *et al.*, 2007
5	Ag with sodium nitrate	3	Spherical	Kim *et al.*, 2008
6	Ag with hydrazine hydrate	9, 11, 24, 30	Nil	Guzman *et al.*, 2008
7	Ag with gallic acid	7, 29, 89	Nil	Castanon *et al.*, 2008
8	Ag with ethanol	16	Spherical	Das *et al.*, 2009

(b) Physical Methods

Physical approaches to synthesize metallic nanoparticles include ultraviolet irradiation (Kundu *et al.*, 2007), sonochemistry (Okitsu *et al.*, 2007), radiolysis (Meyre *et al.*, 2008), laser ablation (Tsuji *et al.*, 2003) and so forth. In this method metallic atoms are vaporized followed by condensation on various supports, where metallic atoms are rearranged and assembled as small cluster of metallic nanoparticles (Egorova and Revina, 2000). The main advantage of the physical approach is that nanoparticles with high purity and desired size can be synthesized.

However, these processes usually require complicated instruments, electrical and radiative heating as well as high power consumption, which results in high operating cost.

(c) Biological Method

The available findings indicate the growing need to develop clean, non-toxic and eco-friendly methods for the synthesis of nanoparticles (Sastry *et al.*, 2003). Hence, biological route appears to be safe for the synthesis of nanoparticles (Husseiny *et al.*, 2007). The particles thus synthesized are defined, require mild conditions for the reaction to occur, take less time, are cost effective and wide availability of natural biological materials and do not involve toxic chemicals (Li *et al.*, 2007).

Conditions needed for synthesis

Synthesis demands various physiochemical and biochemical properties. The size and stability of nanoparticles are affected by several factors including temperature, pH, agitation speed, etc. These properties influence strong entity, biocatalyst, biosensing recording media and sensing. Low temperature promotes stable and enhanced quantity of nanoparticles which are well defined in size and shape. Any increase in temperature diminishes the size and particles become small in size. Another predominant factor is pH which affects the synthesis of the stable nanoparticles. In fact pH also regulates the size of the nanoparticles. Thus, the presence of chloride, iodide and bromide also affect the formation of nanoparticles (Rodriguez *et al.*, 2007). Synthesized nanoparticles are characterized by several techniques and table 2 brings out the details.

Table 2 : Characterization of Silver nanoparticles

S.No.	Name of Instruments	Characterization parameters
1	Particle Size Analyzer: Dynamic Light Scaterring	Measuring the size distribution
2	Transmission Electron Microscopy	To measure the size and shape of entire particle
3	Scanning Electron Microscopy	To measure the surface image of the particle
4	Atomic Force Microscopy	To measure the size, morphology, surface texture and roughness of the particles. It is capable of 3D visualization
5	Fourior Transform Infrared Spectroscopy	To measure the chemical functional groups of the particle
6	X-ray Diffraction	To determine the arrangement of atoms within a particle, their chemical bonds and their disorder
7	Lithography	To transfer previously determined patterns on a substrate
8	Inductively Coupled Plasma Mass Spectroscopy	Capable of sub ppt (parts per trillion) detection limit, which help to measure the nanoparticle movement
9	Energy Dispersive X-ray Spectroscopy	To identify the purity of the particle

Biogenesis of Nanoparticles

Biogenic nanoparticles are made from the extracts of different living systems. In the following we shall briefly discuss their source as plants and micro-organisms.

Plants

Globally, the synthesis of nanoparticles, using plant extracts has attracted the attention of several research groups.

Indium oxide Nanoparticles

Aloe vera appears to be a favourable source of synthesizing indium oxide (In_2O_3) having nanoparticles with particle size ranging from 5-50 nm. The method is simple, cost effective and eco-friendly. The morphology and size of In_2O_3 materials are affected by the calcinations temperature (Meansiri *et al.*, 2008).

Platinum nanoparticles

In the synthesis of platinum nanoparticles *Diopyros kaki* and aqueous $H_2PtCl_6 \cdot 6H_2O$ solution has been used by Song *et al.* (2010). Ideal temperature to obtain 90% conversion was achieved at 95° C with leaf broth of >10%.

Zinc nanoparticles

Here milky latex of *Calotropis procera* has been used as a reducing material and consequently spherical-shaped ZnO-NPs were obtained. In another study Sangeetha *et al.* (2011) produced highly stable and spherical ZnO NPs using *Aloe vera* extract.

Gold nanoparticles

Au particles have been extensively used in several organisms because of their biocompatibility and being considered biologically inert (Bhattacharya and Mukherjee, 2008). The available information reveals that NIR irradiation of Au-based nanomaterials, Au nanospheres, Au nanocages, and Au nanorods can destroy cancer cells and bacteria via photo thermal heating. Moreover, these particles can be combined with photo sensitizers for photodynamic antimicrobial chemotherapy (Kuo *et al.*, 2009 and Pissuwan *et al.*, 2009).

Gold nanotriangles have been reported in the extracts of lemon grass as well as tamarind leaf (Ankamwar *et al.*, 2005) and dead biomass of *Humulus lupulus* (Lopez *et al.*, 2005) and *Emblica officinalis* fruit. In most of these cases the extracts act as reducing agent. Similarly Au nanoparticles have also been obtained from Alfa alfa. In fact the latter plants appear to be cost effective means of phytomining gold from soils (Gardea-Torresdey *et al.*, 2002). In the same vein cumin seeds are powerful antioxidants capable of reducing sodium tetrachloroaurate. In brief, the characteristic chemical, biological, photophysical and magnetic properties of gold nanoparticles furnish commercially viable products for different fields. In recent years concerted efforts are being made to develop non-toxic nanoparticles for wider use. Unarguably, more efforts should be made to use plants containing phytochemicles which are anti-cancerous, tumour specific gold nanoparticles with specific functions (Kattia *et al.*, 2009). Table 3 shows various plant species which have been used to synthesize Au nanoparticles with varying shape.

Table 3 : Biogenesis of Gold Nanoparticles from Plants

S.no.	Species Name	Product	Size (nm)	Shape	Location	References
1	Wheat	Au	10 – 30	Rod shaped	Extra.	Armendariz *et al.*, 2002
2	*Avena sativa* (oat)	Au	5 nm -3 μm	Triangular, hexagonal, irregular	Extra.	Armendraiz *et al.*, 2004
3	*Cicer arietinum* (Bengal green gram)	Au	25 thick	Triangle gold prism	Extra.	Ghule *et al.*, 2006
4	*Aloe vera*	Au	15.2	Spherical, nanotriangle	Extra.	Chandran *et al.*, 2006
5	*Sesbania* seedlings	Au	6 – 20	Spherical	Intra.	Sharma *et al.*, 2007
6	Barbeted skull cup herb	Au	5 to 30	Triangular	Extra.	Wang *et al.*, 2009
7	*Musa pradisiaca*	Au	10 - 20 μm	Nanoplates, microtubules, triangular, hexagonal	Extra.	Banker *et al.*, 2010
8	Sugar beet pulp	Au	20 - 60	Irregular	Extra.	Castro *et al.*, 2010
9	Pear friut extract	Au	200 – 500	Triangular, hexagonal	Extra.	Ghodake *et al.*, 2010
10	*Acanthella elongate*	Au	7.2	Spherical	Extra.	Inbakandan *et al.*, 2010
11	*Coleus amboinicus*	Au	4.6 - 55.1	Spherical, triangular, turncated, hexagonal, decahedral	Extra.	Narayana and Sakthivel , 2010b
12	Clove	Au & Ag	Variable size	Variable shape	Extra.	Singh *et al.*, 2010
13	*Nyctanthus arbortristis* flower	Au	19.5±5	Spherical	Extra.	Das *et al.*, 2011a
14	Rosa Hybrid petal	Au	10	Polydispersed	Extra.	Noruzi *et al.*, 2011
15	Korean red ginseng root (*Panace ginseng* CA Meyer)	Au	5 nm - 3 μm	Spherical	Extra.	Leonad *et al.*, 2011
16	*Medhuca longifolia*	Au	5 nm - 3 μm	Triangular, spherical	Extra.	Fayaz *et al.*, 2011
17	*Cassia fistula*	Au	Nil	Nil	Extra.	Daisy and Saipriya, 2012

Silver nanoparticles

Silver nanoparticles have received intensive research interest because of their wide applications in antimicrobial, catalysis, and surface-enhanced Raman scattering effect (Smith *et al.*, 2006). There are several hypotheses to explain the antibacterial activity of silver nanoparticles which are discussed subsequently in the application part. Silver nanoparticles can be synthesized in a number of ways and stabilization is achieved using capping agents that bind to the nanoparticles surface and improve their stability and water solubility, which are essential for preventing aggregation.

Why Silver

Silver is a naturally occurring pre-historic element used for centuries to treat burns and chronic wounds but its use for medicine or local antibacterial agents was not recognized until the 19thcentury (Oloffs *et al.*, 1994). Since knowing the antimicrobial property of it is also known to attack a broad range of biological processes in microbes including the alteration of cell membrane structure and functions (Pal *et al.*, 2007). Although the specific antimicrobial mechanism of Ag is still not completely understood yet it inhibits the expression of proteins associated with ATP production (Yamanaka *et al.*, 2005). Micro-molar doses of Ag are sufficient to kill bacteria in water (Liu *et al.*, 1994) while Ag can be toxic at high doses to mammals (Conrad *et al.*, 1999), fresh water and marine organisms (Morgan *et al.*, 2005). Such micro-molar concentrations of Ag have no harmful effects on humans (Berger *et al.*, 1976). This makes the wide acceptability of Ag in developing biological and pharmaceutical products (Raad *et al.*, 2002). With the emergence and increase of microbial organisms resistant to multiple antibiotics, and the continuing emphasis on health care costs, many researchers have tried to develop new, effective anti-microbial reagents free of resistance and cost. Silver nanoparticles show efficient antimicrobial property compared to other salts due to their extremely large surface area (Song *et al.*, 2008), which provides better contact with microorganisms.

Properties of Silver

1) Silver as a biocide

It acts as "biocide" and is applied as wood preservatives, to control the infectious agents (for example, *Legionella*). They are also used in water usage, silver and copper based disinfectants in hospital and hotel distribution systems. Silver along with copper, is commonly used to inhibit bacterial and fungal growth in chicken farms and in post harvested cleaning of oysters.

2) Colloidal silver

It was already discovered that the body's most important fluids are colloidal in nature with suspended ultra-fine particles for example blood carries nutrition and oxygen to the body cells. An electro-colloidal method (electrical silver atoms) is used for the manufacture of colloidal silver which appears to be a powerful, natural antibiotic and protective agent against infections. Acting as a catalyst, it disables the enzyme of one-celled bacteria, viruses and fungi needed for their oxygen metabolism. They suffocate them without causing harm to human enzymes or parts of the human body. This results in the destruction of disease-causing organisms in the body as well as in the food.

3) Silver nanoparticles as a catalyst

Silver nanoparticles act as a catalyst and showed high efficiency towards the degradation of non-biodegradable azo dyes and the killing of *Escherichia coli* under visible light irradiation (> 420 nm). The catalyst activity was maintained effectively even after successive cyclic experiments under UV or visible light irradiation without causing any destruction of AgBr. On the basis of instrumental studies the surface Ag species mainly exist as Ag^0 in the structure of all samples before and after reaction, inhibiting the decomposition of AgBr. Silica spheres with immobilized silver nanoparticles have been tested for their ability to catalyse the reduction of dyes by sodium borohydride ($NaBH_4$). Silver nanoparticles have a strong tendency to agglomerate. This reduces the surface to volume ratio and thereby the catalytic effect. To prevent this, a stabilizing agent is often used. However, the agent is adsorbed on the surface of the nanoparticles, shielding them from the oxidant and reductant and thereby inhibiting the catalysis. Instead a new method for stabilizing the nanoparticles is used in which the intermolecular forces which keep the nanoparticles immobilized of the silica spheres, has proven to be strong enough to prevent the process of forming aggregates.

4) Silver as an antimicrobial agent

Interestingly, nanoparticles synthesized through green route are shown to be antibacterial and anti fungal in nature. The toxicity was authenticated through food poisoning and disc diffusion methods.

Ag ions and Ag salts have been used since decades as antimicrobial agents in divergent fields due to their growth-inhibitory capability against microorganisms. They have been used in the medical field for antimicrobial applications specially for preventing HIV binding to host cells (Nino-Martinez *et al.*, 2008). In fact silver nanoparticles and Ag-based compounds having antimicrobial effects are prepared through variety of methods and shown to have effective antimicrobial activity (Lee *et al.*, 2008). When the two are compared, the inorganic nanoparticles synthesized via green route are shown to have distinct advantage over conventional chemical antimicrobial agents.

It is generally suggested that the antibacterial effect of Ag nanoparticles is associated with the peptidoglycan layer, though the precise mechanism of its inhibitory effect on microorganisms is not clear. Some workers have suggested that the positive charge on the Ag ion is crucial for antibacterial activity due to the electrostatic attraction between negative charged cell membrane of microorganism and positive charged nanoparticles (Dibrov *et al.*, 2002). It is also reported that the concentration of Ag nanoparticle affect gram-negative bacteria through the formation of pits in the cell wall (Sondi and Sondi, 2004). Suggestively there is accumulation of Ag nanoparticles in the bacterial membrane causing alteration in permeability and resulting in cell death.

According to Amro *et al.* (2000) another reason for cell death could be attributed to metal depletion resulting in the formation of irregularly shaped pits in the outer membrane. Consequently, membrane permeability is distorted, due to progressive release of lipopolysaccharide molecules and membrane proteins.

Danilczuk and co-workers (2006) through the ESR study of Ag nanoparticles have reported that the antimicrobial mechanism of Ag nanoparticles is somewhat associated to the formation of free radicals which damage the membrane.

Yet another possible mechanism suggested is that the free uptake of free silver ions cause disruption of ATP production and DNA replication (Lok *et al.*, 2006)

It is also reported that synthesized biogenic nanoparticles also exhibited antifungal effects, which was attributed to the destruction of their membrane integrity (Kim *et al.*, 2009) and inhibiting the normal budding process by destroying the cell membrane (Praveen *et al.*, 2012).

From this study the possibility of synthesizing quantum dots by living plants appears to be very promising (Gardea-Torresdey *et al.*, 2003). Table 4 includes the synthesis of silver nanoparticles using different plant extracts.

Table 4 : Biogenesis of Silver nanoparticls from plants

S.no.	Species Name	Products	Size (nm)	Shape	Location	References
1	*Aloe vera*	Ag	15.2 nm ± 4.2 nm	Spherical	Extra.	Chandran *et al.*, 2006
2	*Capsicum annum L.*	Ag	nil	Nil	Extra.	Li *et al.*, 2007
3	*Helianthus annus, Baselia alba, Oryza sativa, Saccharum officinarum, Sorghum bicolour and Zea mays* **(Leaves)**	Ag	nil	Nil	Extra.	Leela and Munusamy, 2008
4	**Xerophytic extract** *(Bryophyllum sps.)*, **Mesophytic extract** *(Cyprus sps.)* and **Hydrophytic extract** *(Hydrilla sps.)*	Ag	2 to 5	Fcc.unit cell	Extra.	Jha *et al.* , 2009
5	**Black Tea leaf extracts**	Ag-Au	nil	Nil	Extra.	Begum *et al.*, 2009
6	*Parthenium hysterophorus* **(Leaves)**	Ag	30 to 80	Nil	Extra.	Parashar *et al.*, 2009
7	*Mentha piperita* **leaf**	Ag	10 to 25	Nil	Extra.	Parashar *et al.*, 2009
8	*Jatropha curcas* **(Latex)**	Ag	30 to 40	Uneven and Large	Extra.	Bar *et al.*, 2009
9	*Volvaricella volvacae*	Au	20 to 150	Triangular, nanoprism, spherical, hexagonal	Extra.	Philip, 2009

S.no.	Species Name	Products	Size (nm)	Shape	Location	References
10		Ag	15	Spherical	Extra.	Philip, 2009
11	**Polyphenolic extracts**	Ag	3.3 to 22.1	Nil	Extra.	Sivaraman *et al.*, 2009
12	***Carica papaya*** **fruit extract**	Ag	15	Nil	Extra.	Jain *et al.*, 2009
13	***Coriandrum sativum*** **(Leaves)**	Ag	26	Nil	Extra.	Satyavathi *et al.*, 2010
14	***Argemone mexicana***	Ag	10 to 50	Nil	Extra.	Singh *et al.*, 2010
15	***Clerodendrum inerme***	Ag	nil	Nil	Extra.	Farooqui *et al.*, 2010
16	***Ocimum sanctum*** **and** ***Vitex negundo***	Ag	nil	Nil	Extra.	Prabhu *et al.*, 2010
17	***Psidium guajava***	Ag	26 ± 5	Nil	Extra.	Raghunandan et al., 2010
18	**Honey**	Ag	4	Nil	Extra.	Philip, 2010
19	***Syzygium aromaticum*** **(leaves and seeds)**	Ag	30 to 92	Nil	Extra.	Kumar *et al.*, 2010b
20	***Lippia citriodora***	Ag	15 to 20	Spherical	Extra.	Cruz *et al.*, 2010
21	***Curcuma longa***	Ag	71 to 80	Decahedral, triangular, spherical and ellipsoidal	Extra.	Satishkumar *et al.*, 2010
22	**Black pepper**	Ag	20 to 50	Spherical	Extra.	Shukla *et al.*, 2010
23	***Sorbus aucuparia*** **leaf extract**	Ag	16 to 18	Spherical, triangular & hexagonal	Extra.	Dubey *et al.*, 2010
24	***Chenopodium album*** **(leaf extract)**	Au - Ag	10 to 30	Quasi-spherical	Extra.	Dwivedi and Gopal, 2010
25	***Allium cepa,***	Ag	33.6	Nil	Extra.	Saxena *et al.*,2010
26	***Azadirachta indica***	Ag, Au & Bio-metallic	nil	Nil	Extra.	Shankar *et al.*, 2004, Tripathy *et al.*, 2008, 2009 & Thirumurugan *et al.*, 2010
27	***Nicotiana tobaccum***	Ag	8	Nil	Extra.	Prasad *et al.*, 2011
28	***Ocimum sanctum***	Ag	3 to 20	Nil	Extra.	Mallikarjuna *et al.*, 2011
29	***Cassia auriculata*** **leaf extract**	Ag	20 to 40	Nil	Extra.	Udaysoorian *et al.*, 2011
30	**Pomegranate fruit**	Ag	10 to 50	Nil	Extra.	Chauhan et al., 2011
31	***Citrullus colocynthis***	Ag	75	Nil	Extra.	Satyavani *et al.*, 2011
32	**Latex of plants**	Ag	nil	Nil	Extra.	Mondal *et al.*, 2011

S.no.	Species Name	Products	Size (nm)	Shape	Location	References
33	*Catharanthus roseus*	Ag	48 to 47	Nil	Extra.	Mukunthan , 2011
34	*Boswellia ovalofoliolata*	Ag	30 to 40	Nil	Extra.	Savithramma *et al.*, 2011
35	*Fenugreek* leaf	Ag	26	Nil	Extra.	Singh *et al.*, 2011
36	*Medicago sativa* seed exudates	Ag	5 to 51	Nil	Extra.	Lukman *et al.*, 2011
37	*Euphorbia hirta , Neriun indicum*	Ag	29	Nil	Extra.	Mano Priya *et al.*, 2011
38	Leaf bud extract of a mangrove plant	Ag	4 to 46	Nil	Extra.	Umashankari *et al.*, 2012
39	*Euphorbia hirta*	Ag	30 to 60	Nil	Extra.	Priyadarshini *et al.*, 2012
40	*Coleus aromaticus* (Leaves)	Ag	44	Nil	Extra.	Vanaja and Annadurai, 2012
41	*Elettaria cardamomom*	Ag	nil	Nil	Extra.	Gnana *et al.*, 2012
42	*Bougainvillea spectabilis* (pink bract)	Ag	nil	Cubic , hexagonal	Extra.	Pareekh *et al.*, 2012
43	*Annona squamosa*	Ag	28.47	Spherical	Extra.	Jagtapand Bapat, 2012
44	*Piper nigrum*	Ag	5 to 50	Crystalline	Extra.	Garg, 2012
45	Walnut	Ag	10 to 50	Monodispersed, spherical	Extra.	Korbekandi *et al.*, 2013
46	*Pedilanthus tithymaloides*	Ag	nil	Nil	Extra.	Sundaravadivelan *et al.*, 2013
47	Pine cone extract	Ag	nil	Traingular & Hexagonal	Extra.	Velmurugan *et al.*, 2013
48	*Artocarpus heterophyllus*	Ag	10.78	Irregular	Extra.	Jagtap and Bapat, 2013
49	Alfa alfa	Ag	Variable size	Spherical, triangular & hexagonal	Extra.	Gardea-Torresdey *et al.*, 2003

Renuga Devi *et al.* (2012) reported the biosynthesis of silver nanoparticles using *C. quadrangularis* and evaluation of its antimicrobial and anticancer potential. The silver nanoparticles have shown antibacterial activity against *E. coli., E. faecalis, B. subtilis, K. pneumonia, S. aureus, S. typhi, V. cholera* and the cytotoxicity of the nanoparticles was studied against HeP 2 and Vero cell lines. Thirunavukkarasu *et al.* (2012) reported the antiparasitic activities of synthesized AgNP using stem aqueous extract of *C. quadrangularis* against the adult *Hippobosca maculata* and the larvae of *Rhipicephalus microplus*. The methanolic extract of the plant was also reported to promot the free radical scavenging activity.

Bacteria

Table 5 below details the biogenesis of nanoparticles from different bacteria.

Table 5 :Biogenesis of nanoparticles from Bacteria

S.no	Microorganisms	Products	Culture Temp. (°C)	Size (nm)	Shape	Location	References
	Bacteria (Extracellular)						
1	*Shewanella oneidensis*	Au	30	12±5	Spherical	Extra.	Suresh *et al.*, 2011
2	*Plectonema boryanum UTEX 485*	Au	25	10 nm - 6 ìm	Octahedral	Extra.	Lengke *et al* ., 2006
3	*Escherichia coli*	Au	37	20 - 30	Triangles, hexagons	Extra.	Du *et al.*, 2007
4	*Pseudomonas aeruginosa*	Au	37	15 -30	Nil	Extra.	Husseiny *et al.*,
5	*Rhodopseudomonas capsulate*	Au	30	10 - 20	Spherical	Extra.	Guo *et al.*, 2007
6	*Escherichia coli*	Ag	37	50	Nil	Extra.	Gurunathan et al., 2009
7	*Corynebacterium glutamicum*	Ag	30	5 - 50	Irregular	Extra.	Sneha *et al.*, 2010
8	*Ureibacillus thermo sphaericus*	Au	60 to 80	50 - 70	Nil	Extra.	Juibari *et al.*, 2011
9	*Escherichia coli*	CdTe	37	2.2 - 3.2	Spherical	Extra.	Bao *et al.*, 2010
10	*DesulfovibrioDesulfuricans*	Pd	25	50	Spherical	Extra.	Lloyd *et al.*, 1998
11	*Lactobacillus sp.*	$BaTiO_3$	25	20 - 80	Tetragonal	Extra.	Jha *et al.*, 2010
12	*Lactobacillus sp.*	TiO_2	25	8 - 35	Spherical	Extra.	Jha *et al.*, 2009
13	*Uncultured Magnetotactic Bacterium*	Probably Poly-phosphates	nil	nil	Rectangular	Extra.	Arakaki *et al.*, 2009
14	*Coriolus versicolor*	CdS	25	100-200	Spherical	Extra.	Sanghi and Verma, 2009
15	*Bacillus Sps*	Ag	nil	5 - 15	Nil	Extra.	Pugazhanthrian *et al.*, 2009
16	*Metal Resistant Bactaria*	Ag	nil	4 and 5	Nil	Extra.	Babu & Gunasekaran, 2009
17	*Bacillus thurengenesis*	Ag	nil	15	Nil	Extra.	Jain *et al.*, 2010

S.no	Microorganisms	Products	Culture Temp. (°C)	Size (nm)	Shape	Location	References
	Bacteria (Extracellular)						
18	*Magnetotactic bacteria*	Gypsum & $CaCO_3$ Layer	nil	nil	Nil	Extra.	Shanker *et al.*, 2004
19	*Magnetotactic bacteria*	Metal nanoparticles	nil	nil	Nil	Extra.	Bazlinski & Frankel, 2004
20	*Pseudomonas stutzeri AG259*	Ag	nil	200	Hexagonal & Pyramidal	Extra.	Klaus *et al.*, 1999
21	*Sulphate Reducing Bacteria*	FeS	nil	2	Spherical	Extra.	Watson *et al.*, 1999
22	*Klebsiella pneumoneae*	Ag	nil	52.5	Nil	Extra.	Shahverdi *et al.*, 2007 and Mokhtari *et al.*, 2009
23	*Marinobactor Pelagius*	Au	nil	Â 10	Varied Shape	Extra.	Sharma et al., 2012
24	*Proteus mirabilis PTCC 1710*	Ag	nil	10 - 20	Spherical	Extra.	Samadi *et al.*, 2009
25	*E.coli*	Ag	nil	nil	Nil	Intra./ Extra.	Natarajan *et al.*, 2010
26	*Bacillus subtilis*	Ag	nil	5 - 60	Nil	Extra.	Saifuddin *et al.*, 2009
27	*Pseudomonas capeutate*	Au	nil	nil	Nil	Extra.	Shiying *et al.*, 2007
28	*Thermomonospora capeutate*	Au	nil	nil	Nil	Extra.	Ahmad *et al.*, 2003
29	*Bacillus lichiniformis*	Au	nil	50	Nil	Extra.	Kalimuthu *et al.*, 2008
30	*Bacillus megaterium* DO1 (dried powder)	Ag	nil	nil	Nil	Extra.	Fu *et al.*, 2006
31	*Lactobacillus* sps. (dried powder)	Ag	nil	nil	Nil	Extra.	Fu *et al.*, 2006
32	*Cornybacterium* species (dried powder)	Ag	nil	10 -15	Nil	Extra.	Fu *et al.*, 2006
33	*Bacillus subtilis*	Ag	nil	5-60 nm	Nil	Extra.	Saifuddin *et al.*, 2009
34	*P. infestans*	Ag	nil	Nil	**Nil**	Extra.	Thirumurugan *et al.*, (2009)

S.no	Microorganisms	Products	Culture Temp. (°C)	Size (nm)	Shape	Location	References
	Bacteria (Intracellular)						
35	*Brevibacterium casei*	PHB	37	100 - 125	Nil	Intra.	Pandian *et al.*, 2009
36	*Rhodococcus sp.*	Au	37	5 -15	Spherical	Intra.	Ahmad *et al.*, 2003
37	*Plectonemaboryanum*	Au	25–100	<10 to 25	Cubic	Intra.	Lengke *et al* ., 2006
38	*Brevibacterium casei*	Au, Ag	37	10 - 50	Spherical	Intra.	Kaliskwaralal *et al.*, 2010
39	*Bacillus cereus*	Ag	37	4 - 5	Spherical	Intra.	Babu & Gunasekaran, 2010
40	*Enterobacter sp*	Hg	30	2 - 5	Spherical	Intra.	Sinha and Khare, 2011
41	*Multicellular Prokaryotes*	Fe_3SO_4	25	nil	Nil	Intra.	Lef'evre *et al.*, 2010
42	*Rhodopseudomonas palustris*	CdS	30	8	Cubic	Intra.	Bai *et al.*, 2009
43	*Lactobacillus*	CdS	25 - 60	4.9±.2	Spherical	Intra.	Prasad and Jha, 2010
44	*Rhodobacter sphareoides*	CdS	nil	8	Hexagonal lattice	Intra.	Bai *et al.*, 2006
45	*Desulfobacteraceae*	CdS	nil	2 - 5	Hexagonal lattice	Intra.	Labrenz *et al.*, 2000
46	*Pseudomonas stutzeri AG259*	Magne-tosome	nil	nil	Nil	Intra.	Schuler and Frankel 1999
47	*E.coli*	$CaSO_4$	nil	nil	Nil	Intra.	Sweeney *et al.*, 2004
48	*Acronomas sps.*	Ag	nil	nil	Nil	Intra.	Mouxing *et al.*, 2006
49	*Lactobacillus strains*	Ag	nil	nil	Nil	Intra.	Nair & Pradeep, 2002
50	*E.coli DH5 Alpha*	Au	nil	nil	Nil	Intra.	Du *et al.*, 2007

Actinomycetes

Actinomycets are the organisms which show an interphase between fungi and bacteria Extracellular gold nanoparticles synthesis is reported in alkalothermophilic actinomycetes - *Thermomonospora* species (Sastry *et al.*, 2003). This species results in monodisperse gol nanoparticles. The biological conditions like alkaline and slightly elevated temperature favou the reaction to occur. Accordingly intracellular monodisperse gold nanoparticles are als reported in alkalotolerant actinomycetes – *Rhodococcus* species (Ahmad *et al.*, 2003).

Algae

Using algae as a source of nanoparticles synthesis is still an unexplored area. Stable gol nanoparticles are reported in *Saragassum wightii* which synthesizes the nanoparticles of averag size 11 nm. The incubation period extended up to 12 hrs, with continuous agitatio (Singaravelu *et al.*, 2007). *Chlorella vulgaris* is also found to be a potent source of gol nanoparticles (Jianping *et al.*, 2007). Ali *et al.* (2011) have fabricated Silicon-Germaniun nanoparticles using freshwater diatom. *Stauroneis sp.* Jeffryes *et al.* (2008) used silica fron cells of *Nitzschia frustulum.*

Fungi

In table 6 biogenesis of nanoparticles using various fungi are shown.

Table 6 : Biogenesis of nanoparticles from Fungi

S.no.	Species	Product	Size (nm)	Shape	Mode of Synthesis	Reference
	Fungi (Extracellular synthesis)					
1	*A. alternata*	Ag	20 - 60	Spherical	Extra.	Gajbhiye *et al.*, (2009)
2	*A. alternata*	Se	30±5	Spherical	Extra.	Sarkar *et al.*, (2011b)
3	*A. alternata*	Au	12±5	Spherical	Extra.	Sarkar *et al.*, (2011a)
4	*A. alternata*	Ag	35-90	ND	Extra.	Acharya *et al.*, (2011)
5	*A. clavatus*	Ag	10 - 25	Spherical, Hexagonal	Extra.	Verma *et al.*, (2010)
6	*A. flavus NJPO8*	Ag	17±5.9	Spherical	Extra.	Jain *et al.*, (2010)
7	*A. flavus*	Ag	7	Spherical	Extra.	Moharrer *et al.*, (2012)
8	*A. fumigatus*	Ag	5 - 25	Spherical, Triangular	Extra.	Bhainsa and D'Souza (2006)
9	*A. fumigatus*	Ag	7 - 19	Variable Shape	Extra.	Navazi *et al.*, (2010)
10	*A. niger*	Ag	15 - 20	ND	Extra.	Gade *et al.*, (2008)

S.no.	Species	Product	Size (nm)	Shape	Mode of Synthesis	Reference
1	*A. oryzae var.viridis*	Au	10 - 60	Triangular, Pentagon, Hexagon	Extra.	Binupriya *et al.*, (2010)
2	*A. terreus*	Ag	1 - 20	Spherical	Extra.	Li *et al.*, (2012)
3	*Bipolaris nodlosa*	Ag	10 - 60	Spherical, Semi-pentagonal, hexahedral	Extra.	Saha *et al.*, (2010)
4	*C. cladosporioides*	Ag	10 - 100	Spherical rods, Triangular	Extra.	Balaji *et al.*, (2009)
5	*Cochliobolus lunatus*	Ag	3 -21	Spherical	Extra.	Salunkhee *et al.*, (2011)
6	*Colletotrichum sp.*	Au	20 - 40	Spherical	Extra.	Shankar *et al.*, (2004a)
7	*C. versicolor*	Ag	25 - 75	Spherical	Extra.	Sanghi & Verma (2009)
8	*C. versicolor*	CdS	100	Spherical	Extra.	Chen *et al.*, (2011)
9	*F.acuminatum EII. and Ev.*	Ag	4 - 50	Spherical	Extra.	Ingle *et al.*, (2008)
	F. oxysporum 284	Au	20 - 40	Multishaped	Extra.	Anitha and Palanivelu, (2011)
	F. oxysporum	Au	20 - 40	Spherical, Triangular	Extra.	Mukherjee *et al.*, (2002)
	F. oxysporum	Ag	5 - 50	Nil	Extra.	Senapati et al (2004)
	F. oxysporum	CdS	5 - 20	Spherical	Extra.	Ahmad *et al.*, (2002)
	F. oxysporum	Ag	5 - 15	Quasi-spherical	Extra.	Ahmad *et al.*, (2003)
	F. oxysporum	SrCO3	Nil	Middle shaped	Extra.	Rautaray *et al.*, (2004)
	F. oxysporum	Zirconium	3 - 11	Quasi-spherical	Extra.	Bansal *et al.*, (2005)
	F. oxysporum	Si	5 - 15	Quasi-spherical	Extra.	Bansal *et al.*, (2005)
	F. oxysporum	Ti	6 - 13	Spherical	Extra.	Bansal *et al.*, (2005)
	F. oxysporum	Ag	20 - 50	Spherical	Extra.	Duran *et al.*, (2005)
	F. oxysporum	Magnetite	20 - 50	Quasi-spherical	Extra.	Bharde *et al.*, (2006)
	F. oxysporum	$BaTiO_3$	4 - 5	Quasi-spherical	Extra.	Bansal *et al.*, (2006)
	F. oxysporum	Ag	10 - 25	Multishaped	Extra.	Kumar *et al.*, (2007a)
	F. oxysporum	CdSe	9 - 15	Spherical	Extra.	Kumar et al (2007b)
	F. oxysporum	Ag	5 - 60	Spherical	Extra.	Mohammadian *et al.*, (2007)
	F. oxysporum	Bi_2O_3	5 - 8	Quasi-spherical	Extra.	Uddin *et al.*, (2008)
	F. oxysporum	Ag	50	Spherical	Extra.	Karbasian *et al.*, (2008)

S.no.	Species	Product	Size (nm)	Shape	Mode of Synthesis	Reference
37	*F. oxysporum*	Pt	20 - 60	Triangular	Extra.	Govender *et al.*, (2009)
38	*F ,oxysporum*	Ag	20 - 70	Multishaped	Extra.	Pandiarajan *et al.*, (2010)
39	*Fusarium semitectum*	Au, Au-Ag	18 - 80	Multishaped	Extra.	Salwe *et al.*, (2008)
40	*F. semitectum*	Ag	10 - 60	Spherical	Extra.	Basavaraja *et al.*, (2008)
41	*Fusarium solani*	Ag	3 - 8	Spherical	Extra.	El-Rafie *et al.*, (2010)
42	*F. solani USM 3799*	Ag	16.23	Spherical	Extra.	Ingle *et al.*, (2009)
43	*Helminthosporum solani*	Au	2 - 70	Rods, Triangle, Pentagons, Stars and Pyramids	Extra.	Kumar *et al.*, (2008)
44	*L. lecanii*	Ag	45 - 100	Spherical	Extra.	Namasivayam and Avimanyu, (2011)
45	*Penicillium sp.*	Ag	16 - 40	Multishaped	Extra.	Sadowski *et al.*, (2008)
46	*Penicillium brevi-compactum WA2315*	Ag	58.35±17.88	Spherical	Extra.	Shaligram *et al.*, (2009)
47	*P. fellutanum*	Ag	5 - 25	Spherical	Extra.	Kathiresan *et al.*, (2009)
48	*Penicillium sp.*	Ag	10 - 100	Spherical	Extra.	Maliszewska *et al.*, (2009)
49	*Penicillium sp.*	Ag	52 - 104	Multishaped	Extra.	Hemath *et al.*, (2010)
50	*Pestalotia sp.*	Ag	10 - 40	Spherical	Extra.	Raheman *et al.*, (2011)
51	*P. chrysosporium*	Ag	5 - 200	Pyramidal	Extra.	Vigneshwaran *et al.*, (2006)
52	*Phanerochaete chrysosporrium*	Au	10 - 100	Spherical	Extra.	Sanghi *et al.*, (2011)
53	*Phoma glomerata*	Ag	60 - 80	Spherical	Extra.	Birla *et al.*, (2009)
54	*Phoma sorghina*	Ag	30 - 40	Rods	Extra.	Gade *et al.*, (2011)
55	*Phoma sp. 3.2883*	Ag	71.60±3.46	Spherical	Extra.	Chen *et al.*, (2003)
56	*Pleurotus sajor caju*	Ag	5 - 50	Spherical	Extra.	Nithya and Ragunathan, (2009)
57	*R .oryzae*	Au	10	Multishaped	Extra.	Das *et al.*, (2009)
58	*R. stolonifer*	Ag	5 - 50	Spherical	Extra.	Afreen and Ranganathan, (2011)
59	*S. rolfsii*	Au	25	Triangular, decahedral and spherical	Extra.	Narayanan and Sakthivel, (2010)
60	*T.asperellum*	Ag	13 - 18	Nil	Extra.	Mukherjee *et al.*, (2008)
61	*Trichoderma harzianum*	Ag	30 - 50	Spherical	Extra.	Singh and Balaji, (2011

S.no.	Species	Product	Size (nm)	Shape	Mode of Synthesis	Reference
62	*Trichoderma reesei*	Ag	5 - 50	Multishaped	Extra.	Vahabi *et al.*, (2011)
63	*T. viride*	Ag	5 - 40	Spherical, rod like	Extra.	Fayaz *et al.*, (2010)
64	*Tricholoma crassum*	Ag	5 - 50	Spherical, Hexagonal	Extra.	Ray *et al.*, (2011)
65	*Trichothecium sp.*	Au	5 - 200	Triangular, Hexagonal	Extra.	Ahmad *et al.*, (2005)
66	*Verticillium sp.*	Magnetite	100 – 400	Cubo-octahedral	Extra.	Bharde *et al.*, (2006)
67	*V. volvacae*	Au. Ag., Au-Ag	20 - 150	Spherical, Hexagonal		Extra. Philip (2009)
	Fungi (Intracellular synthesis)					
68	*A. clavatus*	Au	20 - 35	Triangular	Intra.	Verma *et al.*, (2011)
69	*A. flavus*	Ag	8.92±1.62	Nil	Intra.	Vigneshwaran *et al.*, (2007)
70	*Penicillium sp.*	Au	nil	Spherical	Intra.	Zhang *et al.*, (2009)
71	*V .luteoalbum*	Au	>-10	Spherical, Rods and Spheres	Intra.	Gerike and Pinches, (2006)
72	*Verticillium sp. AAT-TS-4*	Ag	25± 12	Spherical	Intar.	Mukherjee *et al.*, (2001a), Senapati *et al.*, (2004)
73	*Verticillium sp. AAT-TS-4*	Au	20±8	Spherical, Quasi-hexagonal	Intra	Mukherjee *et al.*, (2001a)
	Fungi (Intra/Extra-cellular synthesis)					
74	*F.oxysporum f. sp. lycopersici*	Pt	10 - 50	Triangle, hexagons, square, rectangles	Intra/ Extra-cellular	Riddin *et al.*, (2006)
75	*Neurospora crassa*	Ag	11	Spherical	Intra/ Extra-cellular	Castro-Longoria *et al.*, (2011)
76	*N. crassa*	Au	32	Spherical	Intra/ Extra-cellular	Castro-Longoria *et al.*, (2011)

S.no.	Species	Product	Size (nm)	Shape	Mode of Synthesis	Reference
77	*N. oryzae*	Ag	30 - 90	Spherical	Intra/ Extra-cellular	Saha *et al.*, (2011)
78	*Penicillium sp.*	Au	30 - 50	Spherical	Intra/ Extra-cellular	Du *et al.*, (2011)
79	*Trichothecium sp.*	Au	nil	Spherical, Rods and Triangular	Intra/ Extra-cellular	Ahmad *et al.*, (2005)

Yeast

Table 7 represents various species of yeast used for the synthesis of nanoparticles.

Table 7. Biogenesis of nanoparticles from Yeast

S.no.	Species	Products	Size (nm)	Shape	Location	References
1	*Schizosaccharomyces pombe*	CdS	1 - 1.5	Monodisperse, spherical	Intra.	Kowshik *et al.*, 2002
2	*Torulopsis* species	PbS	2 - 5	Nil	Extra.	Kowshik *et al.*, 2002
3	yeast strains MKY3	Ag	2 - 5	Hexagonal	Extra.	Kowshik *et al*, 2003
4	*Yarrowia lipolytica* NCM 3589	Au	Nil	Triangular	Extra.	Agnihotri *et al.*, 2009
5	Extremophilic yeast	Au	30 - 100	Nil	Extra.	Mourato *et al.*, 2011
6	Extremophilic yeast	Ag	< 20	Nil	Extar.	Mourato *et al.*, 2011
7	MKY3, *Yarrawia lipolytica* NCIM 3589 and *Candida* species	Au-Ag	Nil	Nil	Nil	Kannan *et al*, 2011

Viruses

Intact biological particles have gained much importance for the synthesis of nanoparticles in an eco-friendly manner. Biological molecules such as fatty acids, amino acids , proteins act as a template for the synthesis of CdSe, CdS and CdTe nanocrystals. Other biological molecules like DNA, proteins, biolipid cylinders, viriod capsule S-layer and multicellular super structures act as template for the synthesis of inorganic nanomaterials (Doug!es and Young, 1998).

Mechanism of action

The exact mechanism of action of silver nanoparticles on the microbes is still not vivid but the possible mechanism of action has been suggested according to the morphological and structural changes found in the bacterial cells. These silver nanoparticles have the ability to anchor to the bacterial cell wall and subsequently penetrate it, thereby causing structural changes in the cell membrane like the permeability of the cell membrane and death of the cell causing the formation of 'pits' on the cell surface, and there is accumulation of the nanoparticles on the cell surface (Sondi and Sondi, 2004).

Secondly, the formation of free radicals by the silver nanoparticles may be one of the possible mechanisms causing cell death. Electron spin resonance spectroscopy revealed the formation of free radicals by the AgNP interacting with the bacterial cell membrane, making it porous leading cell death (Kim *et al.,* 2007). It is also proposed that there can be release of silver ions by the nanoparticles (Feng *et al.,* 2008), and these ions can interact with the thiol groups of many vital enzymes and inactivate them (Matsumura *et al.,* 2003). The bacterial cells when come in contact with silver, take in silver ions, which inhibit several functions in the cell and damage the cells leading to the generation of reactive oxygen species, which are possibly produced through the inhibition of a respiratory enzyme by silver ions and attack the cell itself.

Alternatively, silver ions may also interfere with the activity of the DNA, as DNA has sulphur and phosphorus as its major components. These interactions become problematic and thus terminate the microbes. The silver nanoparticles are also found effective in modulating the process of signal transduction (Shrivastava *et al.,* 2007).

Antimicrobial activity of nanoparticles

The size of the nanoparticle implies that it has a large surface area to come in contact with the bacterial cell and hence, it will have a higher percentage of interaction than bigger particles (Pal *et al.,* 2007). The nanoparticles smaller than 10 nm interact with bacteria and produce electronic effects, which enhance the reactivity of nanoparticles. Thus, it is corroborated that the bactericidal effect of silver nanoparticles is size dependent. The antimicrobial efficacy of the nanoparticle depends on the shapes of the nanoparticles also, this can be confirmed by studying the inhibition of bacterial growth by differentially shaped nanoparticles (Morones *et al.,* 2005). According to Pal *et al.,* (2007) truncated triangular nanoparticles show bacterial inhibition with silver content of 1 μg. While, in case of spherical nanoparticles total silver content of 12.5 μg is needed. The rod shaped particles need a total of 50 to 100 μg of silver content. Thus, the silver nanoparticles with different shapes have different effects on bacterial cell.

Application of Nanoparticles

Nanoparticles have applications in many areas, including biomedical, materials sciences, and catalysis due to their unique properties when compared with their bulk solid. In the following a brief description of catalysis action is given.

Catalytic Action

High surface area and energy predetermine metal nanoparticles for being effective catalytic medium. Growing small particles of silver have been observed to be more effective catalysts than stable colloidal particles. These growing particles catalyzed the borohydride reduction of several organic dyes. The reduction rate catalyzed by growing particles is distinctly fast, stable and large silver particles, which are the final products of growing particles. Catalysis is due to efficient particle-mediated electron transfer from the BH_4^- to the dye. The catalytic activity of the particles depends on their size, activity of the dye, and their interaction.

Antimicrobial

Silver has been in use since time immemorial. However the use of silver for medicine or local antibacterial agent was not recognized until the 19th century. Since then the antimicrobial property of silver has been investigated and employed more extensively than any other inorganic antibacterial agent. In the 19th century granulation tissues were removed using silver nitrate to allow epithelization and promote crust formation on the surface of wounds. Due to the increasing bacterial resistance to classic antibiotics, the investigations on the antibacterial activity of silver nanoparticles have increased. It has a significant potential for a wide range of biological application such as antibacterial agents for antibiotic resistant bacteria, preventing infections, healing wounds and anti-inflammatory (Taylor *et al.*, 2009). Bactericidal behaviour of nanoparticles is attributed to the presence of electronic effects that are brought about as a result of change in local electronic structure of the surface due to smaller sizes. These effects are considered to be contributing towards enhancement of reactivity of silver nanoparticles surface. Silver in ionic form strongly interacts with thiol groups of vital enzymes and inactivates them. It has been suggested that DNA loses its replication ability once the bacteria are treated with silver ions (Morones *et al.*, 2005). Silver nanoparticles destabilize plasma membrane potential and cause depletion of levels of intracellular adenosine triphosphate (ATP) by targeting bacterial membrane resulting in bacterial cell death. Convention compounds of silver such as silver nitrate and silver sulfadiazine are used to prevent bacterial growth in drinking water, sterilization and burn care.

Anti Fungal Properties

Nanosilver is an effective and a fast-acting fungicide against a broad spectrum of common fungi including species of *Aspergillus, Candida* and *Saccharomyces* (Wright *et al.*, 1999).

Antiviral Properties

Silver nanoparticles (diameter 5-20 nm, average diameter 10 nm) inhibit HIV -1 (Lago *et al.*, 2011). An interesting observation was that gold nanoparticles (average diameter 10 nm) showed relatively a low anti HIV 1 activity (6-20%) when compared to silver nanoparticles of (98%) (Elechiguerra *et al.*, 2005).

Anti-inflammatory Properties

In animal models, nanosilver alters the expression of matrix metallo-proteinases and suppresses the expression of tumor necrosis factor (TNF-α), interleukin (IL)-12 and IL 1α and induces apoptosis of inflammatory cells (Bhol and Schechter, 2005).

Human Health

Nanoparticles have many different effects on human health relative to bulk material from which they are produced (Albrecht, 2006). Increase the biological activity of nanoparticles can be beneficial, deterimental or both (Kushwaha and Malik, 2012).

Drug Imaging and Cancer treatment

Investigations are going where radio and chemotherapy can be replaced by attaching the nanoparticles to the tumerous cells and the "cook" the tumors inside the body with the help of radio waves and destroys only tumour cells not the healthy cells (Allen *et al.*, 2004). Positively charged gold nanoparticles show their grater affinity towards kidney whereas negatively charged gold nanoparticles show their affinity toward liver (Minchin *et al.*, 2008). Nanoparticles are also successfully used for the treatment of brain tumors and also helps in the removal of even the traces that left behind after the surgical removal (Kircher *et al.*, 2012)

Advances made in the encapsulation using suitable animal model have demonstrated that micro and nanoparticles are capable of enhancing immunization by selective activation of a T cell-mediated immune response (Lutsiak *et al.*, 2002).

Nano vaccines

New devices are discovered for the safe and efficient delivery of the vaccines to cure diseases like HIV and malaria. New particles are devised which consist of fatty acid sphere which are basically the synthetic protein versions produced by the virus. These synthetic particles are responsible for the onset of strong immune responses (Darrell, 2011).

Nano- drugs

Research is carried on the rabbit having congenital CP (cerebral Palsy Disorder), as it replicates the type of neuroinflamation found in human brain and leads to motor deficits in children (Kannan *et al.*, 2012). Nanodrugs are considered to be the most effective method based on nanoparticle delivery having targeted cell action with low side effects. Most commonly used molecules for nano drug preparation are dendrimers, polymer , liposome, micelles, inorganic and organic nanoparticles, etc (Ali *et al.*, 2011).

Through nanodrug it is now possible to make insoluble medicine orally available in nano crystal form. It is now possible to make nano crystals of pharmacologically important molecules which can be easily administered orally and absorbed by the gut even though they are insoluble in water. Several approaches are also used for the preparation of the nano delivery system which will be sophisticated, multifunctional and can be adopted for the multiple drugs and multiple targets (Jocelyn, 2008).

Nanoparticle in Molecular Imaging

Magnetic nanoparticles, which are bound to a suitable antibody, are used successfully to label specific molecules, structures, or microorganisms. Magnetic immunoassay techniques have also been developed in which the magnetic field generated by the magnetically labeled targets is directly detected with the help of sensitive magnetometer. For detection of genetic sequence in a sample gold nanoparticles tagged with short segments of DNA can be used. Multicolor optical coding for biological assays has been achieved by embedding different-sized QDs (nanocrystals of cadmium selenide) into polymeric microbeads. This spectral coding technology is expected to open new opportunities for medical diagnostics.

Nanoparticles also helps to stop bleeding faster but now nanoparticles are tailored to fastener onto blood platelets rapidly and to create a healthy clot which nearly double the survival rate in the critical first hour after injury. These platelet-like nanoparticles are made of biodegradable polymers and are designed to work on injuries after injection. This will even find effective in military treatments as the military has no equivalent of a tourniquet, pressure dressing or other easily transportable technology to stem bleeding from internal injuries.

These nanoparticles also help to overcome ROS stress by releasing their contents at the site of inflammation and get degraded in the presence of hydrogen peroxide produced by these cells. As these capsules are tailored to biodegrade after releasing their consignment on encountering hydrogen peroxide, they prove to be effective in for targeted drug delivery to diseased tissue (Lux *et al.*, 2012).

Bioengineers have designed nanoparticles that can safely and predictably infiltrate deep into the brain to prevent the drug-embedded particles from sticking to their surroundings and can be spread once they are in the brain. These nanoparticles are made by entrapping drug molecules together with microscopic, string-like molecules in a tight ball, which slowly breaks down when it comes in contact with water. These nanoparticles were prepared by nano-sized plastic beads of various sizes with a clinically tested molecule called PEG, or poly (ethylene glycol), that had been shown as stabilizing agent to protect nanoparticles from the body's defense mechanisms (Nance *et al.*, 2012). Combined polyethylene glycol-hydrophilic carbon clusters (PEG-HCC), already being tested and proves to be useful to enhance cancer treatment, are also adept antioxidants. In a traumatic brain injury, brain cells release an excessive amount of an ROS known as superoxide (SO) into the blood. Superoxides are the toxic free radicals used by the immune system to kill invading microorganisms. Usually healthy organisms balance SO with superoxide dismutase (SOD) enzyme,that neutralizes it. But when mild brain trauma occurs it releases superoxides at levels that overwhelm the brain's natural defenses and have very deleterious effects. SO also affects the autoregulatory mechanism that manages the sensitive circulation system in the brain. PEG-HCC nanoparticles immediately and completely quenched superoxide activity and allowed the autoregulatory system to quickly regain its balance. These nano-sized package is doing something that no small drug or protein could do, underscoring the efficacy of active nano-based drugs (Bitner *et al.*, 2012). For details of the use of nanoparticles in human health one may refer to the reference Kushwaha and Malik (2012).

Nanoparticles in Agriculture

Plant pathology is one of the foremost area of studying in agriculture and nanotechnology has immense potential to overcome several plant diseases thus nanoparticles and quantom dots have proves to be potent tool for treating plant disease. In pertinent to plants nanotechnology has been shown to be beneficial specially to enhance crop productivity through the use of nanomaterials, carbon nanotubes, nanoporous zeolites, nano-encapsulation, nanotubes etc.

Silver nanoparticles

Nanoparticles have been successfully used for bioremediation of the contaminated environment, biocides and antifungal on textiles (Barik *et al.*, 2008). As discussed earlier silver nanoparticles have antibacterial and antimicrobial properties hence effectively used to cure microbial plant diseases. Further colloidal nano silver solution is used as reducing agents and stabilizers against antifungal and antibacterial.

Nano silica- silver composite

Silicon is essential for plant to increase disease resistance as well as stress resistance (Brecht *et al.*, 2003). Silicate aqueous solution has affinity against certain fungi. In ionic state it exhibit antimicrobial activity (Mc Cubin, 2003). For details and efficiency against different fungi and bacteria please refer to the table (Fungi and bacteria)

Nano Alumino Silicate

These are very active pesticides which are actively absorbed by the hairs present on the body of the insects. Some insects ingest them as well (Wang *et al.*, 2002).

Nanoparticles in Environmental Control

Iron nanoparticles are used to treat the soil contaminated with heavy metal. Such nanoparticles are sprayed on the soil and encourage bringing about decontamination. This is a preferred process which is less time consuming, cost effective and ecofirendly. Some workers have used for treating waste water to check the microbial growth. Moreover the nitrifying bacteria are inhibited as well which has a detrimental effect on the microorganism (Jha *et al.*, 2011).

For details of the use of nanoparticles in agriculture one may refer to the reference Kushwaha and Malik (2012).

Conclusion

The new developments reviewed here should encourage the preparation of novel materials with different architectures, in order to respond quickly to the demands of Nanotechnology and Materials Science.

Living organisms have huge potential for the production of nanoparticles/nanodevices of wide applications. Organisms from simple bacteria to highly complex eukaryotes in the reaction mixture lead to the production of nanoparticles/nanodevices with desired shape and size. Though nano-biotechnology is at its infancy but various examples through which this technology and their use have been explained here would attract the attention of peoples towards its applications. It is hoped that this chapter is useful in exploring the method for nanoparticle synthesis, as well as for setting the newest trends in this fascinating field. Although a profound concern from a mechanistic view, one aspect that was clear from this review is that, in some of the biological systems, more detailed studies and also the use of comparative methods with different strains in all the organisms discussed may lead in achieving the better understanding of this fascinating topic.

REFERENCES

Abu Bakar NHH, Ismail J and Abu Bakar M (2007) Synthesis and characterization of silver nanoparticles in natural rubber. *Mater Chem Phys* 104(2–3): 276–283

Acharya K, Chattopadhyay D, Patra S, Sarkar J and Saha S (2011) Mycosynthesis of silver nanoparticles and investigation of their antimicrobial activity. *JoNSNEA* 1(1): 17–26

Adlim M, Abu Bakar M, Liew KY and Ismail J (2004) Synthesis of chitosan-stabilized platinum and palladium nanoparticles and their hydrogenation activity. *Jour Mol Catal A Chem* 212 (1–2): 141–149

Afreen RV and Ranganath E (2011) Synthesis of monodispersed silver nanoparticles by *Rhizopus stolonifer* and its antibacterial activity against MDR strains of *Pseudomonas aeruginosa* from burnt patients. *Int Jour Environ Sci* **1**(7): 1582–1592

Agnihotri M, Joshi S, Kumar AR, Zinjarde S and Kulkarni S (2009) Biosynthesis of gold nanoparticles by the tropical marine yeast *Yarrowia lipolytica* NCIM 3589. *Mate Lett.* **63:** 1231-1234

Ahmad A, Senapati S, Khan MI, Kumar R and Sastry M (2003) Extracellular biosynthesis of monodisperse gold nanoparticles by a novel extremophilic actinomycete, *Thermomonospora* sp. *Langmuir* **19**(8) : 3550–3553

Ahmad A, Mukherjee P, Mandal D, Senapati S, Khan MI, Kumar R and Sastry M (2002) Enzyme mediated extracellular synthesis of CdS nanoparticles by the fungus *Fusarium oxysporum. Jour Amri Chem Soc.* 124(1): 12108–12109

Ahmad A, Mukherjee P, Senapati S, Mandal D, Khan MI, Kumar R and Sastry M (2003) Extracellular biosynthesis of silver nanoparticles using the fungus *Fusarium oxysporum. Colloids Surf B* 28(4): 313–318

Ahmad A, Senapati S and Khan MI *et al.,* (2003) Intracellular synthesis of gold nanoparticles by a novel alkalotolerant actinomycete, *Rhodococcus* species," *Nanotechnology* **14(7)** : 824–828

Ahmad A, Senapati S, Khan MI, Kumar R and Sastry M (2005) Extra-/intracellular biosynthesis of gold nanoparticles by an alkalotolerant fungus *Trichothecium* sp. *Jour Biomed Nanotechnol* 1(1): 47–53

Albrecht MA, Evans CW and Raston CL (2006) Green chemistry and the health implications of nanoparticles. *Green Chem* 8: 417–432

Ali MD, Divya C, Gunasekaran M and Thajuddin N (2011) Biosynthesis and characterization of Silicon – Germanium oxide nanocomposite by Diatom, *Dig Jour Nanomater Bios* **6** : 117 – 120

Allen TM and Cullis PR (2004) Drug delivery system: Entering the mainstream. *Science* **303**(5665): 1818-1822

Andreescu D, Eastman C, Balantrapu K and Goia DV (2007) A simple route for manufacturing highly dispersed silver nanoparticles. *Jour Mater Res.* **22**(9): 2488-2496

Anitha TS and Palanivelu P (2011) Synthesis and structural characterization of polydisperse silver and multishaped gold nanoparticles using *Fusarium oxysporum* MTCC 284. *Dig Jour Nanomater Bios* 6(4): 1587–1595

Ankamwar B, Chaudhary M and Sastry M (2005) Gold nanotriangles biologically synthesized using tamarind leaf extract and potential application in vapour sensing., *Synth React Inorg Metal-Org Nano-Metal Chem.*, **35**: 19–26

Amro NA, Kotra LP, Wadu-Mesthrige K, Bulychev A, Mobashery S and Liu G (2000) High-resolution atomic force microscopy studies of the *Escherichia coli* outer membrane: structural basis for permeability. *Langmuir* 16 : 2789- 2796

Arakaki A, Shibusawa M, Hosokawa M and Matsunaga T (2010) Preparation of genomic DNA from a single species of uncultured magnetotactic bacterium by multiple-displacement amplification *Applied and Environmental Microbiology* **76**(5): 1480–1485

Armendariz V, Herrera I, Peralta-Videa JR, Jose-Yacaman M, Troiani H, Santiago P and Cardea-Torresdey JL (2004) Size controlled gold nanoparticle formation by *Avena sativa* biomass: use of plants in nanobiotechnology. *Jour Nanopart Res* 6(4): 377–382

Babu MMG and Gunasekaran P (2009) Production and structural characterization of crystalline silver nanoparticles from *Bacillus cereus* isolate *Colloids and Surfaces B* 74(1) : 191–195

Bai HJ, Zhang ZM and Gong J (2006) Biological synthesis of semiconductor zinc sulphide nanoparticles by immobilized *Rhodobacter sphaeroides Biotechnology Letters* 28(14) : 1135–1139

Bai HJ, Zhang ZM, Guo Y and Yang YE (2009) Biosynthesis of cadmium sulfide nanoparticles by photosynthetic bacteria *Rhodopseudomonas palustris Colloids and Surfaces B* 70(1) : 142–146

Balaji DS, Basavaraja S, Deshpande R, Mahesh DB, Prabhakar BK and Venkataraman A (2009) Extracellular biosynthesis of functionalized silver nanoparticles by strains of *Cladosporium cladosporioides* fungus. *Colloids Surfaces B* 68(1): 88–92

Bankar A, Joshi B, Ravi Kumar A and Zinjarde S (2010) Banana peel extract mediated synthesis of gold nanoparticles. *Colloids Surfaces B* 80(1): 45–50

Bansal V, Poddar P, Ahmad A and Sastry M (2006) Room-temperature biosynthesis of ferroelectric barium titanate nanoparticles. *Jour Am Chem Soc* 128(36): 11958–11963

Bansal V, Rautaray D, Bharde A, Ahire K, Sanyal A, Ahmad A and Sastry M (2005) Fungus-mediated biosynthesis of silica and titania particles. *Jour Mater Chem* 15(26): 2583–2589

Bao H, Lu Z and Cui X ***et al.***, (2010) Extracellular microbial synthesis of biocompatible CdTe quantum dots *Acta Biomaterialia* 6(9) : 3534–3541

Bar H, Bhui DK, Sahoo GP, Sarkar P, De SP and Misra A (2009) Green synthesis of silver nanoparticles using latex of *Jatropha curcas. Colloids and surfaces A: Physicochemical and Engineering aspects.* 339 : 134- 139

Barik TK, Sahu B and Swain V (2008) Nanosilica-from medicine to pest control. *Parasitol Res.* 103(2): 253-258

Basavaraja S, Balaji SD, Legashetty A, Rasab AH and Venkatraman A (2008) Extracellular biosynthesis of silver nanoparticles using the fungus *Fusarium semitectum*. *Mater Res Bull* 43(5): 1164–1170

Bazylinski DA, Garratt-Reed AJ and Frankel RB (1994) Electron microscopic studies of magnetosomes in magnetotactic bacteria *Microscopy Research and Technique* 27(5) : 389–401

Beguma NA, Mondalb S, Basub S, Laskara RA and Mandalb D (2009) Biogenic synthesis of Au and Ag nanoparticles using aqueous solutions of Black Tea leaf extracts. *Colloids and Surfaces B: Biointerfaces* 71 : 113–118

Berger TJ, Spadaro JA, Chapin SE and Becker RO (19.'6) Electrically generated silver ions: Quantitative effects on bacterial and mammalian cells. *Antimicrob. Agents Chemother* **9**: 357-358

Bhainsa KC and D'Souza SF (2006) Extracellular biosynthesis of silver nanoparticles using the fungus Aspergillus fumigatus. *Coll Surf B Bioint.* 47: 160-164

Bharde A, Rautray D, Bansal V, Ahmad A, Sarkar I, Yusuf SM, Sanyal M and Sastry M (2006) Extracellular biosynthesis of magnetite using fungi. *Small* 2(1): 135–141

Bhattacharya R and Mukherjee P (2008) Biological properties of "naked" metal nanoparticles *Advanced Drug Delivery Reviews* 60(11) : 1289–1306

Bhol KC and Schechter PJ (2005) Topical nanocrystalline silver cream suppresses inflammatory cytokines and induces apoptosis of inflammatory cells in a murine model of allergic contact dermatitis. *Br J Dermatol.* 152(6): 1235-1242

Binupriya AR, Sathishkumarb M, Vijayaraghavanb K and Yuna SI (2010) Bioreduction of trivalent aurum to nano-crystalline gold particles by active and inactive cells and cell-free extract of *Aspergillus oryzae* var. viridis. *Jour Hazard Mater* 177: 539–545

Birla SS, Tiwari VV, Gade AK, Ingle AP, Yadav AP and Rai MK (2009) Fabrication of silver nanoparticles by Phoma glomerata and its combined effect against *Escherichia coli, Pseudomonas aeruginosa and Staphylococcus aureus. Lett Appl Microbiol* 48(2): 173–179

Bitner BR, Marcano DC, Berlin JM, Fabian RH, Cherian L, Culver JC, Dickinson ME, Robertson CS, Pautler RG, Kent TA and Tour JM (2012) Antioxidant Carbon Particles Improve Cerebrovascular Dysfunction Following Traumatic Brain Injury. *ACS Nano.* 120815105406004 DOI: 10.1021/nn302615f

Brecht M, Datnoff L, Nagata R and Kucharek T (2003) The role of silicon in suppressing tray leaf spot development in St. Augustine grass. *Publication in University of Florida* 1-4

Castanon GAM, Martinez NN, Gutierrez FM, Mendaza JRM and Ruiz F (2008) Synthesis and activity of silver nanoparticles with different sizes. *Jour Nanoparticle Res.* 10: 1342-1348

Castro L, Bla´zquez ML, Gonza´lez F, Mun˜oz JA and Ballester A (2010) Extracellular biosynthesis of gold nanoparticles using sugar beet pulp. *Chem Eng Jour* 164(1): 92–97

Castro-Longoria E, Vilchis-Nestor AR and Avalos-Borja M (2011) Biosynthesis of silver, gold and bimetallic nanoparticles using the filamentous fungus *Neurospora Crassa Colloids Surf,* 83: 42–48

Chandran PS, Chaudhary M, Pasricha R, Ahmad A and Sastry M (2006) Synthesis of gold triangles and silver nanoparticles using *Aloe vera* plant extract. *Biotechnol Prog* 27: 577-583

Chauhan S Upadhyay MK Rishi N Rishi S (2011) Phytofabrication of silver nanoparticles using pomegranate fruit seeds .*Intern. Jour of Nanomaterials and Biostructures.* 1(2) : 17-21

Chen GQ, Zou ZJ, Zeng GM, Yan M, Fan JQ, Chen AW, Yang F, Zhang WJ and Wang L (2011) Coarsening of extracellularly biosynthesized cadmium crystal particles induced by thioacetamide in solution. *Chemosphere* 83(9): 1201–1207

Chen JC, Lin ZH and Ma XX (2003) Evidence of the production of silver nanoparticles via pretreatment of Phoma sp.3.2883 with silver nitrate. *Lett Appl Microbiol* 37(2): 105–108

Conrad AH, Tramp CR, Long CJ, Wells DC, Paulsen AQ and Conrad GW (1999) Ag+ alters cell growth, neurite extension, cardiomyocyte beating, and fertilized egg constriction. *Aviat Space Environ Med* **70**: 1096-110

Cruz D, Fale´ PL, Mourato A, Vaz PD, Luisa Serralheiro M and Lino ARL (2010) Preparation and physicochemical characterization of Ag nanoparticles biosynthesized by *Lippia citriodora* (Lemon Verbena). *Colloids Surf B* 81(1) : 67–73

Daisy P and Saipriya K (2012) Biochemical analysis of *Cassia fistula* aqueous extract and phytochemically synthesized gold nanoparticles as hypoglycemic treatment for diabetes mellitus. *Int Jour Nanomedicine.* **7**: 1189–1202

Danilczuk M, Lund A, Saldo J, Yamada H and Michalik J (2006) Conduction electron spin resonance of small silver particles. *Spectrochim Acta A* **63** : 189 – 91

Das R, Nath SS, Chakdar D, Gope G and Bhattacharjee R (2009) Prepration of silver nanoparticles and their characterization. *Jour of Nanotechnology* 5: 1-6

Das RK, Gogoi N and Bora U (2011a) Green synthesis of gold nanoparticles using *Nyctanthes arbortristis* flower extract. *Bioprocess Biosyst Eng* 34(5): 615–619

Das SK, Das AR and Guha AK (2009) Gold nanoparticles: microbial synthesis and application in water hygiene management. *Langmuir* 25: 8192–8199

Destrée C and Nagy JB (2006) Mechanism of formation of inorganic and organic nanoparticles from microemulsions. *Adv. Colloid Interface Sci* **123-126** : 353-367

Dibrov P, Dzioba J and Gosink KK (2002) Chemiosmotic mechanism of antimicrobial activity of Ag(+) in *Vibrio cholerae. Antimicrob Agents Chemotherapy* **46** : 2668- 2670

Douglas T and Young M (1998) Host-guest encapsulation of materials by assembled virus protein cages. *Nature* 393: 152–155

Du L, Xian L and Feng J-X (2011) Rapid extra-/intracellular biosynthesis of gold nanoparticles by the fungus Penicillium sp. *Jour Nanopart Res.* 13: 921–930

Du L, Jiang H, Liu X and Wang E (2007) Biosynthesis of gold nanoparticles assisted by *Escherichia coli DH5á* and its application on direct electrochemistry of haemoglobin *Electrochemistry Communications* 9(5) : 1165– 1170

Dubey SP, Lahtinen M, Sarkka H and Sillanpaa M (2010) Bioprospective of *Sorbus aucuparia* leaf extract in development of silver and gold nanocolloids. *Colloids Surf B* 80(1): **26–33**

Duran N, Marcato PD, Alves OL, Desouza G and Esposito E (2005) Mechanistic aspects of biosynthesis of silver nanoparticles by several *Fusarium oxysporum* strains. *Jour Nanobiotechnol* 3: 1–8

Dwivedi AD and Gopal K (2010) Biosynthesis of silver and gold nanoparticles using *Chenopodium album* leaf extract. *Colloids Surf A* 369(1–3): 27–33

Egorova EM and Revina AA (2000) Synthesis of metallic nanoparticles in reverse micelles in the presence of quercetin. *Colloids Surf A* 168(1): 87–96

El-Rafie MH, Mohamed AA, Shaheen THI and Hebeish A (2010) Antimicrobial effect of silver nanoparticles produced by fungal process on cotton fabrics. *Carbohydr Polym* 80: 779–782

Elechiguerra JL, Burt JL, Morones JR, Camacho-Bragado A, Gao X, Lara HH and Yacaman MJ (2005) Interaction of silver nanoparticles with HIV-1 *Journal of Nanobiotechnology* 3: 6

Farooqui MD, Chauhan PS, Krishnamoorthy P and Shaik J (2010) Extraction of Silver nanoparticles from the leaf extract of *Clerodendrum inerme. Digest Jour of Nanomaterials and Biostructures.* 5(1) : 577-583

Fayaz M, Balaji K, Girilal M, Yadav R, Kalaichelvan PT and Venketesan R (2010) Biogenic synthesis of silver nanoparticles and their synergistic effect with antibiotics: a study against gram-positive and gram-negative bacteria. *Nanomed Nanotechnol Biol Med* 6: 103–109

Feng QL, Wu J, Chen GQ, Cui FZ, Kim TN and Kim JO (2008) A mechanistic study of the antibacterial effect of silver ions on *Escherichia coli* and *Staphylococcus aureus. J Biomed Mater Res.* 52: 662–668

Fu M, Li Q, Sun D, Lu Y, He N, Deng X, Wang H and Huang J (2006) Rapid preparation process of silver nanoparticles by bioreduction and their characterizations. *Chinese J Chem Eng.* 14(1): 114-117

Gade A, Bonde PP, Ingle AP, Marcato P, Duran N and Rai MK (2008) Exploitation of *Aspergillus niger* for synthesis of silver nanoparticles. *Jour Biobased Mater Bioenergy* 2(3): 1–5

Gade A, Rai M and Kulkarni S (2011) *Phoma sorghina,* a phytopathogen mediated synthesis of unique silver rods. *Int Jour Green Nanotechnol* 3: 153–159

Gajbhiye M, Kesharwani J, Ingle A, Gade A and Rai M (2009) Fungus mediated synthesis of silver nanoparticles and their activity against pathogenic fungi in combination with fluconazole. *Nanomedicine NBM* 5: 382–386

Gardea-Torresdey JL, Gomez E, Peralta-Videa JR, Parsons JG, Troiani H and Jose-Yacaman M (2003) *Alfalfa* Sprouts: A Natural Source for the Synthesis of Silver Nanoparticles. *Langmuir* 19 (4): 1357-1361

Gardea-Torresdey JL, Parsons LG, Gomez E, Peralta-Videa J, Troiani HE, Santiago P and Yacaman MJ (2002) Formation and Growth of Au Nanoparticles inside Live Alfalfa Plants *Nano Lett.* 2 (4): 397-401

Garg S (2012) Rapid biogenic synthesis of silver nanoparticles using black pepper (*Piper nigrum*) corn extract. *International Jour of Innovations in Biological and Chemical Sciences* 3:

5-10

Gericke M and Pinches A (2006) Biological synthesis of metal nanoparticles. *Hydrometallurgy* 83: 132–140

Ghodake GS, Deshpande NG, Lee YP and Jin ES (2010) Pear fruit extract-assisted room-temperature biosynthesis of gold nanoplates. *Colloids Surf B* 75(2): 584–589

Ghule K, Ghule AV, Liu JY and Ling YCJ (2006) Microscale size triangular gold prisms synthesized using Bengal gram beans (*Cicer arietinum* L.) extract and HAuCl4x3H20: a green biogenic approach. *Nanosci Nanotechnol.* 6(12): 3746-51

Gnana D, Gnana J, Annadurai G and Kannan C (2012) Green synthesis of Silver Nanoparticle using *Elettaria Cardamomom* and Assessment of its Antimicrobial Activity *International Jour of Pharma Sciences and Research.* 2-3(3): 323-330

Govender Y, Riddin T, Gericke M and Whiteley CG (2009) Bioreduction of platinum salts into nanoparticles: a mechanistic perspective. *Biotechnol Lett* 31: 95–100

Guo Z, Kim TY, Lei K, Pereira T, Sugar JG and Hahn HT (2008) Strengthening and thermal stabilization of polyurethane nanocomposites with silicon carbide nanoparticles by a surface-initiated-polymerization approach. *Composites Science and Technology* 68 : 164–170

Gurunathan S, Kalishwaralal K and Vaidyanathan R *et al.*, (2009) Biosynthesis, purification and characterization of silver nanoparticles using *Escherichia coli Colloids and Surfaces B.*74 (1) : 328–335

Guzman MG, Dille J and Godet S (2008) Synthesis of silver nanoparticles by chemical reduction method and their antibacterial activity. *World academy of Science, Engineering and Technology* 43: 357-364

Hemath NKS, Kumar G, Karthik L and Bhaskara Rao KV (2010) Extracellular biosynthesis of silver nanoparticles using the filamentous fungus Penicillium sp. *Arch Appl Sci Res* 2: 161–167

Husseiny MI, El-Aziz MA and Badr Y Mahmoud MA (2007) Biosynthesis of gold nanoparticles using *Pseudomonas aeruginosa. Spectrochimica Acta Part A.* **67**(3-4) : 1003–1006

Inbakandan ID, Ajithkumar TT and Umashankari BT *et al.*, (2012) Saline Systems Mangrove plant, *Rhizophora mucronata* (Lamk, 1804) mediated one pot green synthesis of silver nanoparticles and its antibacterial activity against aquatic pathogens, *Jaganathan Umashankari1* **8**: 11

Ingle A, Gade A, Pierrat S, Sonnichsen C and Rai M (2008) Myco-synthesis of silver nanoparticles using the fungus *Fusarium acuminatum* and its activity against some human pathogenic bacteria. *Curr Nanosci* **4**: 141–144

Ingle A, Gade A, Bawaskar M and Rai M (2009) *Fusarium solani*: a novel biological agent for the extracellular synthesis of silver nanoparticles .*Jour Nanopart Res* 11: 2079–2085

Jagtap UB and Bapat V (2012) Biosynthesis, characterization and antibacterial activity of silver nanoparticles by aqueous *Annona squamosa* L. leaf extract at room temperature. *Jour of Plant Biochemistry and Biotechnology* DOI: 10.1007/s13562-012-0172-8

Jagtap UB and Bapat V (2013) Green synthesis of silver nanoparticles using *Artocarpus heterophyllus* Lam. seed extract and its antibacterial activity. *Industrial Crop and Products* 46: 132-137

Jain D, Daima K, Kachhwaha S and Kothari SL (2009) Synthesis of plant-mediated silver nanoparticles using Papaya fruit extract and evaluation of their antimicrobial activities. *Digest Jour of Nanomaterials and Biostructures* 4(3) : 557-563

Jain D, Kachhawa S, Jain R, Shrivastav S and Kothari SL (2010) Novel microbial route to syntheisze silver nanoparticles using spore crystal mixture of *B. thuringiensis. Indian Jour of Experimental Biology* 48 : 1152-1156

Jain N, Bhargava A, Majumdar S, Tarafdar JC and Panwar J (2010) Extracellular biosynthesis and characterization of silver nanoparticles using *Aspergillus flavus* NJP08: a mechanism perspective. *Nanoscale* 3: 635–641

Jain PK, Huang X, El Sayed LH and El Sayed MA (2007) Review of some interesting Surface Plasmon Resonance- enhanced properties of noble metal Nanoparticles and their applications to biosystems. *Plasmonics* **2**: 107- 118

Jeffryes C, Gutu T, Jiao J and Rorrer GL (2008) Two-stage photobioreactor process for the metabolic insertion of nanostructured germanium into the silica microstructure of the *Diatom Pinnularia sp, Mater Sci Eng C: Biomimetic Supramol Syst.* 28 : 107-118

Jha AK, Prasad K and Kulkarni AR (2009) Synthesis of TiO_2 nanoparticles using microorganisms *Colloids and Surfaces B*, vol. 71(2) : 226–229

Jha AK, Prasad K and Prasad K (2009) A green low- cost biosynthesis of Sb_2O_3 nanoparticles *Biochemical engineering Jour* 43: 303-306

Jha AK, Prasad K, Prasad K and Kulkarni AR (2009) Plant system: Nature's nanofactory *Colloids and Surfaces B: Biointerfaces* (in press) doi: 10.1016/j.colsurfb.2009.05.018

Jha Z, Behar N, Sharma NS, Chandel G, Sharma DK and Pandey MP (2011) Nanotechnology Prospects of Agricultural Advancement. *Nano Vision* **1** (2): 88-100

Jianping X, Jim YL, Daniel ICQ and Yen PI (2007) Identification of active biomolecules in the high- yield synthesis of single–crystalline gold nanoplates in algal solutions. *Int Jour of Applied Biol and Pharma Tech.* 3 (4): 668 – 72

Jocelyn R (2008) Nano Drugs to Starve Tumors: A new delivery system chokes a tumor blood supply while sparing healthy tissues. *Biomedicine.*

Juibari MM, Abbasalizadeh S, Jouzani GS and Noruzi M (2011) Intensified biosynthesis of silver nanoparticles using a native extremophilic *Ureibacillus thermosphaericus* strain *Materials Lett.* 65(6) : 1014–1017

Kalimuthu K, Suresh Babu R, Venkataraman D, Bilal M and Gurunathan S (200 Biosynthesis of silver nanocrystals by *Bacillus licheniformis. Colloids and Surfaces B.* 65(: 150–153

Kalishwaralal K, Deepak V, Pandian SRK *et al.*, (2010) Biosynthesis of silver and go nanoparticles using *Brevibacterium casei Colloids and Surfaces B* 77(2) : 257–262

Kannan N and Subbalaxmi S (2011) Iogenesis of Nanoparticles-A current prespsctive. *R*

Adv Mater Sci. 27: 99-114

Kannan S, Dai H, Navath RS, Balakrishnan B, Jyoti A, Janisse J, Romero R and Kannan RM (2012) Dendrimer-Based Postnatal Therapy for Neuroinflammation and Cerebral Palsy in a Rabbit Model. *Science Translational Medicine* 4 (130): 130-146

Karbasian M, Atyabi SM, Siadat SD, Momem SB and Norouzian D (2008) Optimizing nano-silver formation by *Fusarium oxysporum* (PTCC 5115) employing response surface methodology. *Am Jour Agric Biol Sci* 3: 433–437

Kathiresan K, Manivannan S, Nabeel AM, Dhivya B (2009) Studies on silver nanoparticles synthesized by a marine fungus *Penicillum fellutanum* isolated from coastal mangrove sediment. *Colloids Surf B* 71: 133–137

Kattia K, Chandaa N, Shuklaa R, Zambrea A, Suibramaniana T, Kulkarnia RR, Kannana R and Kattia KV (2009) Green Nanotechnology from Cumin Phytochemicals: Generation of Biocompatible Gold Nanoparticles *Int Jour Green Nanotechnol Biomed*1(1): B39–B52

Kazachenko AS, Legler TAV, Peryanova TOV and Vstavskaya YA (2000) Synthesis and antimicrobial activity of silver complexes with histidine and tryptophan. *Pharm Chem Jour* 34(5): 34–35

Kim JS, Kuk E, Yu K, Kim JH, Park SJ, Lee HJ, Kim SH, Park YK, Park YH, Hwang CY, Kim YK, Lee YS, Jeong DH and Cho MH (2007) Antimicrobial effects of silver nanoparticles. *Nanomedicine* 3 : 95–101

Kim KJ, Sung WS, Moon SK, Choi JS, Kim JG and Lee DG (2008) Antifungal effect of silver nanoparticles on dermatophytes. *Jour of Microbiology and Biotechnology.* 18: 1482–1484

Kim KJ, Sung WS, Suh BK, Seok KM, Jung SC, Jung GK and Dong GL (2009) Antifungal activity and mode of action of silver nano-particles on *Candida albicans. Biometals* **22**(2) : 235-242

Kircher MF, Zerda A, Jokerst JV, Zavaleta CL, Kempen PJ, Mittra E, Pitter K, Huang R, Campos C, Habte F, Sinclair R, Brennan CW, Mellinghoff IK, Holland EC and Gambhir SS (2012) A brain tumor molecular imaging strategy using a new triple-modality MRI-photoacoustic-Raman nanoparticle. *Nature* 18(5): 829-34

Klaus T, Joerger R, Olsson E and Granqvist CG (1999) Silver based crystalline nanoparticles, microbially fabricated *Proceedings of the National Academy of Sciences of the United States of America.* 96(24) : 13611–13614

Korbekandi H, Asghari G, Jalayer SS, Jalayer MS and Bandegani M (2013) Nanosilver Particle Production Using *Juglans Regia* L. (Walnut) Leaf Extract. *Jundishapur Jour Nat Pharm Prod.* : 8(1): in press. DOI: 10.5812/jjnpp.7433

Kotelnikova NE, Wegener G, Stoll M and Demidov VN (2003) Comparative study of intercalation of zero-valent silver into the cellulose matrix by raster and transmission microscopy. *Russ Jour Appl Chem.* 76(1): 117–123

Kowshik M, Ashtaputre S, Kharrazi S, Vogel W, Urban J, Kulkarni SK and Paknikar KM (2003) Extracellular synthesis of silver nanoparticles by a silver- tolerant yeast strain MKY3. *Nanotechnology* 14: 95-100

Kowshik M, Vogel W, Urban J, Kulkarni SK and Paknikar KM (2002) Microbial synthesis of semiconductor PbS nanocrystallites. *Biotechnol Bioeng Adv Mater* 14: 815-818

Kowshik M, Deshmukh N, Vogel W, Urban J, Kulkarni SK and Paknikar KM (2002) Microbial synthesis of semiconductor CdS nanoparticles, their characterization and their use in the fabrication of an ideal diode. *Biotechnology and Bioengineering* 78: 583-588

Kumar AS, Ansary AA, Ahmad A and Khan MI (2007a) Extracellular biosynthesis of CdSe quantum dots by the fungus *Fusarium oxysporum*. *Jour Biomed Nanotechnol* **3**: 190–194

Kumar SA, Abyaneh MK, Gosavi SW, Kulkarni SK, Pasricha R, Ahmad A and Khan MI (2007b) Nitrate reductase-mediated synthesis of silver nanoparticles from $AgNO_3$. *Biotechnol Lett* 29: 439–445

Kumar SA, Peter YA and Nadaeu JL (2008) Facile biosynthesis, separation, conjugation of gold nanoparticles to doxorubicin. *Nanotechnology* 19: 495101

Kumar V, Yadav SC and Yadav SK (2010) *Syzygium cumini* leaf and seed extract mediated biosynthesis of silver nanoparticles and their characterization. *Jour Chem Technol Biotechnol* 85(10): 1301–1309

Kumar DS, Karthik L, Kumar G and Rao BKV (2011) Biosynthesis of Silver anoparticles from Marine Yeast and Their Antimicrobial Activity Against Multidrug Resistant Pathogens *Pharmacologyonline* 3: 1100-1111

Kundu S, Panigrahi S, Praharaj , Basu S, Ghosh SK, Pal A and Pal T (2007) Anisotropic growth of gold clusters to gold nanocubes under UV irradiation. *Nanotechnology* 18(7): 075712

Kuo WS, Chang CN, Chang YT and Yeh CS (2009) Antimicrobial gold nanorods with dual-modality photodynamic inactivation and hyperthermia, *Chem Commun Camb* 32 : 4853–4855

Kushwaha HB and Malik CP (2012) Nanotechnology –A Hope of New Revolution in Agriculture, In *Crop improvement–Physiological, Biotechnological and Nanotechnological Approach,*(Edited) Aavishkar Publishers and Printers, Jaipur, pp-266-285

Labrenz M, Druschel GK, Thomsen-Ebert T, *et al.*, (2000) Formation of sphalerite (ZnS) deposits in natural biofilms of sulfate-reducing bacteria *Science* 290(5497) : 1744–1747

Lago DL, de Oliveira LF, de Almeida Gonçalves K, Kobarg J and Cardoso MB (2011) Nanotechnology against bacteria and HIV. *Materials Chemistry* 21: 12267

Lee BU, Yun SH, Ji JH and Bae GN (2008) Inactivation of *S. epidermidis*, *B. subtilis*, and *E. coli* bacteria bioaerosols deposited on a filter nutilizing airborne silver nanoparticles. *J Microbiol Biotechnol* **18**: 176–182

Leela A and Vivekanandan M (2008) Tapping the unexploited plant resources for the synthesis of silver nanoparticles *African Journal of Biotechnology* **7**(17) : 3162–3165

Lef'ever CT, Abreu F, Lins U and Bazylinski DA (2010) Non magnetotactic multicellular prokaryotes from lowsaline, nonmarine aquatic environments and their unusual negative phototactic behaviour *Applied and Environmental Microbiology* 76(10) : 3220–3227

Leff DV, Ohara PC, Heath JR and Gelbart WM (1995) Thermodynamic control of gold nanocrystal size: experiment and theory. *Jour Phys Chem* 99(18): 7036–7041

Lengke MF, Ravel B, Fleet ME, Wanger G, Gordon RA and Southam G (2006) Mechanisms of gold bioaccumulation by *filamentous cyanobacteria* from gold(III)-chloride complex *Environmental Science & Technology*. 40(20) : 6304–6309

Leonard K, Ahmmad B, Okamura H and Kurawaki J (2011) In situ green synthesis of biocompatible ginseng capped gold nanoparticles with remarkable stability. *Colloids Surf B* 82(2): 391–396

Li G, He D, Qian Y, Guan B, Gao S, Cui Y, Yokoyama K and Wang L (2012) Fungus-mediated green synthesis of silver nanoparticles using *Aspergillus terreus*. *Int Jour Mol Sci* 13: 466– 476

Li S, Shen Y, Xi A, Yu X, Qui L, Zhang L and Zhang Q (2007) Green synthesis of silver nanoparticles using *Capsicum annum* L. extract. *Green Chemistry* 9: 852-858

Lin L, Wang W, Huang J, Li Q, Sun D, Yang X, Wang H, He N and Wang Y (2010) Nature factory of silver nanowires: plant mediated synthesis using broth *of Cassia fistula* leaf. *Chem Eng Jour* 162(2): 852–858

Lin D and Xing B (2007) Phytotoxicity of nanoparti-cles: inhibition of seed germination and root growth. *Environmental Pollution* **150**: 243-250

Liu Z, Stout JE, Tedesco L, Boldin M, Hwang C, Diven WF and Yu VL (1994) Controlled evaluation of copper-silver ionization in eradicating *Legionella pneumophila* from a hospital water distribution system. *J.Infect. Dis* **169**: 919-922

Lloyd JR, Yong P and Macaskie LE (1998) Enzymatic recovery of elemental palladium by using sulfate-reducing bacteria *Applied and Environmental Microbiology* 64(11) : 4607–4609

Lok CN, Ho CM, Chen R, He QY, Yu WY, Sun H, Tam PK, Chiu JF and Che CM (2006) Proteomic analysis of the mode of antibacterial action of silver nanoparticles. *J Proteome Res* **5** : 916–924

Lopez PJ, Gautier C, Livage J and Coradin T (2005) Mimicking biogenic silica nanostructures formation. *Curr Nanosci* 1 : 73-83

Lukman AI, Gong B, Marjo CE, Roessner U and Harris AT (2011) Facile synthesis, stabilization, and anti-bacterial performance of discrete Ag nanoparticles using *Medicago sativa* seed exudates. *Jour Colloid Interface Sci* 353: 433–444

Lutsiak ME, Robinson DR, Coester C, Kwon GS and Samuel J (2002) Analysis of poly(D,L-lactic-co-glycolic acid) nanosphere uptake by human dendritic cells and macrophages in vitro. *Pharm Res.* 19(10): 1480-1487

Lux CG, Barr SJ, Nguyen T, Mahmoud E, Schopf E, Fomina N and Almutairi A (2012) Biocompatible Polymeric Nanoparticles Degrade and Release Cargo in Response to Biologically Relevant Levels of Hydrogen Peroxide. *Journal of the American Chemical Society* 120

Maensiri S, Laokul P, Klinkaewnarong J, Phokha S, Promarak V and Seraphin S (2008) Indium oxide (In 2O3) nanoparticles using *Aloe vera* plant extract: Synthesis and optical properties. *Jour of Optoelectronica and Adv Mater* 10 (3) : 161-165

Mafune´ F, Kohno JY, Takeda Y and Kondow T (2002) Full physical preparation of size-selected gold nanoparticles in solution: Laser ablation and laser-induced size control. *Jour Phys Chem B* 106(31): 7575–7577

Maliszewska I, Szewczyk K and Waszak K (2009) Biological synthesis of silver nanoparticles. *Jour Phys Conf Ser* 146: 2025

Mallikarjuna M, Narasimha G, Dillip R, Praveen B, Shreedhar B, Shree Lakshmi C, Reddy BVS, Deva P and Babu B (2011) Green synthesis of silver nanoparticles using Ocimum leaf extract and their characterization. *Digest Jour of Nanomaterial and Biostructures* 6 : 181-186

Mano Priya P, Karunai Selvi V and John Paul JA (2011) Green synthesis of silver nanoparticles from the leaf extract of *Euphorbia hirta* and *Nerium indicum. Digest Jour of Nanomaterials and Biostructures.* 6(2) : 869-877

Matsumura Y, Yoshikata K, Kunisaki S and Tsuchido T (2003) Mode of bacterial action of silver zeolite and its comparison with that of silver nitrate. *Appl Environ Microbiol* 69 : 4278–4281

McCubin T, Thomas S and Mc Cubin P (2003) A comparison of the antimicrobial effects of four silvercontaining dressings on three organisms. *J Wound Care*. 12: 101-107

Meyre ME, Tre´guer-Delapierre M and Faure C (2008) Radiation induced synthesis of gold nanoparticles within lamellar phases. Formation of aligned colloidal gold by radiolysis. *Langmuir* 24(9): 4421–4425

Minchin R (2008) Sizing up target with nanoparticles. *Nature Nanotechnology* 3(1): 12-13

Mohammadian A, Shaojaosadati SA and Rezee MH (2007) *Fusarium oxysporum* mediates photogeneration of silver nanoparticles. *Sci Iran* **14**: 323–326

Moharrer S, Mohammad B, Gharamohammad RA and Yargol M (2012) Biological synthesis of silver nanoparticles by *Aspergillus flavus*, isolated from soil of Ahar copper mine. *Indian Jour Sci Technol* 5: 2443–2447

Mokhtari M, Deneshpojouh S, Seyedbagheri S, Atashdehghan R, Abdi K, Sarkar S, Minaian S, Shahverdi RH and Shahverdi RA (2009) Biological synthesis of very small silver nanoparticles by culture suspernatant of *Klebsiella pneumonia*. The effects of visible-light irradiation and the liquid mixing process, *Materials Research Bull* 44 : 1415-1421

Mondal AK, Mondal S, Samanta S and Mallick S (2011) Syntheiss of ecofriendly silver nanoparticles from the plant latex used as an important Taxonimic Tool for Phylogenetic Interrelationship. *Advances in Bioresearch.* 2(1): 122-133

Morgan TP, Guadagnolo CM, Grosell M and Wood CM (2005) Effects of water hardness on toxicological responses to chronic waterborne silver exposure in early life stages of rainbow trout (Oncorhynchus mykiss) *Environ Toxicol Chem* **24**: 1642-1647

Morones JR, Elechiguera JL, Camacho A, Holt K, Kouri JB, Ramfrez JT and Yacaman MJ (2005) The bactericidal effect of silver nanoparticles. *Nanotechnology* 16: 2346-2353

Mourato A, Gadanho M, Lino AR, Tenreiro R (2011) Biosynthesis of Crystalline Silver and Gold Nanoparticles by Extremophilic Yeasts, Hindawi Publishing Corporation, *Bioinorganic Chemistry and Applications*, **8** pp doi: 10.1155/2011/546074

Mouxing F, Qingbiao L, Daohua S, Yinghua L, Ning H, Xu D, Huixuan W and Jiale H (2006) Rapid Preparation Process of Silver Nanoparticles by Bioreduction and Their Characterizations. *Chin. J Chem Eng* 14(1): 114-117

Mu L and Feng SS (2003) A novel controlled release formulation for the anticancer drug paclitaxel (Taxol(R)): PLGA nanoparticles containing vitamin E TPGS. *J Control Release* **86**: 33-48

Mukherjee P, Ahmad A, Senapati S, Khan MI, Kumar R, Ramani R and Srinivas V, Sastri M, Ajay Kumar PV, Alam M, Parischa R (2002) Extracellular synthesis of gold nanoparticles by the fungus *Fusarium oxysporum*. *Chem BioChem* 3: 461–463

Mukherjee P, Roy M, Mandal BP, Dey GK, Mukherjee PK, Ghatak J, Tyagi AK and Kale SP (2008) Green synthesis of highly stabilized nanocrystalline silver particles by a non-pathogenic and agriculturally important fungus *T. asperellum*. *Nanotechnology* 19: 075103

Mukherjee P, Ahmad A, Mandal D, Senapati S, Sainkar SR, Khan MI, Parishcha R, Ajayakumar PV, Alam M, Kumar R and Sastry M (2001) Fungus-mediated synthesis of silver nanoparticles and their immobilization in the mycelial matrix: a novel biological approach to nanoparticle synthesis. *Nano Lett* 1: 515–519

Mukunthan KS, Elumalai EK, Patel TN and Murty VR (2011) *Catharanthus roseus*: a natural source for the synthesis of silver nanoparticles. *Asian Pacific Jour of Tropical Biomedecine* 1691(11) : 60041-60045

Nair B and Pradeep T (2002) Coalescence of nanoclusters and formation of submicron crystallites assisted by *lactobacillus strains*. *Crystal Growth & Design*. 2(4) : 293–298

Namasivayam SKR and Avimanyu (2011) Silver nanoparticle synthesis from *Lecanicillium lecanii* and evalutionary treatment on cotton fabrics by measuring their improved antibacterial activity with antibiotics against *Staphylococcus aureus* (ATCC 29213) and E. coli (ATCC 25922) strains. *Int Jour Pharm Pharm Sci* 3: 190–195

Nance EA, Woodworth GF, Sailor KA, Shih TY, Xu Q, Swaminathan G, Xiang D, Eberhart C and Hanes J (2012) A. Dense Poly(Ethylene Glycol) Coating Improves Penetration of Large Polymeric Nanoparticles Within Brain Tissue. *Science Translational Medicine* 4 (149): 119

Narayanan KB and Sakthivel N (2010b) Phytosynthesis of gold nanoparticles using leaf extract of *Coleus amboinicus Lour. Mater Charact* 61(11): 1232–1238

Narayanan KB and Sakthivel N (2010a) Facile green synthesis of gold nanostructures by NADPH-dependent enzyme from the extract of *Sclerotium rolfsii*. *Colloids Surf A* 380: 156- 161

Natarajan V, Veugopal PV and Menon T (2003) Effect of *Azadirachta indica* (neem) on the growth pattern of der-matophytes. *Indian J Med Microbiol*. 21: 98-101

Navazi ZR, Pazouki M and Halek FS (2010) Investigation of culture conditions for biosynthesis of silver nanoparticles using *Aspergillus fumigates*. *Iranian J Biotechnol* 8: 61

Nino-Martinez N, Martinez-Castanon GA, Aragon-Pina A, Martinez-Gutierrez F, Martinez-Mendoza JR and Ruiz F (2008) Characterization of silver nanoparticles synthesized on titanium dioxidefine particles. *Nanotech* **19**: 065711

Nithya R and Ragunathan R (2009) Synthesis of silver nanoparticle using pleurotus Pleurotus sajor caju and its antimicrobial study. *Dig Jour Nanomater Bios* 4: 623–629

Noruzi M, Zare D, Khoshnevisan K and Davoodi D (2011) Rapid green synthesis of gold nanoparticles using Rosa hybrid petal extract at room temperature. *Spectrochim Acta Part A Mol Biomol Spectrosc* 79(5): 1461–1465

Nowack B and Bucheli TD (2007) Occurrence, be-haviour and effects of nanoparticles in the environ-ment. *Environmental Pollution* **150**: 5-22

Okitsu K, Mizukoshi Y, Yamamoto TA, Maeda Y and Nagata Y (2007) Sonochemical synthesis of gold nanoparticles on chitosan. *Mater Lett.* 61(16): 3429–3431

Pal S, Tak YK and Song JM (2007) Does the antibacterial activity of silver nanoparticles depend on the shape of the nanoparticle? A study of the Gram-negative bacterium *Escherichia coli. Appl. Environ. Microbiol* 73: 1712-1720

Pandiarajan G, Govindaraj R, Kumar M and Ganesan V (2010) Biosynthesis of silver nanoparticles using silver nitrate through biotransformation. *Jour Ecobiotechnol* 2: 13–18

Panigrahi S, Kundu S, Ghosh SK, Nath S and Pal T (2004) General method of synthesis for metal nanoparticles. *Jour of Nanoparticle Research* 6: 411-414

Parashar UK, Saxena PS and Shrivastava A (2009) Bioinspired Synthesis of silver nanoparticles. *Digest Jour of Nanomaterials and Biostructures.* 4(1) : 159 – 166

Parashar V, Parashar R, Sharma B and Pandey AC (2009) *Parthenium* mediated synthesis of silver nanoparticles: A novel approach towards weed utilization. *Digest Jour of Nanomaterials and Biostructures* 4(1): 45-50

Pareek N, Singh A and Malik CP (2012) Biogenic Synthesis of Silver Nanoparticles, Using Bougainvillea spectabilis Willd. Bract Extract *National Academy Science Letters* 34 : 1-6

Philip D (2009) Biosynthesis of Au, Ag and Au-Ag nanoparticles using edible mushroom extract. *Spectrochim Acta Part A* 73: 374–381

Philip D (2010) Honey mediated green synthesis of silver nanoparticles. *Spectrochim Acta A.* 75(3): 1078-1081

Pileni MP (1997) Nanosized particles made in colloidal assemblies. *Langmuir* 13(13): 3266–3276

Pissuwan D, Cortie CH, Valenzuela SM and Cortie MB (2009) Functionalised gold nanoparticles for controlling pathogenic bacteria, *Trends in Biotechnol* 28 : 207-213

Prabhu N, Raj DT, Gowri Y, Siddiqua A and Innocent JP (2010) Synthesis of Silver phytonanoparticles and their antibacterial efficacy. *Digest Jour of Nanomaterials and Biostructures* 5(1): 185-189

Prasad K and Jha AK (2010) Biosynthesis of CdS nanoparticles: an improved green and rapid procedure *Journal of Colloid and Interface Science* 342(1) : 68–72

Prasad K, Pathak D, Patel A, Dalwadi P, Prasad R and Patel P *et al.*, (2011) Biogenic synthesis of silver nanoparticles using *Nicotiana tobaccum* leaf extract and study of their antibacterial effect. *Afr J Biotechnol* 10(41): 8122-30

Prathna TC, Chandrasekaran N, Raichur AM and Mukherjee A (2011) Biomimetic synthesis of silver nanoparticles by Citrus limon (lemon) aqueous extract and theoretical prediction of particle size. *Colloids Surf B* 82(1): 152–159

Praveen A, Roy AS and Rao S (2012) Biosynthesis and characterization of silver nanoparticles from *Cassia auriculata* leaf extract and in vitro evaluation of antimicrobial activity. *Inter J Appl Biol and Pharm Tech* **3**(2) : 222-228

Priyadarshani KA, Murugan K, Panneerselvum C, Ponarulselvum S, Hwang JS and Nicoletti M (2012) Biolarvicidal and pupicidal potential of silver nanoparticles synthesized using *Euphorbia hirta* against *Anopheles stephensi* Liston (Diptera: Culicidae). *Parasitology Reserach* 111(3) : 997-1006

Pugazhenthiran N, Anandan S, Kathiravan G, Prakash NKU, Crawford S and Ashok Kumar M (2009) Microbial synthesis of silver nanoparticles by *Bacillus* sp. *Jour Nanopart Res* 11(7): 1811–1815

Raad II and Hanna HA (2002) Intravascular catheter-related infections: New horizons and recent advances. *Arch Intern Med* **162**: 871-878

Raghunandan D, Mahesh BD, Basavaraja S, Balaji SD, Manjunath SY and Venkataraman A (2011) Microwave-assisted rapid extracellular synthesis of stable bio-functionalized silver nanoparticles from guava (*Psidium guajava*) leaf extract *Jour Nanopart Res* 13: 2021–2028

Raheman F, Deshmukh S, Ingle A, Gade A and Rai M (2011) Silver nanoparticles: novel antimicrobial agent synthesized from an endophytic fungus *Pestalotia* sp. isolated from leaves of *Syzygium cumini* (L). *Nano Biomed Eng* 3: 174–178

Rautaray D, Sanyal A, Adyanthaya SD, Ahmad A and Sastry M (2004) Biological synthesis of strontium carbonates crystals using the fungus *Fusarium oxysporum*. *Langmuir* 20: 6827–6833

Ray S, Sarkar S and Kundu S (2011) Extracellular biosynthesis of silver nanoparticles using the mycorrhizal mushroom *Tricholoma crassum* (BERK.) Sacc.: its antimicrobial activity against

Renugadevi K, Inbakandan D, Bavanilatha M and Poornima V (2012) *Cissus quadrangularis* assisted biosynthesis of silver nanoparticles with antimicrobial and anticancer potentials. *Int Jour Pharm Bio Sci* 3(3) : 437 – 445

Riddin T, Gericke M and Whiteley C (2006) Analysis of inter and extracellular formation of platinum nanoparticles by *Fusarium oxysporum* f. sp. lycopersici using response surface methodology. *Nanotechnology* 17: 3482–3489

Roco MC (2003b) Nanotechnology: convergence with modern biology and medicine. *Current Opinion in Biotechnology*. **14**: 337-346

Rodriguez E, Parsons JG, Peralta-Videa JR, Cruz-Jiménez G, Romero-González J, Sánchez Salcido BE, Saupe GB, Duarte-Gardea M and Gardea-Torresdey JL (2007) Potential of

Chilopsis linearis for gold phytomining: Using XAS to determine gold reduction and nanoparticle formation within plant tissues. *Int J Phytorem* 9-2 : 133-147

Sadowski Z, Maliszewska GB, PolowczykI and Kozlecki T (2008) Synthesis of silver nanoparticles using microorganisms. *Mat Sci Poland* 26: 419–425

Saha S, Sarkar J, Chattopadhyay D, Patra S, Chakraborty A, Acharya K (2010) Production of silver nanoparticles by a phytopathogenic fungus *Bipolaris nodulosa* and its antimicrobial activity. *Dig Jour Nanomater Bios* 5: 887–895

Saha S, Chattopadhyay D and Acharya K (2011) Preparation of silver nanoparticles by bio-reduction using *Nigrospora oryzae* culture filtrate and its antimicrobial activity. *Dig Jour Nanomater Bios* 6: 1519–1528

Saifuddin N, Wong CW and Nuryasumira AA (2009) Rapid biosynthesis of silver nanoparticles using culture supernatant of bacteria with microwave irradiation. *E-Jour Chem* 6: 61– 70

Salunkhe RB, Patil SV, Salunke BK, Patil CD and Sonawane AM (2011) Studies on silver accumulation and nanoparticle synthesis by *Cochliobolus lunatus*. *Appl Biochem Biotechnol* 165: 221–234

Samadi N, Golkaran D, Eslamifar A, Jamalifar H, Fazeli MR and Mohseni FA (2009) Intra/ extracellular biosynthesis of silver nanoparticles by an autochthonous strain of Proteus mirabilis isolated from photographic waste. *Jour Biomed Nanotechnol.* 5(3) : 247-53

Sangeetha G, Rajeshwari S and Venckatesh R (2011) Green synthesis of zinc oxide nanoparticles by *Aloe barbadensis miller* leaf extract: Structure and optical properties, *Mater Res Bull* 46 : 2560–2566

Sanghi R and Verma P (2009a) Biomimetic synthesis and characterization of protein capped silver nanoparticles. *Bioresour Technol* 100: 501–504

Sanghi R and Verma P (2009b) A facile green extracellular biosynthesis of CdS nanoparticles by immobilized fungus. *Chemical Engineering Journal* 155(3) : 886–891

Sanghi R, Verma P and Puri S (2011) Enzymatic formation of gold nanoparticles using *Phanerochaete chrysosporium*. *Adv Chem Eng Sci* 1: 154–162

Sarkar J, Chattopadhyay D, Patra S, Deo SS, Sinha S, Ghosh M, Mukherjee A and Acharya K (2011a) *Alternaria alternata* mediated synthesis of protein capped silver nanoparticles and their genotoxic activity. *Dig Jour Nanomater Bios* 6: 563–573

Sarkar J, Dey P, Saha S and Acharya K (2011b) Mycosynthesis of selenium nanoparticles. *IET Micro Nano Lett* 6: 599–602

Sastry M, Ahmad A, Khan MI and Kumar R (2003) Biosynthesis of metal nanoparticles using fungi and actinomycete. *Current science* 85(2) : 162- 170

Sathishkumar M, Sneha K and Yun YS (2010) Immobilization of silver nanoparticles synthesized using *Curcuma longa* tuber powder and extract on cotton cloth for bactericidal activity. *Bioresour Technol* 101(20): 7958–7965

Sathyavani R, Balamuali KM, Rao SV, Saritha R and Narayan Rao (2010) Biosynthesis of silver nanoparticles using *Corriandrum sativum* leaf extract and their application in nonlinear optics. *Adv Sci Letters* 3: 1

Satyavani K, Ramanathan T and Gurudeeban S (2011) Green synthesis of silver nanoparticles by using stem derived callus extract of bitter apple (*Citrullus colocynthis*). *Digest Jour of Nanomaterials and Biostructures* 6 : 1019-1024

Savithramma N, Rao ML and Devi PS (2011) Evaluation of antibacterial efficacy of biologically synthesized silver nanoparticles using bark of *Boswellia ovalofoliolata* Bal. and Henry and Shorea tumbuggaia Roxb. *Jour of Biological Sciences* DOI: 10.3923/jbs.2011

Salwe BD, Salimath B, Deshpande R, Bedre MD, Prabhakar KB and Venkataraman A (2008) Biosynthesis and stabilization of Au and Au–Ag alloy nanoparticles by fungus, *Fusarium semitectum. Sci Technol Adv Mater* 9: 9035012

Saxena A, Tripathi RM and Singh RP (2010) Biological Synthesis of silver nanoparticles by using Onion (*Allium cepa*) extract and their antibacterial activity. *Digest J Nanomater. Biostruct* 5 : 427-432

Schuler D and Frankel RB (1999) Bacterial magnetosomes: microbiology, biomineralization and biotechnological applications. *Appl Microbiol Biotechnol* 52 : 464–473

Senapati S, Mandal D, Ahmad A, Khan MI, Sastry M and Kumar R (2004) Fungus mediated synthesis of silver nanoparticles: anovel biological approach. *Indian Jour Phys A* 78: 101- 105

Shahverdi AR, Minaeian S, Shahverdi HR, Jamalifar H and Nohi AA (2007) Rapid synthesis of silver nanoparticles using culture supernatants of Enterobacteria: a novel biological approach. *Process Biochem* 42: 919–923

Shaligram NS, Bule M, Bhambure R, Singhal RS, Singh SK, Szakac SG and Pandey A (2009) Biosynthesis of silver nanoparticles using aqueous extract from the compactin producing fungal strain. *Process Biochem* 44: 939–943

Shankar SS, Rai A, Ankamwar B, Singh A, Ahmad A and Sastry M (2004a) Biological Synthesis of Triangular Gold Nanoprisms. *Nature Materials* 3: 482-488

Shankar SS, Ahmad A, Rai A, Sastry M (2004b) Rapid synthesis of Au, Ag and bimetallic Au core–Ag shell nanoparticles by using neem (*Azadirachta indica*) leaf broth. *Jour Colloid Interface Sci* 275: 496–502

Sharma N, Pinnaka AK, Raje M, Ashish FNU, Bhattacharyya MS and Choudhury AR (2012) Exploitation of marine bacteria for production of gold nanoparticles. *Microb Cell Fact* 11: 86

Sharma NC, Sahi SV, Nath S, Parsons JG, Gardea-Torresdey JL and Tarasankar P (2007) Synthesis of plant-mediated gold nanoparticles and catalytic role of biomatrix-embedded nanomaterials. *Environ Sci Technol* 41(14): 5137–5142

Shiying H, Zhirui S, Zhanga Y, Zhanga S, Wanga J and Ning G (2007) Biosythesis of nano gold by Pseudomonas capsulata. *Mater Lett* 61: 3984 – 3987

Shrivastava S, Bera T, Roy A, Singh G, Ramachandrarao P and Dash D (2007) Characterisation of enhanced antibacterial effects of novel silver nanoparticles. *Nanotechnology* 18 : 1–9

Shukla VK, Singh RP and Pandey AC (2010) Black pepper assisted biomimetic synthesis of silver nanoparticles. *Jour Alloy Compd* 507(1) : L13–L16

Singaravelu G, Arockiamary JS, Kumar VG and Govindaraju K (2007) A novel extracellular synthesis of monodisperse gold nanoparticles using marine alga, *Sargassum wightii Greville. Colloids and Surfaces B: Biointerfaces* 57 : 97- 101

Singh AK, Talat M, Singh DP and Srivastava ON (2010) Biosynthesis of gold and silver nanoparticles by natural precursor clove and their functionalization with amine group. *Jour Nanopart Res* 12 : 1667–1675

Singh A, Jain D, Upadhaya M K, Khandelwal N and Verma HN (2010) Green synthesis of silver nanoparticles using Argemone Mexicana leaf extract and evaluation of their antimicrobial activities *Digest Journal of Nanomaterials and Biostructures* 5(2) : 483-489

Singh P and Balaji R (2011) Biological synthesis and characterization of silver nanoparticles using the fungus *Trichoderma harzianum. Asian Jour Exp Biol Sci* 2: 600–605

Singh RP, Magesh S, Rakkiyappan C (2011) Formation of Fenugreek (*Trigonella Foenum-Graecum*) extract mediated Ag nanoparticles: Mechanisn and Applications. *International Jour of Bio-Engineering Sciences & Technology* 2(3) : 70-80

Sinha A and Khare SK (2011) Mercury bioaccumulation and simultaneous nanoparticle synthesis by *Enterobacter sp.* cell *Bioresource Technology* 102 : 4281–4284

Smith AM, Duan H, Rhyner MN, Ruan G and Nie SA (2006) Synthesis of Gold Nanoparticles Bearing the Bioconjugation *Phys Chem* 8: 3895

Sneha K, Sathishkumar M, Mao J, Kwak IS and Yun YS (2010) Corynebacterium glutamicum-mediated crystallization of silver ions through sorption and reduction processes. *Chemical Engineering Jour* 162(3) : 989–996

Sondi I and Salopek-Sondi B (2004) Silver nanoparticles as antimicrobial agent: a case study on *E coli* as a model for Gram-negative bacteria. *Jour Colloid Interf Sci* 275 : 177-182

Song HY, Ko KK, Oh LH and Lee BT (2006) Fabrication of silver nanoparticles and their antimicrobial mechanisms. *Eur Cells Mater* 11: 58

Song JY and Kim BS (2008) Rapid biological synthesis of silver nanoparticles using plant leaf extracts. *Bioprocess Biosystem Engineering* **32**: 79-84

Song JY and Kim BS (2009) Rapid biological synthesis of silver nanoparticles using plant leaf extracts. *Bioprocess Biosystem Engineering* **32**: 79–84

Song YJ, Jang HK and Kim SB (2009) Biological synthesis of gold nanopaticles using Magnolia kobus and Diopyros kaki leaf extract *Process Biochemistry* 44: 1133-1138

Sugibayashi K, Morimoto Y, Nadai T and Kato Y (1977) Drug-carrier property of albumin microspheres in chemotherapy.I. Tissue distribution of microsphere-entrapped 5-fluorouracil in mice. *Chem Pharm Bull* 25: 3433–3434

Sundaravadivelan C, Padmanabhan MN, Sivaprasath P and Kishmu L (2013) Biosynthesis of silver nanoparticles from *Pedilanthus tithymaloides* leaf extract with anti-developmental activity against larval instas of Aedes aegypti L (Diptheria; Culicidae). *Parasitology Research* 112 (1) : 303-311

Suresh AK, Pelletier DA and Wang W *et al.*, (2011) Biofabrication of discrete spherical gold nanoparticles using the metalreducing bacterium *Shewanella oneidensis Acta Biomaterialia*

7(5) : 2148–2152

Sweeney RY, Mao C and Gao X *et al.*, (2004) Bacterial biosynthesis of cadmium sulfide nanocrystals *Chemistry & Biology* 11(11) : 1553–1559

Taylor EN and Webster TJ (2009) The use of superparamagnetic nanoparticles for prosthetic biofilm prevention. *Int J Nanomedicine* 4: 145–152

Thirumurugan G, Shaheedha SM and Dhanaraju MD (2009) In vitro evaluation of anti-bacterial activity of silver nanoparticles synthesised by using *Phytophthora infestans. Int Jour Chem Tech R* **1**: 714–716

Thirumurugan A, Jiflin GJ, Rajagomathi G, Tomy NA, Ramachandran S and Jaiganesh R (2010) Biotechnological synthesis of gold nanoparticles of *Azardirachta indica* leaf extract. *Inter Jour of Biol. Techn* 1(1): 75-77

Thirunavukkarasu SK, Abdul AR, Asokan B, Sampath M, Chidambaram J, Arivarasan VK, Chinnaperumal K, Govindasamy R, Abdul AZ, Gandhi E, Kanayairam V, Moorthy L, Chinnadurai S, Loganathan K and Kokati VBR (2012) Evaluation of stem aqueous extract and synthesized silver nano particle using *Cissus quadrangularis* against *Hippobosca maculate* and *Rhipicephalus* (Boophilus) *microplus. Experimental parasitology*, article in press, DOI: 10.1007/s00436-011-2473-6

Tripathy A, Raichur AM, Chandrasekaran N, Prathna TC and Mukherjee A (2009) Process variables in biomimetic synthesis of silver nanoparticles by aqueous extract of *Azadirachta indica* (Neem) leaves. *Jour of Nanoparticle Res* Article in press

Tripathy A, Chandrasekaran N, Raichur AM and Mukherjee A (2008) Antibacterial applications of silver nanoparticles synthesized by aqueous extract of *Azadirachta indica* (Neem) leaves. *Jour of Biomedical Nanotechnology* 4(3): 1- 6

Tsuji T, Kakita T and Tsuji M (2003) Preparation of nano-size particles of silver with femtosecond laser ablation in water. *Appl Surf Sci* 206(1–4): 314–320

Udayasoorian C, Kumar KV and Jayabalajrishana R M (2011) Extracellular Synthesis of Silver Nanoparticles using Leaf Extract of *Cassia auriculata. Digest Jour of Nanomaterials and Biostructures* 6(1): 281 - 285

Uddin I, Adyanthaya S, Syed A, Selvaraj K, Ahmad A and Poddar P (2008) Structure and microbial synthesis of sub-10 nm Bi2O3 nanocrystals. *Jour Nanosci Nanotechnol* 8: 3909–3913

Umashankari J, Inbakandan D, Ajithkumar TT and Balasub T (2012) Mangrove plant, Rhizophora mucronata (Lamk, 1804) mediated one pot green synthesis of silver nanoparticles and its antibacterial activity against aquatic pathogens. *Saline Systems* 8: 11

Vahabi KH, Mansoori GA and Karimi S (2011) Biosynthesis of silver nanoparticles by fungus *Trichoderma Reesei. Sciences* 1: 65-79

Vanaja M and Annadurai G (2012) Coleus aromaticus leaf extract mediated synthesis of silver nanoparticles and its bactericidal activity. *Appl Nanosci* DOI 10.1007/s13204-012-0121-9

Velmurugan P, Lee SM, Iydroose M, Lee KJ and Oh BT (2013) Pine cone-mediated green synthesis of silver nanoparticles and their antibacterial activity against agricultural pathogens. *Applied Microbiology and Biotechnology* 97(1): 361-368

Verma VC, Kharwar RN and Gange AC (2010) Biosynthesis of antimicrobial silver nanoparticles by the endophytic fungus *Aspergillus clavatus. Nanomedicine* 5: 33–40

Verma VC, Singh SK, Solanki R and Prakash S (2011) Biofabrication of anisotropic gold nanotriangles using extract of endophytic *Aspergillus clavatus* as a dual functional reductant and stabilizer. *Nanoscale Res Lett* 6: 16

Vila A, Sanchez A, Tobio M, Calvo P and Alonso MJ (2002) Design of biodegradable particles for protein delivery. *J Control Release* **78**: 15-24

Vigneshwaran N, Ashtaputre NM, Varadarajan PV, Nachane RP, Paralikar KM and Balasubramanya RH (2007) Biological synthesis of silver nanoparticles using the fungus *Aspergillus flavus. Mater Lett* 61: 1413–1418

Vigneshwaran N, Kathe AA, Varadarajan PV, Nachane RP and Balasubramanya RH (2006) Biomimetics of silver nanoparticles by white rot fungus, *Phaenerochaete chrysosporium. Colloids Surf B* 53: 55–59

Vijayaraghavan K and Nalini SPK (2010) Biotemplates in the green synthesis of silver nanoparticles. *Biotechnol Jour* 5(10): 1098–1110

Wang Y, He X, Wang K, Zhang X and Tan W (2009) Barbated Skullcup herb extract-mediated biosynthesis of gold nanoparticles and its primary application in electrochemistry. *Colloids Surf B* 73(1): 75–79

Wang YA, Li JJ, Chen HY and Peng XG (2002) Stabilization of inorganic nanocrystals by organic dendrons. *J Am. Chem Soc* 124: 2293-8

Watson JHP, Ellwood DC, Soper AK and Charnock J (1999) Nanosized strongly-magnetic bacterially-produced iron sulphide materials. *Journal of Magnetism and Magnetic Materials* 203(1–3) : 69–72

Wright JB, Lam K, Hansen D and Burell RE (1999) Efficacy of Topical Silver Against Fungal Burn Wound Pathogens *Am J Infect Control* 27 : 344-350

Xu ZP, Zeng QH, Lu GQ and Yu AB (2006) Inorganic nanoparticles as carriers for efficient cellular delivery. *Chemical Engineering Science* **61**: 1027- 1040

Yamanaka M, Hara K, and Kudo J (2005) Bactericidal actions of a silver ion solution on *Escherichia coli*, studied by energy-filtering transmission electron microscopy and proteomic analysis. *Appl. Environ. Microbiol.* **71**: 7589-7593

Yu-Nam Y and Lead R (2008) Manufactured nano-particles: an overview of their chemistry, interactions and potential environmental implications. *Science of the Total Environment* **400**: 396-414

Zhang X, He X, Wang K, Wang Y, Li H and Tan W (2009) Biosynthesis of size-controlled gold nanoparticles using fungus, *Penicillium* sp. *Jour Nanosci Nanotechnol* 9: 5738–5744

10

Mutations Breeding in Genomics Era

New Opportunities and Challenges

Gulzar S Sanghera and *Subhash C Kashyap*

Advances in molecular genetics and DNA technologies have brought plant breeding, including mutation breeding, into a molecular era. With ever-increasing knowledge of molecular genetics and genomics and rapidly emerging molecular techniques, breeders can now use mutation techniques in breeding new varieties more wisely and efficiently than ever before. Plant molecular mutation breeding is here defined as mutation breeding, in which molecular or genomic information and tools are used in the development of breeding strategies, screening, selection and verification of induced mutants, and in the utilization of mutated genes in the breeding process. It is built upon the science of DNA damage, repair and mutagenesis, plant molecular genetics and genomics of important agronomic traits as well as induced mutations. Mutagenic treatment, super-mutable genetic lines, molecular markers and high throughput DNA technologies for mutation screening such as TILLING (Targeting Induced Local Lesions in Genomes), are the key techniques and resources in molecular mutation breeding. Molecular mutation breeding will significantly increase both the efficiency and efficacy of mutation techniques in crop breeding. A perspective molecular mutation breeding scheme is proposed for discussion.

Introduction

Increasing crop yields to ensure food security is a major challenge. Amongst the obstacles against this are the changing climate (increasing temperatures and more erratic rainfall) which most often compromise crop productivity (Parry *et al.*, 2005) and the need to produce additional food and crops for bio-energy whilst minimizing the carbon costs of production (Powlson *et al.*, 2005). There is therefore an urgent requirement for new higher yielding varieties (Parry *et al.*, 2007; Reynolds *et al.*, 2009) with improved nutrient (Lea and Azevedo, 2006) and water use efficiency (Richards, 2000). An apparent limitation of mutagenesis versus wide crossing or transgenesis methods is that breeders can manipulate only genes already present in the genome. No new genes can be added by this method. Furthermore, nearly all mutations result in a loss of gene function, meaning that mutagenesis is concerned more with reducing the effects of unwanted genes than increasing the expression of desirable genes. At first sight, this might seem like a serious limitation to the creation of useful new agronomic traits. However, recent genomic studies reveal the surprising fact that during the 10,000-year history of agriculture, loss-of-function alleles were associated with nine out of 19 key episodes in crop improvement and/or varietal divergence (Doebley *et al.*, 2006; Burger *et al.*, 2008). Therefore, the past or future potency of mutagenesis for crop improvement cannot be underestimated.

During last two decades, there has been a dramatic increase in the amount of genome sequence data available for world major food crops, their pests and pathogens. Complete genome sequences have been reported for rice (Matsumoto *et al.*, 2005) and sorghum (Paterson *et al.*, 2005) and also for several crop pathogens e.g. *Agrobacterium tumefaciens* (Wood *et al.*, 2001), Phytoplasma (Oshima *et al.*, 2004), *Fusarium graminearum* (Cuomo *et al.*, 2007), *Magnaporthe grisea* (Dean *et al.*, 2005). For the other major global crop, wheat, and other crop pests and pathogens the sequences of expressed sequence tags have become available. The exploitation of these sequence data for crop improvement is limited by the complexity of many of the traits that determine agronomic performance (Parry *et al.*, 2005; Parry and Reynolds, 2007). However, reverse genetics approaches allow progress to be made on the major challenge of linking sequence information to the biological function of genes and on determining their contribution to important characters and traits. Typically, these approaches rely on the disruption of candidate genes by mutagenesis, transposons, and T-DNA tagging or RNA interference (RNAi) (Sanghera *et al.*, 2010). The overall objective of this chapter is to understand the mechanisms of mutagenesis and molecular genetic features of induced mutation in plants, to provide the scientific basis for developing more efficient mutation techniques, and their proper use for crop improvement.

Revival mutation research in genomic era

Mutagenesis is one of the few biotechnologies currently that is used much more in developing countries than elsewhere. Both radiation and chemical mutagenesis have been used for crop improvement since the 1930s. During the 1950s, FAO began working with the International Atomic Energy Agency (IAEA) to make irradiation technology more widely available to developing countries, in a collaboration that is now known as the Atoms for Food global partnership (FAO and IAEA, 2008). More than 200 mutation-derived varieties have been obtained world-wide in the last sixty years, generating benefits worth billions of dollars, mainly in developing countries (Ahloowalia *et al.*, 2004).

Modification of DNA to create novel phenotypes involves the use of mutagenic agents, such as chemicals or radiation (Donini and Sonnino, 1998). It also includes somatic mutagenesis whereby tissue or cell cultures may undergo useful epigenetic modifications, provided that the resultant traits are stable in future generations. Induced mutagenesis has been practised with great success in crop breeding programmes in developing countries since the 1930s (Ahloowalia *et al.*, 2004), but its scope and utility have recently been greatly enhanced and extended by the new molecular-based technology of TILLING (detailed below). Although induced mutations has been widely used in crop breeding and basic research, the fundamental processes that lead to mutations and the molecular nature of mutations induced by physical mutagens remain largely unknown in plants for longer period of time. However, the rapid progress in functional genomics has provided unprecedented opportunities to study the mechanisms underpinning the response of plant genomes to physical and chemical mutagens. Recent advances in genomics have created an opportunity to apply our current knowledge of DNA repair and mutagenesis in model systems to crop plants. Various molecular and fundamental mechanisms broadly encompasses all aspects of research that address the detection of mutations, the mechanisms by which mutations in genes and chromosomes arise, and the modulation of mutagenesis by mutation avoidance pathways such as DNA repair, cell cycle control and apoptosis. It includes the role of genetic variation in the genesis and manifestation of mutations, ranging from the variable manner in which xenobiotics are metabolized to variations in the capacity of cells to replicate and repair damaged DNA. It also includes the contributions of these mechanisms, when perturbed, to animal disease models and to human disease, with particular emphasis on carcinogenic mechanisms. Chromosome stability is paramount for maintaining cellular homeostasis. Therefore, genesis of aneuploidy and isodisomy, including the roles played by cell cycle checkpoints, spindle microtubules, centrosomes and kinetocore proteins, and agents that might disrupt them are of great significance. Since isodisomy can occur as a consequence of recombination, all aspects of homologous recombination and non-homologous end joining are appropriate.

Critically, mutations in important traits or genes (e.g. in nutritional quality, resource use efficiency, architecture or phenology) can be readily exploited by plant breeders without the legislative restrictions, licensing costs, and societal opposition applied to genetically modified (GM) approaches. This is despite the fact that transcriptomic analyses have shown that large-scale plant mutagenesis may induce greater changes in gene expression patterns than transgene insertion (Batista *et al.*, 2008). The rapid development of plant molecular genetics and genomics in areas relevant to mutation breeding has been reinvigorating this breeding method; it is expected that mutation breeding will directly benefit from the rapid scientific and technological advances in molecular genetics and genomics.

New technologies for detection of gene variations

The goal in mutagenic breeding is to cause maximal genomic variation with a minimum decrease in viability. Among the radiation-based methods, γ-ray and fast neutron bombardment are now used in most applications. Of these, γ-ray bombardment is less destructive causing point mutations and small deletions whereas fast neutron bombardment causes translocations, chromosome losses, and large deletions. Compared to chemical mutagens, both types of radiations cause damage on a larger scale and severely reduces viability (Leung *et al.*, 2001, Wu *et al.*, 2005). In general, plant molecular breeding depends on the understanding of the molecular genetic control of target traits of interest. This information is also of great help in developing a proper mutation breeding strategy.

In vitro mutagenesis

It is caused by changes in DNA induced during *in vitro* culture (Durrant, 1962). Somaclonal variation is normally regarded as an undesirable by-product of the stresses imposed on a plant by subjecting it to tissue culture. These stresses include abiotic factors, such as cold, water deficiency, or high salt concentrations; excess or dearth of nutrients; the effects of chemical growth regulators; and infections by pathogens. The stresses of tissue culture can result in single-gene mutations; the deletion or transposition of larger lengths of DNA, including chromosome segments; methylation or de-methylation of genes; and even the duplication or loss of entire chromosomes. Provided they are carefully controlled, somaclonal changes in cultured plant cells can potentially provide a powerful new tool to generate variation for crop breeders (Sala and Labra, 2003). Somaclonal mutagenesis has been used to manipulate traits such as disease resistance, insect resistance, nutritional value, drought and salt tolerance in crops ranging from sugarcane to banana. A somaclonal sugarcane variant for improved yield is reported to be commercially distributed in India (FAO-BioDeC, www.fao.org/biotech/inventory_admin/dep/default.asp).

Site-specific mutagenesis

An interesting example of site-specific mutagenesis is the use of oligonucleotides with mismatches to the specific gene to be mutated. The company, Cibus LLC, is expected to soon release herbicide resistant sorghum (Wenzel, 2007). The company estimates a development cost of about 3-5 million US dollars compared to 30-40 million US dollars for transgenics due to avoiding regulatory hurdles. Such a method of directed mutagenesis is considered to be a non-transgenic approach. This more inexpensive approach may facilitate the generation of new traits in minor crops.

Homoeologous recombination

The *Ph1* locus controls the pairing of the sets of chromosomes in wheat. In crosses with wild relatives, this locus unfortunately prevents the pairing of wheat and the chromosomes of wild relatives making it difficult to introgress new genes into wheat. The ability to alter the control exerted by Ph1 would enable wheat breeders to access a much greater range of genetic diversity. A cdc2 gene complex is thought to be the Ph1 locus responsible for the pairing of homoeologous chromosomes in wheat (Griffith *et al.*, 2006). Wheat has three genomes that are similar but vary in genetic constitution. Recombination between these similar chromosomes can lead to new variation. Control of recombination in crosses of wheat with wild relatives may be possible through the use of okadaic acid, a phophatase inhibitor, and lead to more variation.

Transposable element insertions

The introduction or activation of transposable elements of various sorts provides the possibility of altering genes to generate phenotypically detectable mutants (forward genetics). The presence of such elements in a gene allows one to correlate these changes in a given genetic sequence with a specific phenotype (reverse genetics). Fortunately, many genetic stocks have been produced in a variety of organisms to make this a robust technology. For example, in rice there are about 50,000 lines with Tos17 insertions produced at the National Institute of Agrobiological resources http://tos.nias.affrc.go.jp). In addition, about 150,000 lines of rice possess Ac/Ds, enhancer traps, T-DNA, and activation tags which have been produced by researchers in Korea, Australia, China, Taiwan, France, Singapore, Netherlands, and the U.S. Perhaps a total of a million tagged sites are available in japonica rice (Phillips *et al.*, 2007).

Gene targeting

Targeting specific genes for modification is becoming more and more common. Zinc-finger nucleases can be targeted to specific genes causing a double-stranded break which disables the gene (Porteus and Carroll, 2005, Shukla *et al.*, 2008). Zinc finger nucleases are comprised of a DNA-recognition domain and a cleavage domain. The double-strand breaks at specific locations may disable the targeted allele or even lead to a modified sequence upon repair. The potential exists to insert a gene at the double-strand break.

TILLING

It is important for assessing the feasibility and potential to induce a mutation of interest. Since the mutation frequency for any given fragment of DNA or gene is more or less similar, the opportunity to obtain a mutant of different traits would therefore be dependent on the number of genes that control the trait. For example, many genes can affect the growth duration (i.e. days from sowing to heading for cereals), therefore, the mutation frequency of such trait is often far higher than single-gene controlled traits. Secondly, a mutation may have pleiotropic effects if the gene is at the upstream or at the middle of a long biosynthetic pathway, such as the MIPS gene in phytic acid biosynthesis and precautions should be taken for such a mutation project. Third, knowledge of genes controlling a trait of interest would constitute the very basis of the TILLING method. It can be viewed as an updated high-tech version of mutation breeding (McCallum *et al.*, 2000a; 2000b). First, mutagenic agents, such as alkylating agents or radiation, are used as normal to create a population of thousands of mutagenized plants. Next, the second (or M2) generation of these mutants is screened using a semiautomated high-throughput DNA-based method to detect mutations in genes of interest. Screening involves the use of polymerase chain reaction (PCR) to amplify gene fragments of interest, plus rapid identification of any mutation-induced lesions by looking for mismatches in duplexes with non-mutagenized DNA sequences. The third step is to evaluate the phenotypes of a limited number of selected mutant plants (Fig.1). TILLING is also amenable to automation including high-throughput robotic screening systems, making it especially suitable for large and complex polyploid genomes found in several major crops. As well as screening mutagenized populations, TILLING can be used to screen variation in natural populations, in what has been termed EcoTILLING (Henikoff *et al.*, 2004). As with other technologies, TILLING will eventually get cheaper and more accessible, so it can be applied more readily by developing countries. However, the wider applications of this and other new biotechnologies depend critically on how and where they have been developed e.g. chemical/radiation mutagenesis was pioneered in the public sector, and was subsequently disseminated around the world.

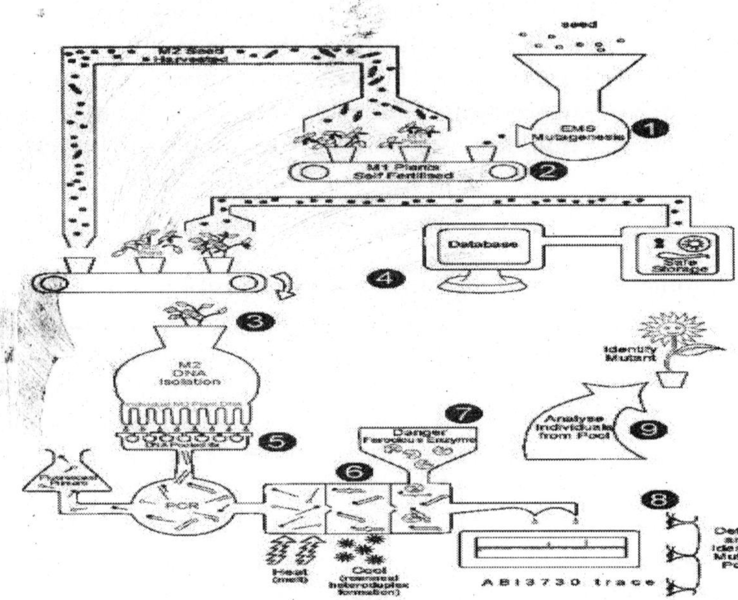

Fig. 1 Schematic representation of Mutation Assisted Breeding programme, 1= seed treatment with mutagen, 2= M1 generation selfed, 3= DNA isolation of M2 plants, 4 = preparing database, 5,6,7 = PCR 8= detection of mutant pool, 9 = analysis of individual from pool to identify mutant.

Source: *http://revgenuk.jic.ac.uk*

In contrast, other technologies, such as maize F1 hybrids and transgenesis, were commercialized by the private sector and, outside the arena of globally traded commodity crops, they have spread more slowly andless widely. In the case of TILLING, it will be important to maintain a balance between protecting the legitimate commercial interests and research investments of the exploiting companies, while making the technology available for non-profit, public-good applications in developing countries, an overview of TILLING approach used among various crops is appended below (Table 1).

Re-sequencing

DNA resequencing involves sequencing an individual's DNA for a specific region and comparing it to a reference sequence that is already available in order to detect mutations. Resequencing the genome of many individuals allows determination of the relationship between sequence variation and specific phenotypes. This substantially increases the ability to detect gene variation. In rice, genome-wide SNP (Single Nucleotide Polymorphism) discovery was attempted across the unique sequence fraction of the Nipponbare rice genome. Twenty diverse varieties were selected for resequencing based on geographic representation, diversity, usage, and traits by a group of rice researchers in the *Oryza* SNP Consortium (McNally *et al.*, 2006) (www.oryzasnp.org). Overall, the consortium (McNally and Leung, personal communication) found 2.6 SNPs per kb (146,576 genic variants and 112,623 intergenic variants). QTL mapping studies are often restricted due to the absence of known polymorphic

Table 1 : Year-wise overview of TILLING approach among various crops

Year	Organism	Genus species	Mutagen applied	Mutagen dose	Estimated Genome	Ploidy	Mutation rate
2000	Arabidopsis	*Arabidopsis thaliana*	EMS	20mM to 40 mM	125 Mbp	2X	1/300 kb
2001	Arabidopsis	*Arabidopsis thaliana*	EMS	20mM to 30mM	125 Mbp	2X	1000/genome
2002	Arabidopsis	*Arabidopsis thaliana*	Fast neutrons	60 Gy	125 Mbp	2X	0.7 to 12 kb
2003	Lotus	*Lotus japonicas*	E MS	0mL/10mL (v/v)	472Mbp	2X	1/154kb
2004	Barley	*Hordeum vulgare*	EMS	20mM to 30 mM	5300 Mbp	2X	1/Mb
200	4 Maize	*Zea mays*	EMS	0.0625%	2500 Mbp	2X	0.93/kb
2005	Rice	*Oryza sativa*	MNU	1mM	430 Mbp	2X	0.80/1kb
2006	Field mustard	*Brassica rapa*	γ radiation	500Gy	500Mbp	2X	1/6190kb
2007	Pea	*Pisum sativum*	EMS	4 mM	4300 Mbp	2X	1/669 kb
2007	Rice	*Oryza sativa*	EMS	1.5%	430 Mbp	2X	1/294 kb
2007	Rice	*Oryza sativa*	Az-MNU	1 to 15 mM	430 Mbp	2X	1/265 kb
2008	Rape seed	*Brassica napus*	EMS	0.4% to 1.2% w/v	1150 Mbp	4X	1/41.5kb
2008	Sorghum	*Sorghum bicolor*	EMS	0.1 to 0.6% (v/v)	7.35 ×108 bp	2X	1/526 kb
2008	Soybean	*Glycine max*	NMU	2.5 mM	1115 Mbp	2X	~1/140 kb
2008	Soybean	*Glycine max*	EMS	50 mM	1115 Mbp	2X	1/250 kb
2009	Barley	*Hordeum vulgare*	Sodium azide	1.5 mM	5500 Mbp	2X	1/ 2.5 Mb
2009	Clover	*Medicago truncatula*	EMS	0.075% to 0.40%	470 Mbp	2X	1/485kb
2009	Common Bean	*Phaseolus vulgaris*	EMS	20mM to 60 mM	625 Mbp	2X	2 to 3/Mb
2009	Peanut	*Arachis hypogaea*	EMS/γ-radiation	(0.5%)/ 20 &30 kr	3.0 x 109 bp	4X	0.5 x10-7
2009	Potato	*Solanum tuberosum*	γ radiation	0.5% to 2.0%	850Mbp	4X	Â1/1810 bp
2009	Wheat	*Triticum turgidum*	EMS	0.7% to-1%	12000 Mbp	4X	1/40 kb
2010	Field mustard	*Brassica rapa*	EMS	0-1%	500Mbp	2X	1/60kb
2010	Oat	*Avina sativa*	EMS	0.9% (v/v)	13000 Mbp	6X	1/ 20-40 kb
2010	Tomato	*Solanum lycopersicum*	EMS	0.7% to 1.0%	950Mbp	2X	1/322kb

(Source: Rashid *et al.*, 2011)

sites between parental lines. Having a large number of SNPs reduces this problem by providing information on nucleotide variation between cultivars. Each SNP reflects gene variation. Directed mutagenesis through regulation of gene expression: RNA interference is a RNA-guided regulation of gene expression utilizing double-stranded ribonucleic acid complementary to the genes for which expression is to be inhibited (Sanghera *et al.*, 2010). Although a relatively new discovery, RNAi has already been demonstrated to be useful in generating variation for important traits. Root-knot nematode resistance has been produced in Arabidopsis (Huang *et al.*, 2006), delayed senescence in wheat (Uauy *et al.*, 2006), gossypol reduction in cotton seed (Sunikumar *et al.*, 2006), and cytoplasmic male sterility in tobacco and tomato (Sandhu *et al.*, 2007).

Applications in crop improvement

Van Harten (1998) described in "Mutation Breeding -Theory and practical application," "An explanation for the decreasing interest in mutation breeding, at least in most "developed" countries, may be that during the past two decades attention has become more and more directed towards studying the possibilities offered to plant breeding by various new molecular technologies. As a result of these developments mutation breeding seems to have lost part of its previous attraction for young researchers." It is not necessary to mention, however, that mutation breeding is still a very interesting and useful technology for isolating genes and for elucidating gene mechanisms and metabolic pathways in various crops. The record has also shown that mutation induction is a very useful conventional breeding tool for developing superior cultivars. Today, site directed mutagenesis in vivo or in vitro cell can be envisioned and many researchers are conducting programmes in this direction. New fields of science and technologies were developed on the basis of achievements of traditional or classic methods in the present genomic era and are explained below:

From Genotype to Phenotype

Contrary to traditional screening methods done by plant breeders, TILLING focuses on first identifying mutations within genes of interest and then linking those mutations to a specific phenotype. However, this approach is only possible when a gene linked to the trait of interest is known and the gene sequence available. Using software and maps of conserved sequences within the gene it is then possible to predict which of the identified mutations that are most likely to cause changes in protein structure or aborted translation resulting in a nonfunctional product. The potential phenotypes identified in this way can then be verified by anatomical, histological, physiological, or biochemical studies. Although theoretically straightforward, there are several problems that might arise during the screening process and subsequent analysis. Since the screening takes place at the DNA-level, enhancer and promotor mutations that are upstream of the gene of interest can be difficult, if not impossible to detect unless a full genomic sequence is available, which is not the case for most non model systems? Another complication stems from the fact that a single mutation, even if predicted to be deleterious does not necessarily affect overall cellular function. Homologs or paralogs of the gene of interest may still be expressed, leading to a low or nonexistent penetration of the mutation. This is especially true for hexaploid plants where a homolog of the gene of interest may exist in all three genomes and when one allele is mutated, two others may compensate for the loss. In practice it is therefore often necessary to identify knockout mutations in all alleles by laborious screenings followed by time-consuming crosses to stack the different mutations in the same genome. This can severely delay the development of the final trait. Despite these drawbacks, several groups have reported successes in linking genotypic change to novel phenotypes in a variety of crops. Most noticeably in wheat, where traits related to the *waxy* phenotype (Slade *et al.*, 2005; Dong *et al.*, 2009; Sestilli *et al.*, 2010) and grain hardness are being developed (Dong *et al.*, 2009), in soybean where TILLING has proven useful in increasing the oleic acid content through the identification of mutations in the *FAD1, 2,* and *3* genes (Dierking *et al.*, 2009) and in *Sorghum* where lignin content has been decreased though mutation of *COMT* (Xin *et al.*, 2008).

From Phenotype to Genotype

To be truly useful, identification of a strong genetic character in a mutagenised population by a phenotypic screening procedure should be followed by a characterization of the molecular event underlying the modified character. In plants with sequenced genomes, that is, where reference sequences are available, novel phenotypes can be characterised using a combination of whole-genome resequencing, linkage maps, and microarrays, providing a comprehensive picture of gene expression changes and newly introduced SNPs compared to wild-type specimens. A classical example is the identification of a GA20 oxidase mutation as a cause for the semidwarf phenotype used in many commercial rice varieties. Using genetic maps, the trait was linked to a region of chromosome 1. Combined with the knowledge that the dwarf phenotype had reduced levels of gibberellic acid (GA), a putative GA gene in that area was identified and sequenced using the rice reference genome as a base. The sequence showed a 280 bp deletion resulting in an inactive protein, explaining the decreased GA-levels (Spielmeyer *et al.*, 2002). Microarray technology has also been successfully applied in rice and *Arabidopsis* to connect genome-wide variations to specific phenotypes (Singer *et al.*, 2006, Edwards *et al.*, 2008) However, next-generation technologies such as Illumina sequencing now outperform the more traditional microarray methods for SNP identification (Huang *et al.*, 2009) In one such approach, EMS-induced *Arabidopsis* Col-0 mutants with slow growth and light green leaves were screened to identify the causative mutations. The recessive mutants were first crossed with the *Landsberg erecta* ecotype. DNA from 500 F2 individuals was then pooled and sequenced using Illumina sequencing to up to 22-fold genome coverage. Software called SHORE map was then developed to identify the mutations in the segregating population. The software detected a mutation causing serine to asparagine nonsynonymous codon change in the AT4G35090 gene (Schneeberger *et al.*, 2009). In yet another approach, Austin *et al.* (2011) identified three genes involved in cell wall biosynthesis. They first screened the *Arabidopsis* EMS treated Col-0 mutants for sensitivity to flupoxam that were previously known to affect cell wall assembly or integrity. The mutants were then crossed to *Landsberg erecta* ecotype. The genomic DNA was extracted from the F_2 population and screened using Illumina GA sequencing. Through an in-house developed statistical approach, they were able to correctly identify the causative mutations and hence the genes responsible for the phenotype (Austin *et al.*, 2011). Since a mutation does not necessarily need to be in an exon of the candidate gene, identifying a mutation may be difficult if a reference genome is unavailable. Mutations such as promotor mutations, mutations changing genome structure, mutations upstream in the regulation pathway, and various micro-RNA mutations can all be responsible for the downstream effect. When a reference genome is not available, these factors can be extremely difficult and time-consuming to evaluate comprehensively. In such cases, an initial approach would be to obtain as many mutants as possible and evaluate each one separately, re-sequencing all genes of interest and performing qPCR experiments to gauge any possible changes in expression among the candidate genes. Although difficult, it is not impossible to obtain a genotype-phenotype association this way. Using EST libraries instead of the fully sequenced genome, Feiz *et al.* (2011) linked wheat grain hardness to *Puroindoline a* and *b* mutations in an EMS-mutagenised population. A major caveat is that a link between genetic maps and genes are unknown in many cases, thus effectively robbing the researchers of a valuable selection tool for limiting the number of candidate genes.

Identification of Novel Traits in Mutated Populations

Biochemical Screening

The main purpose of TILLING is to allow identification of mutations at the genetic level. However, this does not exclude the fact that TILLING populations, as well as other mutagenised populations also can be used for phenotypic screens. Macromolecular composition and quantity of bioactive compounds like lignin and other fibers, lipid, and starch content are all quality characters that cannot be scored in the field. Lignin is found in secondary plant cell walls and provides rigidity to the plant. Lignin is considered a negative component in foragers as it blocks the digestion of cell-wall polysaccharides by microbial enzymes and is itself indigestible. Thus, crop varieties with lower lignin levels in the cell walls are preferred for feed since they aremore energy efficient. A quick and economical assay for visually screening for altered lignin levels in seeds is the phloroglucinol-HCl assay (Wiesner test) (Speer, 1987). We screened seeds from 1824 lines from an oat TILLING population (Caldwell *et al.*, 2004) and identified 17 lines where the seeds had reduced lignin stain intensity. For further confirmation, an acetyl-bromide method was then used for accurate quantification of lignin levels in themutant seeds (Chawade *et al.*, 2010; Iiyama and Wallis, 1988). Increased levels of dietary components that directly interfere with cholesterol absorption or excretion and thereby contribute to lowered plasma cholesterol levels are also very important breeding goals. One example is the mixed-linkage (1→3), (1→4) β-D-glucan soluble fibre which is mainly found in cereals and where oat and barley contain the highest concentrations. Using an assay kit available from Megazyme (McCleary and Codd, 1991) they measured β-glucan content in seeds from 1500 random lines in an oat TILLING population. We identified lines with increased levels of β-glucan as well as lines with levels less than half of what is found in Belinda, the original cultivar. With the rising number of TILLING-populations they anticipated that these populations will be increasingly screened not only by TILLING, that is, genetic screening, but also by various advanced biochemical assays to identify important quality characters. Recently, they set up an GC-MS assay and screened 1000 lines for β-sitosterol content and, in this relatively small sample size, identified lines with almost twice the normal levels. The advantage of a screen at the phenotypic level is that the target character is directly identified. The disadvantage, compared to a genotypic screening is that the specific mutation(s) mediating the phenotype remains undiscovered. There are several other examples from the literature elegantly demonstrating the power of biochemical screens (Barkawi *et al.*, 2010; Reiter *et al.*, 1997; Junhyun *et al.*, 2007).

Physiological screening

Fungal pathogens represent a major threat to global agriculture. Global climate change with mild winters and higher humidity is expected to increase the problem even further. One particularly troublesome pathogen with high relevance in North America and Europe including Sweden is *Fusarium* (Parry *et al.*, 1995). Comprised of more than 1000 different species, *Fusarium* cause diseases in major agricultural crops like wheat, barley, maize and oats. In addition, *Fusarium sp.* also produce a plethora of mycotoxins which accumulate in the grain, enter the food chain, and pose serious threats to human and animal health. A particular challenge is *Fusarium* head blight disease (FHB), for which there are currently no satisfactory management strategies available and where fungicide treatments give mixed and unpredictable results, sometimes even worsening mycotoxin contamination (McMullen *et al.*, 1997). Unfortunately, the variation in the breeding populations does not seem to be high enough to identify and develop lines resistant to the disease. On the other hand, even for characters that vary considerably with environmental factors, like pathogen resistance, mutagenised populations could be used to identify resistant lines with a strong genetic component. The trick is to design an *in vitro* assay with such a stringent selection that single rare lines with strong resistance against the disease can be identified. We tested this concept by designing a petri dish assay to identify *Fusarium*-tolerant oat from a mutated population with a high variety (Chawade *et al.*, 2010). They placed 5 seeds from each line of the oat TILLING population on water agar and inoculated each seed with approximately 5000 spores of *Fusarium culmorum*.

At first sight, this might seem like a serious limitation to the creation of useful new agronomic traits. However, genomic studies reveal the surprising fact that during the 10 000-year history of agriculture, loss-of-function alleles were associated with nine out of 19 key episodes in crop improvement and/or varietal divergence (Doebley *et al.*, 2006; Burger *et al.*, 2008). Therefore, the past or future potency of mutagenesis for crop improvement cannot be underestimated.

Crop varietal development through mutation breeding

Human population growth is projected to reach a staggering nine billion people by 2050. Who will feed the growing billions; how can more food be produced on less land; how can we protect crop yields from climate variability, pests and diseases? These are major issues of present generation. Mutation breeding has played a major role in improving crops and stabilizing food security concerns worldwide. Classic examples are improved yields brought about by semi-dwarf mutant varieties of wheat and rice in the mid-twentieth century, which created the 'Green Revolution'. Mutation induction via irradiation is an established method in providing useful variation for crop improvement, despite this the sought after mutant is a rare event.

Public agencies, including the Joint FAO/IAEA Division and universities have been effective proponents of mutagenesis technology and there are essentially no IPR barriers to its deployment for public-good crop breeding. Today mutants can be detected efficiently using high-throughput phenotyping and genotyping methods. Hence, many mutagenized crop varieties have been produced by and for developing countries. As of early 2009, more than 2770 varieties of mutation-bred crop varieties, including all the major staple species (Ahloowalia *et al.*, 2004), had been released into cultivation in at least 59 developing countries, mostly in Asia. The largest mutation breeding programmes are in China and India, but dozens of other countries are also using the technology (Maluszynski *et al.*, 2003; Kodym and Afza, 2003). Widely used mutagenized crops include: Soghat bread wheat in Pakistan, Zhefu rice in Thailand, Shwewartun rice in Myanmar, and Bajra pearl millet in India. In Vietnam, three new varieties of rice with improved food quality and salt tolerance have been developed since 1996. Since their release in the Mekong Delta region, they have increased smallholder incomes by USD 350/farmer/year and include some of the top export varieties (FAO and IAEA, 2008).

In India, oilseed and legume crops are important food components as they are major contributors to dietary oils and proteins. In order to generate genetic variability in these crops, mutation research is extensively carried out at Bhabha Atomic Research Centre (BARC), Mumbai for the past half a century. Besides cytogenetic studies, the era of direct mutants as crop varieties began in groundnut, mustard, pigeonpea and mungbean. Induction of modified traits and their incorporation in an ideal genotype was achieved by judicious use of induced mutation and hybridization techniques. So far, 110 varieties in oilseed and legumes have been developed in India by incorporating desirable traits like large seed, semi-dwarf habit, high harvest index, better partitioning, fresh seed dormancy, yellow seed color, drought tolerance, resistance to powdery mildew, yellow mosaic virus, and bacterial pustule diseases (Kharakwal and Shu, 2009). Many of the national/state breeding programmes have been utilizing these varieties as parental materials/donors and developed several improved varieties. Several of these varieties have high patronage from the farming community and are extensively grown in the country. Groundnut varieties have made considerable impact by giving record yields across the country. Further, mungbean varieties were also surging ahead by virtue of their resistance to yellow mosaic virus, *Rhizoctonia* root-rot and powdery mildew diseases with suitability to rice fallow situations. Blackgram variety TAU-1 has occupied maximum blackgram area in Maharashtra state. These varieties also facilitated farmers to develop newer cropping systems. Thus, induced mutation research remained in the forefront of Indian agricultural research by developing popular varieties with higher productivity potential in oilseeds and legumes (D'Souza *et al.*, 2009).

Table 1 : Number of released mutant varieties in 57 crop species in India

S. No	Common name	Latin name	No. of varieties
1	Moench Okra	*Ablemoschus esculentus* L.	2
2	Groundnut	*Arachis hypogaea* L.	18
3	Wild Bougainvillea	*Bougainvillea spectabilis*	13
4	Mustard	*Brassica juncea* L.	9
5	Pigeonpea	*Cajanus cajan* L. Mill sp.	5
6	Green pepper	*Capsicum annum* L.	1
7	Papaya	*Carica papaya* L.	1

S. No	Common name	Latin name	No. of varieties
8	Chrysanthemum	*Chrysanthemum* sp.	49
9	Chickpea	*Cicer arietinum* L.	8
10	White jute	*Corchorus capsularis L.*	2
11	Tossa jute	*Corchorus olitorius L.*	3
12	Turmeric	*Curcuma domestica* Val.	2
13	Citronella	*Cymbopogon winterianus* Jowitt.	9
14	Cluster bean	*Cyamopsis tetragonoloba* L.	1
15	Dahlia	*Dahlia* sp.	11
16	Hyacinth bean	Dolichos lablab L.	2
17	Finger millet	*Eleusine coracana* L.	7
18	Gladiolus	*Gladiolus* L.	2
19	Soybean	*Glycine max* L. Merr.	7
20	Desi cotton	*Gossypium arborium* L.	1
21	American cotton	*Gossypium hirsutum* L.	8
22	Sunflower	*Helianthus annus* L.	1
23	Hibiscus	*Hibiscus sinensis* L.	2
24	Barley	*Hordeum vulgare* L.	13
25	Indian henbane	*Hyocyamus niger*	1
26	Wild sage	*Lantana depressa* L.	3
27	Lentil	*Lens culinaris* L. Medik.	3
28	Ridged gourd	*Luffa acutangula* Roxb.	1
29	Tomato	*Lycopersicon esculentum*	4
30	Germen chamomile	*Matricario cammomilla*	1
31	Spearmint	*Mentha spicata*	1
32	Bitter gourd	*Momordica charantia* L.	1
33	Mulberry	*Morus alba* L.	1
34	Tobacco	*Nicotiana tabacum* L.	1
35	Rice	*Oryza sativa* L.	42
36	Opium poppy	*Papaver somniferum* L.	2
37	Pearl millet	*Pennisetum typhoides* L.	5
38	French bean	*Phaseolus vulgaris* L.	1
39	Pea	*Pisum sativum* L.	1

S. No	Common name	Latin name	No. of varieties
40	Isabgol	*Plantago ovata* L.	2
41	Tuberose	*Polyanthus tuberosa* L.	2
42	Portulaca	*Portulaca grandiflora* L.	11
43	Castor	*Ricinus communis* L.	4
44	Rose	*Rosa* sp.	16
45	Sugarcane	*Sachharum officinarum* L.	9
46	Sesame	*Sesamum indicum* L.	5
47	Foxtail millet	*Setaria italica* L.	1
48	Clarke Khasianum	*Solanum khasianum*	1
49	Brinjal	*Solanum melongena* L.	1
50	Coleus	*Solenostemon rotundifolius*	1
51	Snake gourd	*Trichosanthus anguina* L.	1
52	Egyptian clover	*Trifolium alexandrium* L.	1
53	Wheat	*Triticum aestivum* L.	4
54	Moth bean	*Vigna aconitifolia* Jacq. M.	5
55	Blackgram	*Vigna mungo* L. Hepper	9
56	Mungbean	*Vigna radiata* L. Wiczeck	15
57	Cowpea	*Vigna unguiculata* L. Walp.	10
Total			**343**

(Source: Kharakwal and Shu, 2009)

Conclusion and future outlook

Induced mutations are necessary to enhance rate of genetic variability since spontaneous mutation rate is very slow and that prevents breeders to exploit them in plant breeding programs. The major advantage in induced mutations is that multiple trait mutants can be isolated as compared to transgenic approach where single trait can be introduced in the crop and moreover there is a lack of acceptance of genetically modified (GM) food. A specific advantage of mutation induction is to develop a range of mutant lines, and identify trait specific genes in order to set up molecular gene database, study molecular functional genomics and develop bio-informatics for future to develop plant variety to grow the existing arable land under climate change to feed rapid human population growth. Moreover another advantage of mutagenesis is to isolate mutants with multiple traits and hopefully would be ideal to grow under climate change. The use of mutagenesis with plant tissue culture and length of culture period especially in cereals could increase the genetic variability for exploitation by plant breeders. This approach could be utilized more frequently before genetic engineering becomes a regular and a reliable tool in plant breeding. The developing countries with high population growth can't wait before genetic engineering can reap high harvest. Thereby, a combination of tissue culture and mutagenesis could create genetic variability and that is utilized by plant breeders in plant breeding programs.

The new gene discovery with reverse and forward genetics will open the way for developing functional genomic plant breeding (Hirochika *et al.*, 2004). The general strategy for reverse genetics is called TILLING (Targeting Induced Local Lesions in Genomes) or coming together of traditional mutagenesis with functional genomics (Martin *et al.*, 2009). TILLING enables the reverse selection of single point mutations by cleavage of mismatches in heteroduplex DNA with the endonucleases CEL 1. This strategy was first applied in an *Arabidopsis thaliana* mutant collection induced with ethyl methanesulphonate (Bradley *et al.*, 2007). EcoTILLING, a variation of this technique, represents a means to determine the extent of natural variation in selected genes in crops (Martin *et al.*, 2009). It may be a cost effective approach for haplotyping and SNP discovery. Furthermore, DNA sequence information of crop plants facilitates the isolation of cisgenes (Jacobsen and Schouten, 2007), which are genes from crop plants themselves or from crossable species. The increasing number of these isolated genes, and the development of transformation protocols that do not leave marker genes behind, provide an opportunity to improve plant breeding while remaining within the gene pool of the classical breeder or mutation breeding.

REFERENCES

Ahloowalia BS, Maluszynski M and Nichterlein K *(2004)* Global impact of mutation-derived varieties. *Euphytica 135: 187-204*

Anonymous (2012) *Plant Breeding & Genetics Newsletter* 29:1-2

Austin R S, Vidaurre D, Stamatiou G *et al.* (2011) Next generation mapping of *Arabidopsis* genes. *Plant Journal* 67: 715–725

Barkawi L S, Tam Y Y, Tillman J A, Normanly J and Cohen J D (2010) A high-throughput method for the quantitative analysis of auxins. *Nature Protocols* 5: 1609–1618

Batista R, Saibo N, Lourenco T and Oliveira MM (2008) Microarray analysis reveal that plant mutagenesis may induce more transcriptomic changes than transgene insertion. *Proceedings of the National Academy of Sciences,* USA 105: 3640–3645

Bradley T, Till J, Cooper J, Tai T H, Colowit P, Greene E A, Henikoff S and Comai L (2007) Discovery of chemically induced mutations in rice by TILLING. *BMC Plant Biol* 7: 19

Burger JC, Champan MA and Burke JM (2008) Molecular insights into the evolution of crop plants. *Am J Bot* 95: 113–122

Caldwell D G, McCallum N, Shaw P, Muehlbauer G J, Marshal D F and Waugh R (2004) A structured mutant population for forward and reverse genetics in Barley (*Hordeum vulgare* L.). *Plant Journal* 40: 143–150

Chawade A, Sikora P and Br¨autigamet M (2010) Development and characterization of an oat TILLING-population and identification of mutations in lignin and beta-glucan biosynthesis genes. *BMC Plant Biology* 10: 86

Cuomo CA, Gueldener U, Xu JR, *et al.* (2007) The *Fusarium graminearum* genome reveals a link between localized polymorphism and pathogen specialization. *Science* 317: 1400–1402

D'Souza S F, Reddy K S, Badigannavar A M, Manjaya J G and Jambhulkar S J (2009) Mutation Breeding in Oilseeds and Grain Legumes in India: ccomplishments and Socio-Economic Impact in Q.Y. Shu (ed.), Induced Plant Mutations in the Genomics Era. Food and Agriculture Organization of the United Nations, Rome, 2009, 55-57

Dean RA, Talbot NJ, Ebbole DJ, et al. (2005) The genome sequence of the rice blast fungus *Magnaporthe grisea. Nature* 434:980–986

Dierking E C and Bilyeu K D (2009) New sources of soybean seed meal and oil composition traits identified through TILLING. *BMC Plant Biology* 9: 89

Doebley J, Gaut BS and Smith BD (2006) The molecular genetics of crop domestication. *Cell* 127:1309-1321

Dong C, Dalton-Morgan C, Vincent K and Sharp P (2009) A modified TILLING method for wheat breeding. *Plant Genetic* 2: 39–47

Donini P and Sonnino A 91998) Induced mutation in plant breeding: Current status and future outlook In: Somaclonal Variation and Induced Mutations in Crop Improvement, Jain SM, Brar DS, Ahloowalia BS (Ed). Kluwer Academic Publishers, Dordrecht. 255-291

Durrant A (1962) The environmental induction of heritable changes in *Linum. Heredity* 17: 2761-2764

Edwards J D, Janda J, Sweeney M T *et al.* (2008) Development and evaluation of a high-throughput, low-cost genotyping platform based on oligonucleotide microarrays in rice. *Plant Methods* 4:e 13

Feiz L, Martin J M and Giroux M J (2009) Creation and functional analysis of new Puroindoline alleles in *Triticum aestivum. Theoretical and Applied Genetics* 118: 247–257

Griffiths S, Sharp R, Foote TN, Bertin I, Wanous M, Reader S, Colas I and Moore G (2006) Molecular characterization of Ph1 as a major chromosome pairing locus in polyploidy wheat. *Nature* 439: 749-752

Hirochika H; Guiderdon E, Hsing Y, Eun M Y, Han C, Upadhyaya N, Ramachandran S, Zgang Q, Pereira A, Sundaresan V and Leung H (2004) Rice mutant resources for gene discovery. *Plant Mol Biol* 54: 325-334

Huang G, Allen R, Davis EL, Baum TJ and Hussey RS (2006) Engineering broad root-knot resistance in transgenic plants by RNAi silencing of a conserved and essential root-knot nematode parasitism gene. *Proc Natl Acad Sci* USA 103: 14302-14306

Huang X, Feng Q, Qian Q *et al.* (2009) High-throughput genotyping by whole-genome resequencing. *Genome Research* 19: 1068–1076

Iiyama K and Wallis A F A (1988) An improved acetyl bromide procedure for determining lignin in woods and wood pulps. *Wood Science and Technology* 22: 271–280

Jacobsen E and Schouten H J (2007) Cisgenesis strongly improves introgression breeding and induced translocation breeding of plants. *Trends in Biotechnology* 25: 219-223

Junhyun J, Park S-Y, Chi M-H et al. (2007) High throughput phenotype screening pipeline for functional genomics in *Magnaporthe oryzae. Protocol Exchange* 2: 137-142

Kharkwal MC and Shu Q Y (2009) The Role of Induced Mutations in World Food Security. In: Induced Plant Mutations in the Genomics Era, Shu, Q.Y. (Ed.). Food and Agriculture Organization of the United Nations, Rome, Italy, pp: 33-38.

Lea PJ and Azevedo RA (2006) Nitrogen use efficiency 1. Uptake of nitrogen from the soil. *Annals of Applied Biology* 149: 243–247

Leung H, Wu C, Baraoidan M *et al.*. (2001) Deletion mutants for functional genomics: progress in phenotyping, sequence assignment, and database development. In *Rice Genetics*, D. Brar, B. Hardy, and G. Khush, Eds., vol. 4, pp. 239–251, International Rice Research Institute

Martin B, Ramiro M, Martinex-Zapater J M and Alonso-Blanco C (2009) A high-density collection of EMS-induced mutations for TILLING in *Landsberg erecta* genetic background of Arabidopsis. *BMC Plant Biol* 9: 147

McCleary B V and Codd R (1991) Measurement of (1 ’! 3), (1 ’!4)-â-D-glucan in barley and oats: a streamlined enzymic procedure. *Journal of the Science of Food and Agriculture* 55: 303–312

McMullen M, Jones R and Gallenberg D (1997) Scab of wheat and barley: a re-emerging disease of devastating impact. *Plant Disease* 81: 1340–1348

McNally R, Bruskiewich R, Mackill D, Buell CR, Leach JE and Leung H (2006) Sequencing multiple and diverse rice varieties, connecting whole-genome variation with phenotypes. *Plant Physiol* 141: 26-31

Oshima K, Kakizawa S, Nishigawa H, *et al.* (2004) Reductive evolution suggested from the complete genome sequence of a plant pathogenic phytoplasma. *Nature Genetics* 36: 27–29

Parry D W, Jenkinson P and McLeod L (1995) Fusarium ear blight (scab) in small grain cereals—a review," *Plant Pathology* 44: 207–238

Parry MAJ and Reynolds MP (2007) Improving resource use efficiency. *Annals of Applied Biology* 151: 133–135

Parry MAJ, Flexas J and Medrano H (2005) Prospects for crop production under drought: research priorities and future directions. *Annals of Applied Biology* 147: 211–226

Parry MAJ, Madgwick PJ, Carvalho JFC and Andralojc PJ (2007) Prospects for increasing photosynthesis by overcoming the limitations of Rubisco. *Journal of Agricultural Science* 145: 31–43

Paterson AH, Bowers JE, Vandepoele K, Van de Peer Y (2005) *Ancient duplication of cereal genomes. New Phytol* 165: 658–661

Phillips RL, Odland, WE and Kahler AL (2007) Rice as a reference genome and more. In: 5th Intl. Rice Genetics Symp. (Brar, D.S., Mackill, D. and Hardy, B., Eds.), CD,

International Rice Research Institute, Philippines, 3-15

Porteus MH and Carroll D (2005) Gene targeting using zinc finger nucleases. *Nature Biotechnol* **23:** 967-973

Powlson DS, Riche AB and Shield I (2005) Biofuels and other approaches for decreasing fossil fuel emissions from agriculture. *Annals of Applied Biology* 146: 193–201

Rashid M, He G, Guanxiao Y and Khurram Z (2011) Relevance of tilling in plant genomics. *AJCS* 5(4):411-420

Reiter W D, Chapple C and Somerville C R (1997) Mutants of *Arabidopsis thaliana* with altered cell wall polysaccharide composition. *Plant Journal* 12: 335–345

Reynolds M, Foulkes MJ, Slafer GA, Berry P, Parry MAJ, Snape JW and Angus WJ (2009) Raising yield potential in wheat. *Journal of Experimental Botany* 60: 1899–1918

Richards RA (2000) Selectable traits to increase crop photosynthesis and yield of grain crops. *Journal of Experimental Botany* 51: 447–458

Sala F and Labra M (2003) Antibiotic-induced DNA methylation changes in calluses of *Arabidopsis thaliana. Plant Cell Tissue and Organ Culture* 72: 157-162

Sandhu APS, Abdelnoor RV and Mackenzie SA (2007) Transgenic induction of mitochondrial rearrangements for cytoplasmic male sterility in crop plants. *Proc Natl Acad Sci* ***USA*** 104:1766-1770

Sanghera GS, Gill MS, Gosal SS, Wani SH (2010) RNA Interference: Its Concept and Application in Crop Plants In: Biotechnology: Cracking new pastures eds. CP **Malik.** MD Publications New Delhi, India pp. 33-78

Schneeberger K, Ossowski S, Lanz C *et al.* (2009) SHOREmap: simultaneous mapping **and** mutation identification by deep sequencing. *Nature Methods* 6: 550–551

Sestili F, Botticella E, Bedo Z, Phillips A and Lafiandra D (2010) Production of novel allelic variation for genes involved in starch biosynthesis through mutagenesis. *Molecular Breeding* 25: 145–154

Shukla F, Arnold N, Dekelver R, Doyon Y, Mitchell J, Rowland L, Simpson M, Worden S, Urnov F and Gao Z (2008) Zinc finger nuclease-mediated gene targeting in maize. In: 50th Annu. Maize Genet. Conf., Washington, D.C. Abstr. P.37

Singer T, Fan Y, Chang H S, Zhu T and Hazen S P (2006) A high-resolution map of Arabidopsis recombinant inbred lines by whole-genome exon array hybridization. *PLoS Genetics* 2: e144

Slade AJ, Fuerstenberg SI, Loeffler D, Steine MN and Facciotti D (2004) A reverse genetic, nontransgenic approach to wheat crop improvement by TILLING. *Nature Biotechnol* 23: 75-81

Slade J, Fuerstenberg SI, Loeffler D, Steine M N and Facciotti D (2005) A reverse genetic, non transgenic approach to wheat crop improvement by TILLING. *Nature Biotechnology* 23: 75–81

Speer E O (1987) A method of retaining phloroglucinol proof of lignin. *Stain Technology* **62:** 279–280

Spielmeyer W, Ellis M H and Chandler PM (2002) Semidwarf (sd-1), "green revolution" rice, contains a defective gibberellins 20-oxidase gene. *Proceedings of the National Academy of Sciences of the USA* 99: 9043–9048

Sunikumar G, Campbell LM, Puckhaber L, Stipanovic RD and Rathore KS (2006) Engineering cottonseed for use in human nutrition by tissue-specific reduction of toxic gossypol. *Proc Natl Acad Sci USA* 103: 18054-18059

Uauy C, Distelfeld A, Fahima T, Blechl A, Dubcovsky JA (2006) NAC gene regulating senescence improves grain protein, zinc, and iron content in wheat. *Science* 314: 1298-1301

Van Harten AM (1998) Mutation breeding - theory and practical breeding.Cambridge University Press, Cambridge, UK.

Wenzel W (2007) Fast-Track Traits. *Farm Journal,* March 17,

Wood DW, Setubal JC, Kaul R, *et al.* (2001) The genome of the natural genetic engineer *Agrobacterium tumefaciens* C58. *Science* 294: 2317–2323

Wu J L, Wu C, Lei C *et al.* (2005) Chemical- and irradiation induced mutants of *indica* rice IR64 for forward and reverse genetics. *Plant Molecular Biology* 59: 85–97

Xin Z, LiWang M, Barkley N A *et al.*. (2008) Applying genotyping (TILLING) and phenotyping analyses to elucidate gene function in a chemically induced sorghum mutant population. *BMC Plant Biology* 8: 103

Index

B

C

D

E

F

G

H

I